Sulphur in Biology

The Ciba Foundation for the promotion of international cooperation in medical and chemical research is a scientific and educational charity established by CIBA Limited—now CIBA-GEIGY Limited—of Basle. The Foundation operates independently in London under English trust law.
Ciba Foundation Symposia are published in collaboration with Excerpta Medica in Amsterdam.

Excerpta Medica, P.O. Box 211, Amsterdam

Sulphur in Biology

Ciba Foundation Symposium 72 (new series)

1980

Excerpta Medica

Amsterdam • Oxford • New York

ISBN Excerpta Medica 90 219 4078 7
ISBN Elsevier/North-Holland 0 444 90108 6

Published in April 1980 by Excerpta Medica, P.O. Box 211, Amsterdam and
Elsevier/North-Holland, Inc., 52 Vanderbilt Avenue, New York, N.Y. 10017.

Suggested series entry for library catalogues: Ciba Foundation Symposia.
Suggested publisher's entry for library catalogues: Excerpta Medica.

Ciba Foundation Symposium 72 (new series)
324 pages, 66 figures, 43 tables

Library of Congress Cataloging in Publication Data
Symposium on Biology of Sulphur, Ciba Foundation, 1979.
Sulphur in biology.

(Ciba Foundation symposium; 72 (new ser.))
Bibliography: p.
Includes indexes.
1. Sulphur metabolism - Congresses. I. Title.
II. Series: Ciba Foundation. Symposium; new ser., 72.
QP535.S1S93 1979 574.1'9214 79-24939
ISBN 0-444-09108-6

Printed in The Netherlands by Casparie, Amsterdam

Contents

Participants

Symposium on Biology of Sulphur, held at the Ciba Foundation, London, 24th—26th April 1979

J. R. POSTGATE *(Chairman)* Unit of Nitrogen Fixation, University of Sussex, Falmer, Brighton, Sussex, BN1 9RQ, UK

C. BRIERLEY The New Mexico Bureau of Mines, Socorro, New Mexico 87801, USA

S. W. J. BRIGHT Biochemistry Department, Rothamsted Experimental Station, Harpenden, Hertfordshire, AL5 2JQ, UK

F. CHATAGNER Laboratoire de Biochimie, 96 Boulevard Raspail, F-75006 Paris, France

K. S. DODGSON Department of Biochemistry, University College, PO Box 78, Cardiff, CF1 1XL, UK

W. A. HAMILTON Department of Microbiology, Marischal College, Aberdeen University, Aberdeen, Scotland, AB9 1AS, UK

J. R. IDLE Department of Biochemical Pharmacology, St Mary's Hospital Medical School, Norfolk Place, London W2 1PG, UK

J. L. JOHNSON Department of Biochemistry, Duke University Medical Center, Durham, North Carolina 27710, USA

M. C. JONES-MORTIMER Department of Biochemistry, University of Cambridge, Tennis Court Road, Cambridge, CB2 1QW, UK

J. H. R. KÄGI Biochemisches Institut, University of Zürich, Zürichbergstrasse 4, CH 8028 Zürich, Switzerland

D. P. KELLY Department of Environmental Sciences, University of Warwick, Coventry, CV4 7AL, UK

N. M. KREDICH Department of Medicine, Division of Rheumatic & Genetic Diseases, Duke University Medical Center, Durham, North Carolina 27710, USA

J. LE GALL Laboratoire de Chimie Bactérienne, CNRS, 31 Chemin Joseph-Aiguier, 13274 Marseilles – Cedex 2, France

A. MEISTER Department of Biochemistry, Cornell University Medical College, 1300 York Avenue, New York, New York 10021, USA

S. H. MUDD Laboratory of General and Comparative Biochemistry, National Institute of Mental Health, 9000 Rockville Pike, Bethesda, Maryland 20205, USA

D. K. RASSIN Department of Human Development and Nutrition, New York State Institute for Basic Research in Mental Retardation, 1050 Forest Hill Road, Staten Island, New York 10314, USA

M. RECASENS Centre de Neurochimie du CNRS, 11 rue Humann, 67085 Strasbourg — Cedex, France

F. A. ROSE Department of Biochemistry, University College, PO Box 78, Cardiff, CF1 1XL, UK

A. B. ROY Department of Physical Biochemistry, The John Curtin School of Medical Research, Australian National University, PO Box 334, Canberra City, ACT 2601, Australia

J. A. SCHIFF* Institute for Photobiology of Cells and Organelles, Brandeis University, Waltham, Massachusetts 02154, USA

I. H. SEGEL Department of Biochemistry and Biophysics, University of California, Davis, California 95616, USA

H. E. SWAISGOOD Department of Food Science & Biochemistry, North Carolina State University, Raleigh, North Carolina 27650, USA

F. R. WHATLEY Botany School, University of Oxford, South Parks Road, Oxford, OX1 3RA, UK

D. M. ZIEGLER Department of Chemistry, Clayton Foundation Biochemical Institute, The University of Texas, Austin, Texas 78712, USA

Editors: KATHERINE ELLIOTT *(Organizer)* and JULIE WHELAN

* Contributed *in absentia,* because of illness.

Introduction

J. R. POSTGATE

Unit of Nitrogen Fixation, University of Sussex, Brighton, Sussex, BN1 9RQ, UK

I may seem a strange choice for chairman of this symposium since I have not been involved in sulphur research for the past seven years and my contribution, such as it was, was made more than a decade ago. However, I am delighted to be back in the intellectual environment of sulphur.

I would like to make one general point about the biology of sulphur. For the past couple of decades I have been increasingly, and now almost exclusively, involved with another element, nitrogen, and the reason that this happened is in part because of its overriding importance in world food production. One result of this importance is that a specialized international symposium on the subject of nitrogen fixation is held every two years, and almost every international congress of microbiology, biochemistry or botany has sections devoted to nitrogen fixation. There are at least four national meetings every year devoted to nitrogen, with particular emphasis on its fixation. Since 1960, the scientific world seems to have blown its top on the subject of nitrogen. The consequences have been various: sometimes they have been comic, as in some of the ludicrous grant proposals one sees; sometimes they have been sad, because competitiveness in certain areas has become almost counterproductive. But there can be no question that the rate of advance of our knowledge of the nitrogen cycle and of nitrogen fixation in particular has been dramatic, because people realized that if research was not done they would not eat, or else their children and grandchildren would not eat.

In the same period there have been about six meetings devoted to the biology of sulphur. Yet one has known for three decades that the gross national productivity in any industrialized country is directly related to its consumption of the element sulphur; that all the sulphur used is biogenic, having been formed largely in the Permian Era; that it is effectively a reserve of fossil energy that we are now consuming with the same abandon with which we consume oil, coal or natural gas; and that therefore when these energy supplies run out the energy cost of making

Sulphur in biology
(Ciba Foundation Symposium 72) p 1-2

substitutes will go up and up. The analogy I am drawing here is that when something is economically important, funds become available for research. It is not an analogy that I would wish to push too far, and I would be the last to suggest that research on the biology of sulphur should be determined by the international price of sulphur. There was a period when this happened and the results to research were disastrous. I do note with regret that the importance of sulphur, both economically and in medicine and biochemistry, has not had the same pull in terms of research as has the realization of the importance of nitrogen. We are, therefore, fortunate that the Ciba Foundation has considered sulphur to be a subject worth serious attention. This will do something to demonstrate that sulphur is a significant research subject.

Turning to the programme, I am well aware of certain gaps, which is perhaps inevitable with such a wide subject. As we shall hear, the biology of sulphur is entirely dependent on the activities of microbes, but many important types will not be covered. In particular, the photosynthetic sulphur bacteria will not be discussed as such, nor the new species of sulphate-reducing bacteria. The ecology of the sulfuretum (the ecosystem based on the sulphur cycle; see Postgate 1979) is not overtly featured. In the area of biochemistry, the iron-sulphur proteins are major terminal products of sulphur metabolism which do not feature in the formal papers. We hope that participants will bring these topics into the discussions of the subjects selected for major emphasis.

Reference

Postgate JR 1979 The sulphate-reducing bacteria. Cambridge University Press, Cambridge

The sulphur cycle: definitions, mechanisms and dynamics

DONOVAN P. KELLY

Department of Environmental Sciences, University of Warwick, Coventry, CV4 7AL, UK

Abstract The principal biochemical processes of the sulphur cycle are described and the types of organisms known to catalyse the reductive and oxidative phases of the cycle outlined. Attention is drawn to the shortcomings in our current knowledge of the scale of turnover of the sulphur cycle and of our understanding of the microorganisms involved in specialized environments. Examples of some special habitats are used to illustrate these points. The role of sulphate-reducing bacteria and sulphur-oxidizing chemolithotrophs in the formation and recycling of sulphide minerals is described.

Along with carbon, nitrogen, oxygen, hydrogen and phosphorus, sulphur is a major metabolic nutrient element and exhibits perfect biogeochemical cycling with intermediate exchange between the terrestrial, aquatic and atmospheric phases of the environment. Most biologists know the sulphur cycle: the alternation of dissimilatory sulphate reduction and sulphide oxidation with the minor cycle of assimilatory sulphate metabolism and subsequent sulphide release during putrefaction. True, the broad outline of the cycle is easily understood (Trudinger 1969, Junge 1972, Schlegel 1976), but the global kinetics and actual relative contributions of different kinds of microorganisms (even the nature of all the organisms involved) are poorly understood in terms of microbial ecology and global biogeochemistry, even though some organisms of the cycle have been studied biochemically in great detail (Baas Becking & Wood 1955). In the following sections I have outlined the main features of the cycle and dwelt specifically on some of the less well-known aspects of sulphur turnover. Some of the problems in quantifying the operation of the cycle are considered and illustrated with some examples of specific habitats and microbial systems in which sulphur transformations occur.

Sulphur in biology
(Ciba Foundation Symposium 72) p 3-18

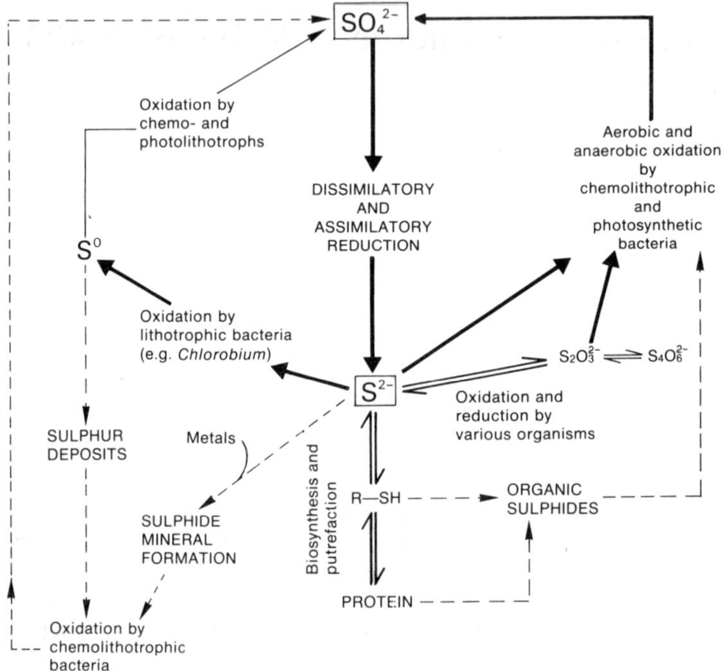

FIG. 1. The essential features of the global biological sulphur cycle.

PRINCIPLES OF THE SULPHUR CYCLE

The cycle as illustrated (Fig. 1) can be considered as four interlinked phases: (a) assimilatory sulphate reduction to sulphide on a scale sufficient only to provide the organism with adequate sulphide to effect the synthesis of sulphur amino acids for protein formation; (b) protein and organic sulphur compound degradation following death (or of excreta) of the organisms, leading to free sulphide release and in part to volatile organic sulphide production: (c) dissimilatory sulphate reduction to free hydrogen sulphide on a large scale, the sulphide production vastly exceeding sulphur requirements for biosynthesis as the process is a respiratory one rather than biosynthetic; (d) sulphide oxidation resulting ultimately in the regeneration of sulphate.

Assimilatory sulphate reduction occurs in very many micro- and macro-organisms, generally by the reactions of Fig. 2. Putrefaction and degradation of organic sulphur compounds to give free sulphide results from microbiological activity, heterotrophic bacteria in particular releasing sulphide in the process of degrading the carbon skeletons for their own biosynthesis. In some cases organic

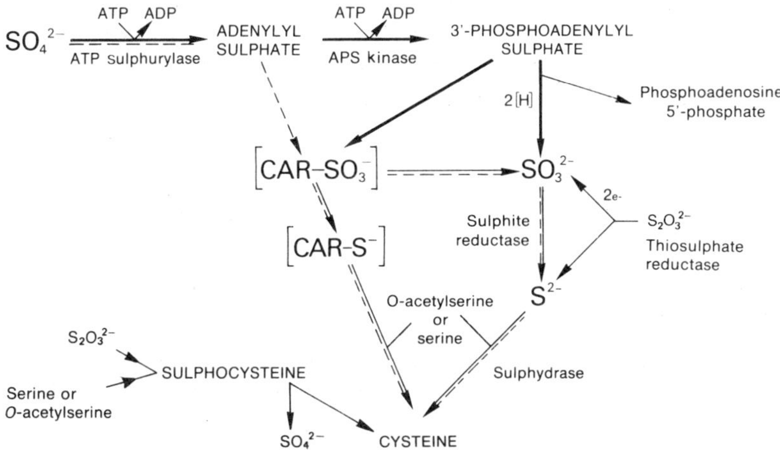

FIG. 2. Assimilatory sulphate reduction and cysteine formation. ———, bacterial and fungal pathways; - - -, pathways in oxygen-evolving eukaryotic phototrophs.

sulphur degradations seem to lead directly to sulphate (Stapley & Starkey 1970, Kelly 1972). Organic sulphides, particularly dimethyl sulphide and methyl mercaptan, are released both from living and decaying tissues and may have a significant role in sulphur cycling (Lovelock et al 1972, Zinder et al 1977). Dissimilatory sulphate reduction is effected only by specialized types of anaerobic bacteria that use sulphate as the terminal electron acceptor for the respiratory oxidation of organic substrates or hydrogen. The best-studied genera are *Desulfovibrio* and *Desulfotomaculum*. Recently a genus, *Desulfuromonas*, has been found that reduces elemental sulphur (Pfennig & Biebl 1976). Fig. 3 illustrates the probable reactions of sulphate reduction in *Desulfovibrio*. Reoxidation of sulphide can be chemical by oxygen or is effected by bacteria in anaerobic and aerobic environments. Aerobically, the thiobacilli and related chemolithotrophic bacteria are best known. These obtain all the energy required for growth from the oxidations (Kelly 1978). Anaerobically, nitrate-reducing thiobacilli oxidize

ATP ADP
SO_4^{2-} → ADENYLYL SULPHATE
ATP sulphurylase
2[H] AMP
→ SO_3^{2-}
APS reductase
2 [H] 2 [H] 2 [H]
[$S_3O_6^{2-}$ → $S_2O_3^{2-}$] → S^{2-}
SO_3^{2-} SO_3^{2-}
Sulphite reductase

FIG. 3. Probable pathway of dissimilatory sulphate reduction in *Desulfovibrio*. The occurrence of trithionate and thiosulphate as intermediates is problematical.

sulphur compounds in the absence of light, but the green and red photosynthetic sulphur bacteria consume sulphide in illuminated anaerobic habitats. The organisms involved in these phases and the reactions effected are summarized in

TABLE 1

Some of the microorganisms of the sulphur cycle and the reactions they catalyse

Organism	*Habitats*	*Reactions catalysed*
Desulfovibrio *Desulfotomaculum*	Anaerobic sediments, muds and soils	Oxidation of organic acids, alcohols and hydrogen $8[H] + SO_4^{2-} = H_2S + 2H_2O + 2OH^-$ $6[H] + SO_3^{2-} = H_2S + H_2O + 2OH^-$ $4CH_3.CO.COOH + H_2SO_4 = H_2S + 4CH_3COOH + 4CO_2$ $2[H] + S_2O_3^{2-} = H_2S + SO_3^{2-}$ Polythionates + x [H] → sulphide
Desulfuromonas	Marine muds and from *'Chloro-pseudomonas'* mixed cultures	Acetate, ethanol or pyruvate oxidized $2[H] + S^0 = H_2S$
Facultative anaerobes (e.g. marine bacteria, *Salmonella, Proteus, Citrobacter)*	Diverse: soil, water, animal gut	$2[H] + S_4O_6^{2-} = 2S_2O_3^{2-} + 2H^+$ Thiosulphate + pyruvate = sulphide Sulphite + pyruvate = sulphide + thiosulphate
Green photosynthetic bacteria	Illuminated anaerobic muds and sulphide-rich waters	$2H_2S + CO_2 = (CH_2O) + H_2O + 2S^0$
Red/purple photosynthetic sulphur bacteria (e.g. *Chromatium*)	As above; some potential for aerobic sulphur oxidation	Sulphide, sulphur, thiosulphate → sulphate
Purple photosynthetic non-sulphur bacteria (e.g. *Rhodopseudomonas*)	As above Some facultative aerobes	Sulphide → sulphur Sulphide → sulphate
Thiobacillus spp. *Thiomicrospira* spp. Probably other lithotrophic genera	Aerobic water or soils; a few species able to reduce nitrate anaerobically	$H_2S + 2O_2 = H_2SO_4$ $Na_2S_2O_3 + 2O_2 + H_2O = Na_2SO_4 + H_2SO_4$ $S^0 + 1.5O_2 + H_2O = H_2SO_4$ $Na_2S_4O_6 + 3.5O_2 + 3H_2O = Na_2SO_4 + 3H_2SO_4$ $5Na_2S_2O_3 + 8NaNO_3 + H_2O = 9Na_2SO_4 + H_2SO_4$
Some heterotrophic bacteria	Soil pseudomonads Some marine bacteria	$4Na_2S_2O_3 + O_2 + 2H_2O = 2Na_2S_4O_6 + 4NaOH$
Beggiatoa	Fresh water; paddy fields	Sulphide → sulphur + sulphate

Table 1. The mechanisms of inorganic sulphur oxidation are reviewed elsewhere (Kelly 1968, 1972, 1978, Trudinger 1969).

We can see that interactions of numerous bacteria are involved in sulphur cycling but our understanding of the dynamic relation between these organisms is rather limited. In some cases there is such intimate association between sulphide-generating and photosynthetic sulphide oxidizers that they are difficult to separate physically from each other. In general, sulphate-reducing bacteria are restricted to dark anaerobic environments (Postgate 1965) such as sediments (although they have also been found in the rumen: Huisingh et al 1974) and the sulphide they generate is reoxidized either anaerobically or aerobically after diffusion from the sediments. Oxidation can thus occur at various depths in the aquatic environment depending on the sulphide tolerance of the bacteria involved, on the availability of light, nitrate or oxygen (depending on the organisms involved), and on any physicochemical limitations of the bacteria present. Assessment of the relative roles of the organisms and of the total sulphide being generated is difficult in such complex environments. Chemical oxidation of sulphide may generate thiosulphate as well as sulphur and sulphate. Similarly polythionates may be formed chemically and by thiobacilli oxidizing sulphur compounds. Thiosulphate and tetrathionate may be reduced by heterotrophic bacteria other than *Desulfovibrio* (Le Minor & Pichinoty 1963, Tuttle & Jannasch 1973a), thus returning sulphur through the reductive phase without first seeing regeneration to sulphate. Similarly, thiosulphate can be oxidized at least to tetrathionate by some heterotrophs (Kelly 1972) and thus influence the oxidative phase of the cycle. Sulphide oxidation by purple non-sulphur bacteria (Hansen & van Gemerden 1972) and algae has also been reported.

CLOSED VERSUS OPEN FLOW OF SULPHUR IN THE GLOBAL BIOGEOCHEMICAL SULPHUR CYCLE

A perfect biogeochemical cycle is characterized by free flow of the element through all phases of the biosphere. This requires a gaseous, water-soluble intermediate that can mediate between the atmosphere and the terrestrial and aquatic environments. For sulphur this role is met by hydrogen sulphide and possibly dimethyl sulphide. Atmospheric H_2S is in dynamic equilibrium with the land and sea and the annual flux of H_2S through the atmosphere is about 10^8 tonnes (Junge 1972). This, however, is no measure of total sulphide production on the earth since we have no reliable estimate of the amount of H_2S oxidized chemically and biologically before it can escape to the atmosphere. It seems unlikely, considering the extent of freshwater and marine sedimentary environments in which sulphate reduction occurs but in which little or no H_2S is detectable, that the flux into the atmosphere represents more than a very small part of the total annual H_2S generation, most of which must consequently be recycled in 'closed' environments.

'Closed' cycling of sulphur is most easily seen in aquatic environments where sulphide generated in sediments diffuses through the water column and is oxidized probably by anaerobic denitrifying bacteria and by photosynthetic bacteria at deep levels in the photic zone. Aerobically chemical and bacterial oxidation would consume residual sulphide reaching oxygenated waters. In soils, sulphate reduction in waterlogged levels produces sulphide that again is oxidized in the aerobic layers: the presence of sulphide in bogs for example can be detected by a blackening of a silver wire inserted vertically through the aerobic and anoxic layers and usually indicates little sulphide diffusion to the soil surface. The well-known 'Winogradsky column' (e.g. Schlegel 1976) is a good way of illustrating the closed sulfuretum system and the stratification of organisms that can occur in it in response to sulphide and oxygen concentration gradients, and the availability of light in the case of phototrophs.

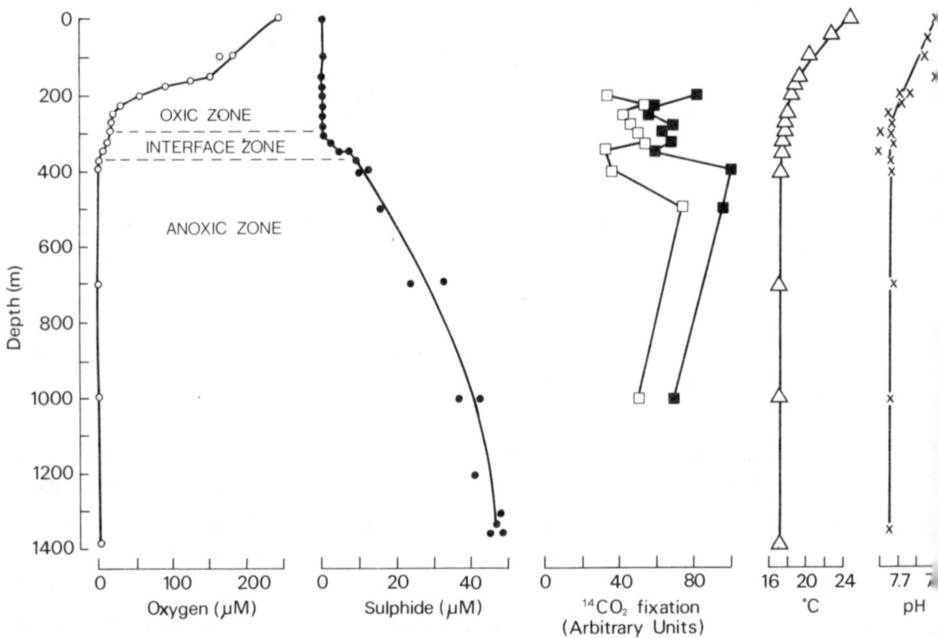

FIG. 4. The Cariaco Trench. Oxygen (○) and sulphide (●) concentrations, water temperature (Δ) and pH (×) at depths from the top to near the bottom of the Trench were measured at a location 10° 31.0′N and 64° 38.2′W between 27th March and 1st April 1979. Fixation of $^{14}CO_2$ by microorganisms filtered from water sampled from different depths and resuspended in sulphide-free seawater was measured aerobically with (■) and without (□) added thiosulphate (0.2mM). (Previously unpublished data of D. P. Kelly, A. L. Smith, W. Kaplan & I. Morris.)

An unusual example of a marine environment in which large-scale sulphide generation occurs is the Cariaco Trench off the coast of Venezuela (Richards & Vaccaro 1956). Here a body of water some 1400 m deep is anoxic and rich in sulphide at depths below about 300 m (Fig. 4). At the O_2/S^{2-} interface, which is far below the photic zone so photosynthetic sulphide oxidation cannot occur, sulphide oxidation must result from combined chemical oxidation and the activity of chemolithotrophic bacteria. The potential for carbon dioxide fixation in these anoxic and low-oxygen waters is significant and is stimulated aerobically by thiosulphate (Fig. 4), which probably indicates the presence of thiobacillus-like bacteria (Tuttle & Jannasch 1973b). Such bacteria are probably important in the oxidation of sulphides throughout the oceans (Tilton 1968, Tilton et al 1967, Adair & Gundersen 1969, Tuttle & Jannasch 1972). Many studies on the oxidation of sulphur compounds by thiobacillus-like bacteria have employed highly aerobic conditions, whereas in nature the bulk of sulphide oxidation probably occurs under low oxygen supply. The demonstration of relatively abundant nitrate-reducing sulphur oxidizers in sea waters (Tuttle & Jannasch 1972) and of sulphide-tolerant bacteria with thiobacillus-like physiology from marine muds (Kuenen & Veldkamp 1972) suggests that organisms with this kind of physiology may be more significant in the oxidative phase at least of the marine sulphur cycle than the better-studied aerobic thiobacilli. Similarly, *Beggiatoa* is of importance in sulphide oxidation in waterlogged rice paddies (Joshi & Hollis 1977).

FORMATION AND BREAKDOWN OF MINERALS IN THE SULPHUR CYCLE

The ocean and soil environments are the locations for transformations of soluble sulphur compounds and sulphur, the most studied substrates being of course sulphate, sulphite, sulphide, thiosulphate and polythionates. Cycling with solutes can be rapid with short turnover times for the substrates. It is demonstrable, however, that sulphide generated by sulphate reduction can be precipitated as metal sulphides in aquatic or soil environments, and many sedimentary mineral sulphide deposits owe their origin to this mechanism (Temple 1964, Bloomfield 1969, Hallberg 1972a, Trudinger 1976, Trudinger et al 1972). Of particular significance are the economically important deposits of copper, zinc and lead sulphides and the formation of pyrite both as a mineral deposit and in acid sulphate soils. The mechanisms of pyrite formation in sulphate-reducing systems have been studied in detail (Hallberg 1972b, Trudinger 1976) and involve the progressive physicochemical transformation of an initial ferrous sulphide precipitate to different sulphide minerals, ultimately producing pyrite (Fig. 5). Pyrite and other sulphide minerals are substrates for oxidation by acidophilic sulphur- and iron-oxidizing bacteria such

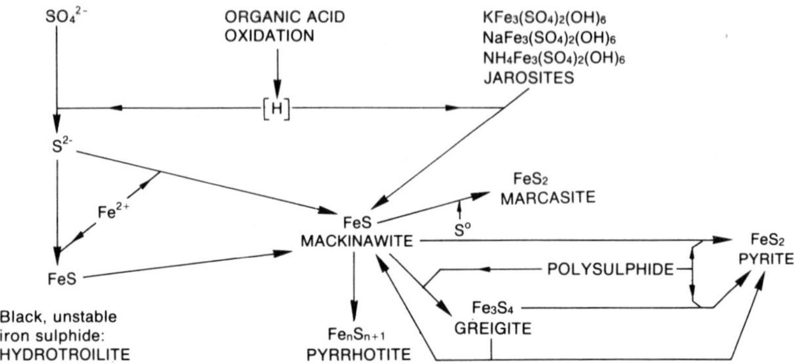

FIG. 5. Iron sulphide mineral formation in cultures of *Desulfovibrio* reducing soluble sulphate or jarosites. (Based on Hallberg 1972b; Ivarson et al 1976, Trudinger 1976.)

as *Thiobacillus ferrooxidans* (Brierley 1978, Kelly et al 1979), regenerating free sulphuric acid and ferric iron:

$$4FeS_2 + 15O_2 + 2H_2O = 2Fe_2(SO_4)_3 + 2H_2SO_4$$

Thus iron and sulphur trapped in mineral synthesis can be remobilized and re-enter the sulphur cycle. This can be of economic benefit, since metals are released (Brierley 1978), or can result in acidification of soils (Bloomfield 1972, Kelly 1972). Typically the ferric iron so generated precipitates as basic ferric sulphates (jarosites) at pH values above pH 2 in the presence of potassium, sodium or ammonium ions. Thus insoluble pyrite can be transformed by these aerobic bacteria to an insoluble jarosite. Equally interesting is that these iron jarosites can be substrates for sulphate reduction by *Desulfovibrio,* eventually regenerating pyrite (Ivarson & Hallberg 1976, Ivarson et al 1976). Cyclic interconversions of pyrite and jarosites can thus occur on a seasonal or geological time scale and represent a longer-term cycling of sulphur through insoluble as well as soluble sulphates and sulphides, as illustrated in Fig. 6.

FUTURE STUDY OF THE SULPHUR CYCLE

Clearly, much detailed biochemical work remains to be done on the well-known sulphate reducers and sulphur oxidizers, but much more study is needed of the bacteriology of the specialized environments in which high sulphide and low oxygen concentrations coexist, the transformation of low concentrations of dissolved reduced sulphur compounds and the origin and destruction of organic sulphides. More detailed understanding of the rates and quantities of sulphate reduction in sediments and sulphide oxidation in aquatic environments could also

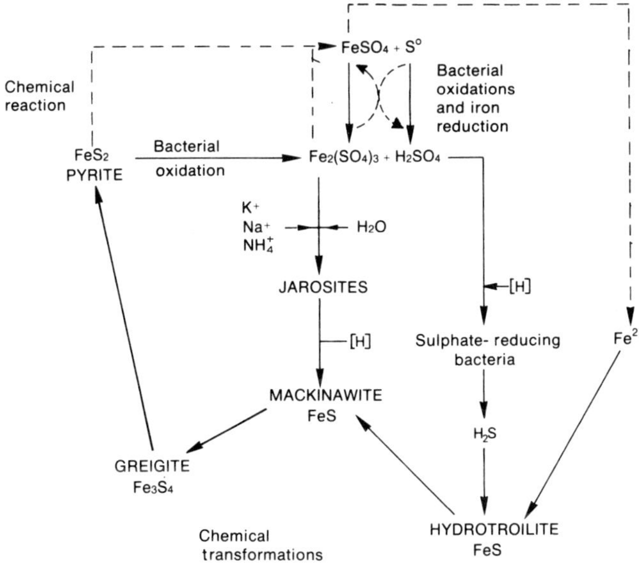

FIG. 6. The iron-sulphur cycle catalysed by the sulphate-reducing bacteria and iron- and sulphide-oxidizing thiobacilli and involving interconversion of jarosites and pyrite. These processes occur for example in sulphate soils subject to periodic waterlogging. On a geological time scale the formation and breakdown of sulphide minerals is also an example of this cycle.

enable some better estimate to be made of the proportion of organic carbon (ultimately derived from primary photosynthetic production) being returned to carbon dioxide through the activity of sulphate-reducing bacteria. Similarly the link between the sulphur and nitrogen cycles could be better quantified if more was known about sulphide-dependent nitrate reduction in the seas and soil environment. Finally, much more needs to be learned about ecological interactions between organisms of the sulphur cycle.

ACKNOWLEDGEMENTS

I am grateful to Dr I. Morris (Bigelow Laboratory for Ocean Sciences) for making it possible for me to work on the Cariaco Trench during a cruise by the R/V Endeavor in March 1979; and The Royal Society and the Natural Environment Research Council for grants in aid of the work.

References

Adair FW, Gundersen K 1969 Chemoautotrophic sulfur bacteria in the marine environment. Can J Microbiol 15:355-359

Baas Becking LGM, Wood EJF 1955 Biological processes in the estuarine environment. Koninkl Nederl Akad van Wetenschappen Ser B Phys Sci Proc B58:160-181; B59:85-108
Bloomfield C 1969 Sulphate reduction in waterlogged soils. J Soil Sci 20:207-221
Bloomfield C 1972 The oxidation of iron sulphides in soils in relation to the formation of acid sulphate soils, and of ochre deposits in field drains. J Soil Sci 23:1-16
Brierley CL 1978 Bacterial leaching. CRC Crit Rev Microbiol 5:207-262
Hallberg RO 1972a Sedimentary sulfide mineral formation — an energy circuit system approach. Mineral Deposita 7:189-201
Hallberg RO 1972b Iron and zinc sulfides formed in a continuous culture of sulfate-reducing bacteria. Neues Jahrb Mineral Monatsh 11:481-500
Hansen TA, van Gemerden H 1972 Sulfide utilization by purple nonsulfur bacteria. Arch Mikrobiol 86:49-56
Huisingh J, McNeill JJ, Matrone G 1974 Sulfate reduction by a *Desulfovibrio* species isolated from sheep rumen. Appl Microbiol 28:489-497
Ivarson KC, Hallberg RO 1976 Formation of mackinawite by the microbial reduction of jarosite and its application to tidal sediments. Geoderma 16:1-7
Ivarson KC, Hallberg RO, Wadsten T 1976 The pyritization of basic ferric sulfates in acid sulfate soils: a microbiological interpretation. Can J Soil Sci 56:393-406
Joshi MM, Hollis JP 1977 Interaction of *Beggiatoa* and rice plant: detoxification of hydrogen sulfide in the rice rhizosphere. Science (Wash DC) 195:179-180
Junge C 1972 Sulphur supplies of atmospheric origin. In: Symposium international sur le Soufre en Agriculture. Ann Agron Numéro hors série: 235-247
Kelly DP 1968 Biochemistry of oxidation of inorganic sulphur compounds by microorganisms. Aust J Sci 31:165-173
Kelly DP 1972 Transformations of sulphur and its compounds in soils. In: Symposium international sur le Soufre en Agriculture. Ann Agron Numéro hors série: 217-232
Kelly DP 1978 Bioenergetics of chemolithotrophic bacteria. In: Bull AT, Meadow PM (eds) Companion to microbiology. Longman, London, p 363-386
Kelly DP, Norris PR, Brierley CL 1979 Microbiological methods for the extraction and recovery of metals. Soc Gen Microbiol Symp 29:263-308
Kuenen JG, Veldkamp H 1972 *Thiomicrospira pelophila*, gen.n., sp.n., a new obligately chemolithotrophic colourless sulfur bacterium. Antonie van Leeuwenhoek J Microbiol Serol 38:241-256
Le Minor L, Pichinoty F 1963 Recherche de la tétrathionate-réductase chez les bactéries gram-négatives anaérobies facultatives. Ann Inst Pasteur (Paris) 104:384-393
Lovelock JE, Maggs RJ, Rasmussen RA 1972 Atmospheric dimethyl sulphide and the natural sulphur cycle. Nature (Lond) 237:452-453
Pfennig N, Biebl H 1976 *Desulfuromonas acetoxidans* gen.nov. and sp.nov., a new anaerobic, sulfur-reducing, acetate-oxidizing bacterium. Arch Microbiol 110:3-12
Postgate JR 1965 Recent advances in the study of the sulfate-reducing bacteria. Bacteriol Rev 29:425-441
Richards FA, Vaccaro RF 1956 The Cariaco Trench, an anaerobic basin in the Caribbean Sea. Deep-Sea Res 3:214-228
Schlegel HG 1976 Allgemeine Mikrobiologie, 4th edn. Thieme-Verlag, Stuttgart
Stapley EO, Starkey RL 1970 Decomposition of cysteic acid and taurine by soil microorganisms. J Gen Microbiol 64:77-84
Temple KL 1964 Syngenesis of sulfide ores: an evaluation of biochemical aspects. Econ Geol 59:1473-1491
Tilton RC 1968 The distribution and characterization of marine sulfur bacteria. Rev Int Oceanogr Med 9:237-253
Tilton RC, Cobet AB, Jones GE 1967 Marine thiobacilli. Can J Microbiol 13:1521-1534
Trudinger PA 1969 Assimilatory and dissimilatory metabolism of inorganic sulphur compounds by microorganisms. Adv Microb Physiol 3:111-158
Trudinger PA 1976 Microbiological processes in relation to ore genesis. In: Wolf KH (ed) Handbook of stratabound and stratiform ore deposits. Elsevier, Amsterdam, p 135-190

Trudinger PA, Lambert IB, Skyring GW 1972 Biogenic sulfide ores: a feasibility study. Econ Geol 67:1114-1127

Tuttle JH, Jannasch HW 1972 Occurrence and types of thiobacillus-like bacteria in the sea. Limnol Oceanogr 17:532-543

Tuttle JH, Jannasch HW 1973a Dissimilatory reduction of inorganic sulfur by facultatively anaerobic marine bacteria. J Bacteriol 115:732-737

Tuttle JH, Jannasch HW 1973b Sulfide and thiosulfate-oxidizing bacteria in anoxic marine basins. Mar Biol (NY) 20:64-70

Zinder SH, Doemel WN, Brock TD 1977 Production of volatile sulfur compounds during the decomposition of algal mats. Appl Environ Microbiol 34:859-860

Discussion

Postgate: You mentioned the associations of bacteria which have considerable ecological significance. They often consist of a sulphate reducer growing physically close to a sulphide oxidizer. These 'consortia' behave like prokaryotic equivalents of a differentiated multicellular organism. The ecological importance of such consortia is being increasingly recognized (e.g. Pfennig 1978).

But turning to the closed sulfuretum, as found in the Cariaco Trench, several large and many small sulfureta are known. When such sulfureta are functioning they are of necessity anaerobic, except at the interface of the sulfuretum with the atmosphere, where oxidative processes occur. Board (1976) has suggested that ecologically, on a global scale, such sulfureta are collectively responsible for maintaining the oxygen tension of the atmosphere at its Po_2 of 0.2, whereas one would predict from rates of photosynthesis and CO_2 fixation that it would rise steadily. If the major oxygen sink is in fact the sulfuretum, or the edge of it, it is an interesting thought that we aerobes live in the 'oxidized' side of an ecosystem of which we don't normally see half.

Dodgson: I wonder whether there is a potential problem arising from the scientific advances currently being made in our understanding and manipulation of the processes involved in nitrogen fixation. Increased productivity; in terms of plant growth and yield, may follow, but will we face a new problem in terms of inadequate supplies of sulphur for plant growth?

Whatley: The New Zealanders have evidence that sulphur is already the principal element limiting the growth of grass in pastures.

Postgate: Yes. The answer is not that use of nitrogen fertilizer may or will do this, but that it is already happening. This was illustrated by the fact that, as a result of the world sulphur shortage in about 1950, agriculturalists throughout the world ceased using ammonium sulphate as the major fertilizer and went over to ammonium or calcium nitrate. As a result, sulphur-deficient agricultural soils, formerly known only in East Africa, have been found all over the world, in both temperate and tropical regions.

Dodgson: Is there any real evidence that sulphur is being steadily transformed into a form or state from which it cannot be recycled naturally?

Kelly: The annual net flux of sulphur from the atmosphere and the land to the sea is of the order of 95 million tonnes a year, made up of 73 million tonnes in river run-off and 22 million tonnes from the atmosphere by gas absorption, precipitation and dry deposition (Junge 1972).

Whatley: Is that comparable to phosphorus leaching?

Kelly: It is the same sort of thing, a one-way flow, except that the sulphur is lost more slowly because phosphorus doesn't return whereas sulphur recycles via biological H_2S generation and anthropogenic SO_2.

The actual turnover of sulphur is vastly greater than the 10^8 tonnes annual flux of H_2S through the atmosphere that I quoted, because huge amounts of H_2S are generated in anoxic environments but reoxidized before ever coming into contact with the atmosphere.

Postgate: For nitrogen, at least, these numerical estimates of fluxes are not very reliable and often not more than inspired guesses!

Kelly: I agree. The same goes for the argument about anoxic basins being the controlling factor for the oxygen balance of the atmosphere; one has no quantitative idea of how important the processes of sulphate reduction and oxygen consumption are over the entire ocean bed.

Rose: Do we know how much sulphur is brought into circulation annually as a result of the utilization of elemental sulphur, sulphide-containing mineral deposits, fossil fuels and so on? We should be able to calculate it. Of course, some of it may be immediately 'locked up' again in an unavailable form, depending on the use to which it is put, but some must come into circulation.

Postgate: The important information is that extremely little of the sulphur consumed by industry actually appears in the industrial product. An exception is the sulphur which is part of vulcanized rubber. Most of the sulphur is sulphuric acid, used for chemical treatment of products, so that it eventually goes down the drain. So the sulphur is originating from a concentrated form and becoming a more dilute form, and therefore difficult to recover.

Rose: We hear a great deal about SO_2 entering the atmosphere as a result of industrial combustion processes. This alone must be a significant contribution.

Whatley: Does that addition alter the argument? Is it only a minor loss?

Kelly: No. It can be calculated from the annual consumption of coal and low- or high-sulphur oil. It gives a very large global output of about 10^8 tonnes of SO_2 (e.g. Junge 1972).

Dodgson: You didn't emphasize the contribution of the sulphur being returned to soils in forms other than sulphides. John Harwood in our laboratory has recently stressed the enormous quantities of plant chloroplast sulpholipid that is returned to the soil each year (Harwood & Nicholls 1979). How the sulphur of this sulpholipid becomes available for a re-cycling is unknown.

Kelly: The only well-studied organic sulphur systems are taurine and cysteine breakdown; these seem to be converted directly to sulphate via sulphite, rather than to sulphides (Stapley & Starkey 1970). I wasn't aware of the sulpholipid contribution.

Whatley: Isn't plant sulpholipid a lipid sulphate? It makes up a high proportion of the chloroplast.

Dodgson: It is a 1,2-diacyl-3-(6-sulpho-α-D-quinovopyranosyl)-L-glycerol in which the sugar moiety, quinovose, carries a sulphonate group at C-6. The linkage is therefore C–S rather than the C–O–S typical of a sulphate ester.

We don't know exactly how or at what stage of biodegradation the C—S bond is broken.

Segel: If the sulphur in sulpholipid is extracted by the type of mechanism operative with cysteic acid, it probably enters the sulphate assimilation pathway as sulphite. (Many microbial mutants blocked at sulphate transport, activation, and reduction to sulphite will grow on cysteic acid, but mutants blocked at sulphite reduction cannot.)

Dodgson: C—S bond rupture must also occur in the biodegradation of some detergents containing sulphonate groups. For example, Thysse & Wanders (1974) have evidence that the initial steps in the biodegradation of alkane sulphonates by certain pseudomonads involve hydroxylation of the carbon atom bearing the sulphonate group followed by hydrolysis to yield bisulphate and the appropriate long-chain aldehyde.

Segel: Dr Kelly, how much tetrathionate exists in ordinary surface soil? *Penicillium notatum,* which is probably typical of aerobic filamentous fungi, possesses a membrane transport system that is specific for tetrathionate and just as active as a sulphate permease. Its K_m is about 10^{-5}M; the V_{max} is 1.5 μmol/g dry weight cells (Tweedie & Segel 1970).

Kelly: I don't know the answer. Some measurements were made of polythionate levels in ocean water samples (Tilton 1968, Tuttle & Jannasch 1972), and it has been demonstrated, more qualitatively than quantitatively, in soils. It is fairly easy to find soil pseudomonads which happily oxidize thiosulphate to tetrathionate, but we don't know why they do this since there is apparently no energy gained metabolically and no effect on growth.

Tetrathionate is commonly used by *Proteus* and some other Enterobacteriaceae as a respiratory electron acceptor, so it must have occurred in the environment for a very long time.

Brierley: Why is aluminium jarosite not reduced by sulphate-reducing microorganisms when ammonium and potassium jarosites are?

Kelly: I don't know. This suggests that the organism *(Desulfovibrio)* recognizes not only that the substrate contains sulphate but contains iron as well, which is odd, and presumably requires a correct spatial relation of the Fe and SO_4 groups.

Again, we don't know how the organism attacks something like a jarosite, which is highly insoluble, and may consequently require bacterial attachment.

Whatley: Has this recognition of iron/sulphate complex in the minerals anything to do with the delocalization of electrons in iron-sulphur compounds?

Postgate: Desulfovibrios have a special problem with their iron nutrition. These organisms are rich in haemoproteins such as cytochrome c_3 as well as non-haem iron-sulphur proteins, all of which contain large amounts of iron. This iron must be assimilated from an environment where there is free H_2S most of the time. In that circumstance the solubility product of FeS gives an iron concentration of less than 10^{-9} M, so there must be a specific physiological mechanism for recognizing that iron, and incorporating it into iron proteins, from extremely low external concentrations.

Kelly: Perhaps the most interesting things for microbiologists and microbial ecologists to look at in the future in relation to the sulphur cycle are these low oxygen and high sulphide environments, where we are finding new organisms and strange associations of them, and where we find the possibility of simultaneous micro-aerophilic metabolism and possibly nitrate reduction as well. It will be interesting to see the relations and associations between these organisms; for example, the competition between different kinds of sulphur oxidizers in these mixed environments and with mixed substrates. The organisms that predominate there may not necessarily be those that have been studied in the laboratory. The more adaptable mixotrophic and facultative heterotrophic organisms may well be doing all the work in sulphur turnover in these unusual environments.

Whatley: In the Cariaco Trench, where there is an anoxic zone containing H_2S and H_2S is diffusing up to the oxygen-mixing area, sulphate is formed. What happens to it? Does it leak out, in which case the system isn't closed, or does it fall back in again, in the form of microorganisms?

Kelly: It may well do! Except that the dissimilatory process will be generating much more sulphide than required for assimilatory processes. I think the system is 'closed' in the sense of sulphur conservation.

Mudd: I have a question relating to the cycling of sulphur by higher green plants. Wilson et al (1978) recently showed that the leaves of a number of species liberate H_2S into the atmosphere; they estimated that in certain special circumstances, at least, the rate was comparable to the total assimilatory capacity. Is it known to what extent this happens normally, in physiological conditions, and the relative contribution such volatilization of sulphur makes to atmospheric sulphur?

Kelly: In assimilatory pathways, some leaf sulphide is free and some is bound, depending on the system. Are there any figures on how closely these processes are regulated? How much overspill is there, releasing free sulphide? Taking the terrestrial primary production as about 10^{11} tonnes of carbon fixed per annum, and assuming a minimum 0.1% sulphur content of the dry weight of plants, all coming from sulphate reduction, then if there is a 10% excess production of sulphide which escapes, some 20–30 million tonnes of H_2S per annum could be entering the atmosphere. So it could be a significant contribution, simply through a 10% failure to regulate the assimilatory sulphate reduction process absolutely precisely.

Mudd: In one species that we have studied, *Lemna* (duckweed), which is a higher aquatic microphyte, we were unable to detect volatile sulphur after feeding $^{35}SO_4$. This would limit sulphur volatilization to less than 1% of the amount of sulphur entering terminal products, such as methionine and cysteine.

Kelly: Among yeasts, some produce sulphide and some don't. Presumably this is by assimilatory sulphate reduction.

Dodgson: It has frequently been pointed out that in many aerobic soils about 50% of the sulphur content is present in sulphate ester form — mostly in the complex soil colloids, humic acid and fulvic acid (Freeney et al 1969).

Dr Rose has attempted to achieve the enzymic desulphation of humic acid but without success. It would be interesting to know the extent to which sulphur, bound in humic acid, is being removed from circulation.

Rose: When we said that sulphate was trapped in humic acid, we meant that we were unable to release sulphate ions from humic acid enzymically (Houghton & Rose 1976). This could simply have reflected a necessity for the 'macromolecule' first to be cleaved in some way to make the sulphate groups accessible. Unfortunately, we were unable to test this because we did not possess an appropriate agent, such as a lignase, which might perform such a function in the soil environment. Presumably humic acid must be broken down in nature.

Whatley: If there wasn't anything to degrade it, the world would be full of humic acid!

Dodgson: This is what worries me about the quoted values for soil sulphur. Apparently up to 50% of that sulphur can be attributed to sulphate ester, yet Houghton & Rose (1976) have shown that a number of different types of sulphate ester are quite rapidly desulphated when they are added to soil samples.

References

Board PA 1976 Bacterial sulphate reduction and the anaerobic regulation of atmospheric oxygen. Atmos Environ 10:339-342

Freeney JR, Melville GE, Williams CH 1969 Extraction, chemical nature and properties of soil organic sulphur. J Sci Food Agric 20:440-445

Harwood JL, Nicholls RG 1979 The plant sulpholipid — a major component of the sulphur cycle. Biochem Soc Trans 7:440-447

Houghton C, Rose FA 1976 Liberation of sulfate from sulfate esters by soils. Appl Environ Microbiol 31:969-976

Junge C 1972 Sulphur supplies of atmospheric origin. Ann Agron (Paris) Numéro hors série 1972:235-255

Pfennig N 1978 Syntrophic associations and consortia with photosynthetic bacteria. Abstracts 12th International Congress of Microbiology, Munich (S11.6) p 16

Stapley EO, Starkey RL 1970 Decomposition of cysteic acid and taurine by soil microorganisms. J Gen Microbiol 64:77-84

Thysse GJE, Wanders TH 1974 Initial steps in the degradation of N-alkane-1-sulphonates by *Pseudomonas*. Antonie van Leeuwenhoek J Microbiol Serol 40:25-37

Tilton RC 1968 The distribution and isolation of marine sulfur bacteria. Rev Int Oceanogr Med 9:237-253

Tuttle JH, Jannasch HW 1972 Occurrence and types of Thiobacillus-like bacteria in the sea. Limnol Oceanogr 17:532-543

Tweedie JW, Segel IH 1970 Specificity of transport process for sulphur, selenium, and molybdenum anions by filamentous fungi. Biochim Biophys Acta 196:95-106

Wilson LG, Bressan RA, Filner P 1978 Light-dependent emission of hydrogen sulfide from plants. Plant Physiol 61:184-189

Kinetic and chemical properties of ATP sulphurylase from *Penicillium chrysogenum*

PETER A. SEUBERT, PAMELA A. GRANT, ELIZABETH A. CHRISTIE, JOHN R. FARLEY and IRWIN H. SEGEL

Department of Biochemistry and Biophysics, University of California, Davis, California 95616, USA

Abstract Adenosine triphosphate sulphurylase (ATP: sulphate adenylyltransferase, EC 2.7.7.4) has been purified from the filamentous fungus, *Penicillium chrysogenum,* and characterized physically, kinetically, and chemically. The *P. chrysogenum* enzyme is an octomer (mol. wt. 440 000) composed of eight identical subunits (mol. wt. 55 000). Some physical constants are $S_{20,w} = 13.0 \times 10^{-13}$ s, $D_{20,w} = 2.94 \times 10^{-7}$ $cm^2 \times s^{-1}$, $\bar{v} = 0.733\ cm^3 \times g^{-1}$, $A^{1\%}_{1\ cm} = 8.71$ at 278 nm.

The enzyme catalyses (a) the synthesis of adenosine 5′-phosphosulphate (APS) and $MgPP_i$ from MgATP and SO_4^{2-}, (b) the hydrolysis of MgATP to AMP and $MgPP_i$ in the absence of SO_4^{2-}, (c) $Mg^{32}PP_i$-MgATP exchange in the absence of SO_4^{2-}, (d) molybdolysis of MgATP to AMP and $MgPP_i$, (e) synthesis of MgATP and SO_4^{2-} from APS and $MgPP_i$, and (f) $Mg^{32}PP_i$-MgATP exchange in the presence of SO_4^{2-}. The V_{max} values of reactions (a)-(c) are about 0.10-0.35 μmole × min^{-1} × mg enzyme^{-1}. The V_{max} values of reactions (d)-(f) are about 12-19 μmole × min^{-1} × mg enzyme^{-1}.

The catalytic activity of the enzyme in the direction of APS synthesis is rather low (0.13 unit × mg protein^{-1}, corresponding to an active site turnover number of 7.15 min^{-1}). However, the ATP sulphurylase content of mycelium growing on excess SO_4^{2-} is 0.22 unit × g dry wt.$^{-1}$, which is sufficient to account for the maximum *in vivo* rate of SO_4^{2-} assimilation.

The normal catalytic reaction is *Ordered Bi Bi* with A = MgATP, B = SO_4^{2-}, P = $MgPP_i$, and Q = APS. Several lines of kinetic evidence suggest that the E·MgATP and E·APS complexes isomerize (to E~AMP·$MgPP_i$ and E~AMP·SO_4, respectively) before the second substrate binds.

Chemical modification studies have disclosed the presence of essential arginine, histidine, carboxyl, and tyrosine residues. The latter is rather acidic (pK_a = 7 or less). Nitration of the tyrosine increases the K_m for MgATP without significantly affecting K_{ia} for MgATP or V_{max_f}. This result, and the fact that MgATP plus nitrate protects the enzyme against inactivation by tetranitromethane while MgATP alone does not, suggests that the essential tyrosine plays a role in nucleotide isomerization (perhaps as an adenylyl acceptor).

Sulphur in biology
(Ciba Foundation Symposium 72) p 19-47

INTRODUCTION: REACTION AND PHYSIOLOGICAL ROLES OF THE SULPHATE-ACTIVATING ENZYMES

All forms of life are capable of incorporating inorganic sulphate into organic molecules. The first step in the metabolism of sulphate is catalysed by the 'sulphate-activating' enzyme, ATP sulphurylase (ATP: sulphate adenylyltransferase, EC 2.7.7.4) which forms adenosine 5′-phosphosulphate (APS). In most organisms, APS is converted to 3′-phosphoadenosine 5′-phosphosulphate (PAPS) by the second 'sulphate-activating' enzyme, APS kinase (ATP: adenylylsulphate 3′-phosphotransferase, EC 2.7.1.25). The equilibrium of the ATP sulphurylase reaction lies far to the left. Nevertheless, the overall production of PAPS *in vivo* is promoted by the thermodynamically favourable APS kinase reaction and the hydrolysis of inorganic pyrophosphate, as shown below.

(1) $SO_4^{2-} + MgATP \leftleftharpoons APS + MgPP_i$	$\Delta G' = +11\,000$ cal/mole
(2) $APS + MgATP \rightleftharpoons PAPS + MgADP$	$\Delta G' = -6000$ cal/mole
(3) $MgPP_i + H_2O \rightleftharpoons 2\,P_i + Mg^{2+}$	$\Delta G' = -5000$ cal/mole
Sum (4) $SO_4^{2-} + 2\,MgATP + H_2O \rightleftharpoons PAPS + MgADP + 2\,P_i + Mg^{2+}$	$\Delta G' = 0$ cal/mole

PAPS, the product of the two-step activation sequence, serves as the sulphate donor for the biosynthesis of all known biological sulphate esters. For example, animals use PAPS to produce the acid mucopolysaccharides (chondroitin sulphates, dermatan sulphate, keratan sulphate) and the structurally related anticoagulant, heparin. PAPS is also involved in the biosynthesis of phenol sulphates, steroid sulphates, cerebroside sulphate, serotonin sulphate, ascorbic acid sulphate, luciferyl sulphate, and many different sulphated glycoproteins. Seaweeds produce a variety of sulphated polysaccharides. Higher plants, algae, filamentous fungi, a few red yeasts, and some bacteria synthesize choline *O*-sulphate. Thus, PAPS plays a role in sulphate ester biosynthesis analogous to that of ATP in phosphate ester biosynthesis. But sulphate activation is more than just a prelude to sulphate ester formation. In microorganisms and plants the 'active sulphates' are the first products of a branched biosynthetic ('sulphate assimilation') pathway leading to the formation of the sulphur amino acids (cysteine and methionine), their derivatives (e.g., cysteamine, taurine, glutathione, *S*-adenosylmethionine), and a variety of specialized sulphur compounds (e.g., lipoic acid, thiamine, chloroplast sulpholipid). In *Desulfovibrio desulfuricans* and other 'dissimilatory' sulphate reducers, APS serves an additional role—that of terminal electron acceptor of anaerobic metabolism.

In spite of the ubiquity and importance of the sulphate-activating enzymes, they have been purified from only a few organisms. This paper describes some of the physical, kinetic, and chemical properties of ATP sulphurylase from *Penicillium chrysogenum*.

PHYSICAL PROPERTIES OF ATP SULPHURYLASE FROM *PENICILLIUM CHRYSOGENUM*

The enzyme from *P. chrysogenum* has a molecular weight of about 440 000, as established from velocity sedimentation and gel filtration studies (Tweedie & Segel 1971a). Sodium dodecyl sulphate (SDS) gel electrophoresis reveals a single protein band of mol.wt. about 55 000. Thus, the fungal enzyme appears to be an octomer composed of eight identical subunits. Other physical properties (Tweedie & Segel 1971b) are $S_{20,w} = 13.0 \times 10^{-13}$ s, $D_{20,w} = 2.94 \times 10^{-7} \text{ cm}^2 \times \text{s}^{-1}$, Stokes radius = 7.2 nm, $\bar{v}$ (based on amino acid analysis) = $0.733 \text{ cm}^3 \times \text{g}^{-1}$, and $A^{1\%}_{1\,cm} = 8.71$ at 278 nm; A_{278}/A_{260} is about 2.

CATALYTIC PROPERTIES OF ATP SULPHURYLASE

Only ATP, and to a lesser extent, dATP, serve as effective substrates for the *P. chrysogenum* ATP sulphurylase (Tweedie & Segel 1971a). The true substrate is MgATP; free ATP is an inhibitor competitive with MgATP.

The K_i for free ATP is about the same as the K_{ia} value for MgATP (0.5–0.8 mM), suggesting that the Mg^{2+} is required for catalysis and not for nucleotide binding (Tweedie & Segel 1971b). Other adenine nucleotides such as 5′-AMP, 5′-ADP, 3′, 5′-ADP, α,β-methylene ATP, NAD^+ and $NADP^+$ are good competitive inhibitors with respect to MgATP, whereas ribose-5-phosphate, adenosine, 2′-AMP, 3′-AMP and 5′-adenosine monosulphate are relatively poor inhibitors (Farley et al 1976, 1978). These results suggest that the nucleotide subsite is designed to bind the AMP portion of ATP. In the reverse direction, $MgPP_i$ is the substrate.

The sulphate subsite has a similar high degree of catalytic specificity and lower degree of binding specificity. Only Group VI anions serve as substrates. Selenate forms a relatively stable APSe complex. Molybdate also reacts (Wilson & Bandurski 1958), but the nucleotide product, APMo, spontaneously breaks down to AMP, regenerating MoO_4^{2-} (reaction 5):

$$(5)\quad MoO_4^{2-} + MgATP \rightleftharpoons APMo + MgPP_i$$

$$APMo + H_2O \longrightarrow AMP + MoO_4^{2-}$$

There is, in fact, good reason to believe that APMo never really forms. The molybdolysis reaction provides a convenient way of assaying ATP sulphurylase in partially purified preparations free of ATPase. (The $MgPP_i$ is converted to $2P_i$ + Mg^{2+} with inorganic pyrophosphatase; the P_i is determined colorimetrically.) A number of anions bind to the sulphate subsite without promoting the cleavage of MgATP. Nitrate and chlorate (Shaw & Anderson 1974, Farley et al 1976) are the most potent of these dead-end inhibitors, but formate, thiosulphate and phosphate also bind (Farley et al 1978).

Besides the normal forward and reverse reaction, ATP sulphurylase also catalyses the hydrolysis of MgATP to AMP and $MgPP_i$ in the absence of sulphate (Farley et al 1978), the hydrolysis of APS to AMP and sulphate in the absence of $MgPP_i$ (Reuveny & Filner 1976, Farley et al 1978), and an isotope exchange between $Mg^{32}PP_i$ and MgATP in the presence of sulphate (Shaw & Anderson 1972, 1974, Farley et al 1978). The *P. chrysogenum* enzyme also catalyses a sulphate-independent $Mg^{32}PP_i$-MgATP exchange (Farley et al 1978). This exchange rate is much lower than that observed in the presence of sulphate, but it is comparable to the rate of the normal forward reaction. Reactions involving the synthesis or cleavage of APS are specific for Mg^{2+}. Mn^{2+} or Co^{2+} can substitute for Mg^{2+} in the molybdolysis, the ATP hydrolysis, and the $Mg^{32}PP_i$-MgATP exchange reactions. These results raise the possibility that the enzyme possesses a high affinity Mg^{2+} site involved in APS synthesis and cleavage that is distinct from the

TABLE 1

Maximum velocities of the reactions catalysed by ATP sulphurylase[a]

Reaction	*Maximum velocity ($\mu mole \times min^{-1} \times mg\ enzyme^{-1}$)*
APS synthesis	0.13[b]
ATP synthesis	12.2[c]
ATP hydrolysis	0.097
APS hydrolysis	13.3
Molybdolysis	18.5
PP_i-ATP exchange (with saturating SO_4^{2-})[d]	17.5
PP_i-ATP exchange (with saturating SeO_4^{2-})[d]	8.0
PP_i-ATP exchange (in the absence of SO_4^{2-})	0.35

[a]All rates were calculated by extrapolation to saturating substrate concentrations, and all were determined with the same homogeneous enzyme preparation at 30 °C.
[b]Corresponds to an active site turnover number of 7.15 min^{-1}.
[c]Corresponds to an active site turnover number of 671 min^{-1}.
[d]Sulphate and selenate were varied from 0.5 to 2.0 mM at 0.48 mM (saturating) $^{32}PP_i$, 15 mM-$MgCl_2$, and 5 mM-ATP. No substrate inhibition was observed.

metal-ATP or metal-PP_i site. Table 1 summarizes the maximal velocities of the reactions catalysed by the ATP sulphurylase of *P. chrysogenum*.

Several observations suggest that the hydrolysis reactions are catalysed by ATP sulphurylase, and not by minor contaminants: (a) the products of MgATP hydrolysis are $MgPP_i$ and AMP. Most ATPases produce ADP and P_i; (b) the K_m for MgATP hydrolysis is essentially the same as the K_m for the normal forward reactions; (c) MgATP hydrolysis is inhibited by sulphate analogues (nitrate, chlorate) and by APS; (d) APS hydrolysis is inhibited by ATP; (e) the ratio of maximal rates for MgATP and APS hydrolysis is about the same as the ratio of V_{max_f}/V_{max_r} for the normal catalytic reactions; and (f) APS hydrolysis shows substrate inhibition at high [APS], just as the normal reverse reaction. It is likely that the hydrolyses are suppressed in the presence of the normal cosubstrates. (In the presence of excess $Mg^{32}PP_i$, APS is stoichiometrically converted to MgATP-^{32}P.)

SULPHATE ACTIVATION RATES *IN VIVO* AND *IN VITRO*

One striking feature of the ATP sulphurylase from *Penicillium chrysogenum* is its relatively low catalytic activity in the physiologically important direction of APS synthesis: 0.13 units per mg protein, corresponding to an active site turnover number of only 7.15 min^{-1}. The question naturally arises: is there enough ATP sulphurylase in *P. chrysogenum* to account for the observed growth rate of the organism in medium containing inorganic sulphate as the only sulphur source? If the answer were 'no', we would be forced to conclude that either (a) the *in vitro* assays do not reflect the true catalytic activity of the enzyme *in vivo* (e.g., either the enzyme requires an unknown activator that is destroyed or diluted during extraction and purification, or the characteristics of the enzyme are altered during purification, or only a fraction of the cellular content of the enzyme is recovered by the extraction procedure used); or (b) ATP sulphurylase is not the only enzyme involved in the assimilation of inorganic sulphate into the metabolic stream. In fact, an ADP sulphurylase has been demonstrated in yeast (Robbins & Lipmann 1958, Adams & Nicholas 1972) and in spinach leaves (Burnell & Anderson 1973). However, we have been unable to demonstrate any such activity in cell-free extracts of *P. chrysogenum*.

Some experiments designed to estimate the *in vivo* rate of sulphate activation in *P. chrysogenum* were performed (Farley et al 1979). The rate was determined to be 0.19 ± 0.09 μmoles × g dry wt.$^{-1}$ × min^{-1} by the following methods. (a) The maximum growth of the organism in synthetic medium was a linear function of the initial Na_2SO_4 concentration between 0 and 8 × 10^{-4} M-Na_2SO_4. The growth yield was 1.64 × 10^{-2} g dry wt. mycelium per μmole of added sulphate, corresponding to a minimum sulphur requirement of 61 μmoles per g dry wt. In these conditions (limiting sulphate) the minimum doubling time of *P. chrysogenum* in submerged culture was about 3.8 hours, corresponding to a maximum exponential growth rate constant of about 3.0×10^{-3} min^{-1}. If all the sulphur in this mycelium passed through APS, the rate of sulphate activation *in vivo* must have been 0.183 μmoles × min^{-1} × g dry $wt.^{-1}$ (b) In the presence of excess $^{35}SO_4^{2-}$, the total organic ^{35}S produced varied with the mycelial growth rate. However, until the culture approached maximum density, the product of (growth rate constant) × (organic ^{35}S content) was nearly constant at 0.24–0.28 μmoles × min^{-1} × g dry $wt.^{-1}$ (c) A sulphur-starved mycelium pulsed with 10^{-4} M $^{35}SO_4^{2-}$ produced organic ^{35}S at a rate of about 0.10–0.20 μmoles × min^{-1} × g dry $wt.^{-1}$ under conditions where the internal concentrations of ATP (Chandler & Segel 1978) would permit ATP sulphurylase to operate at about 70% of its V_{max}.

Cell-free extracts of *P. chrysogenum* growing rapidly on excess sulphate contained 0.22 unit of ATP sulphurylase per g dry wt. Thus, in spite of the relatively low specific activity of homogeneous ATP sulphurylase, the mycelial content of the enzyme is sufficient to account for the observed growth rate of the organism on inorganic sulphate as the sole sulphur source.

KINETIC MECHANISM OF ATP SULPHURYLASE

Plots of $1/v$ versus 1/[MgATP] at several different fixed concentrations of SO_4^{2-}, and of $1/v$ versus $1/[SO_4^{2-}]$ at several different fixed concentrations of MgATP (Fig. 1), were linear (Farley et al 1976). Each family of initial velocity plots intersected to the left of the $1/v$ axis, above the horizontal axis. All slope and $1/v$-axis intercept replots were linear with finite intercepts. Thus, the forward reaction is clearly 'sequential', i.e., ordered or random, but not ping-pong (Segel 1975). The product, APS, was competitive with MgATP and a mixed-type inhibitor with respect to SO_4^{2-}. Nitrate was competitive with SO_4^{2-} and uncompetitive with respect to MgATP. All replots were linear with finite intercepts. The dead-end inhibitor, AMP, behaved qualitatively the same as APS. These results are consistent with a *steady-state ordered* kinetic mechanism in which MgATP adds to the enzyme before SO_4^{2-}, and $MgPP_i$ dissociates before APS.

In the reverse direction, plots of $1/v$ versus 1/[APS] at different fixed $[MgPP_i]$ (Fig. 2a) were linear and intersected to the left of the $1/v$ axis, above the horizontal axis. The $1/v$-axis intercept replot was linear with finite intercepts, but the slope replot had a zero intercept. The plots $1/v$ versus $1/[MgPP_i]$ at different fixed [APS] (Fig. 2b) intersected on the $1/v$-axis. The slope replot was linear. These initial velocity patterns are diagnostic; they characterize a *rapid equilibrium ordered* sequence in which APS adds to the enzyme before $MgPP_i$ (see p 320–329 of Segel 1975). MgATP was competitive with both APS and $MgPP_i$, a pattern consistent

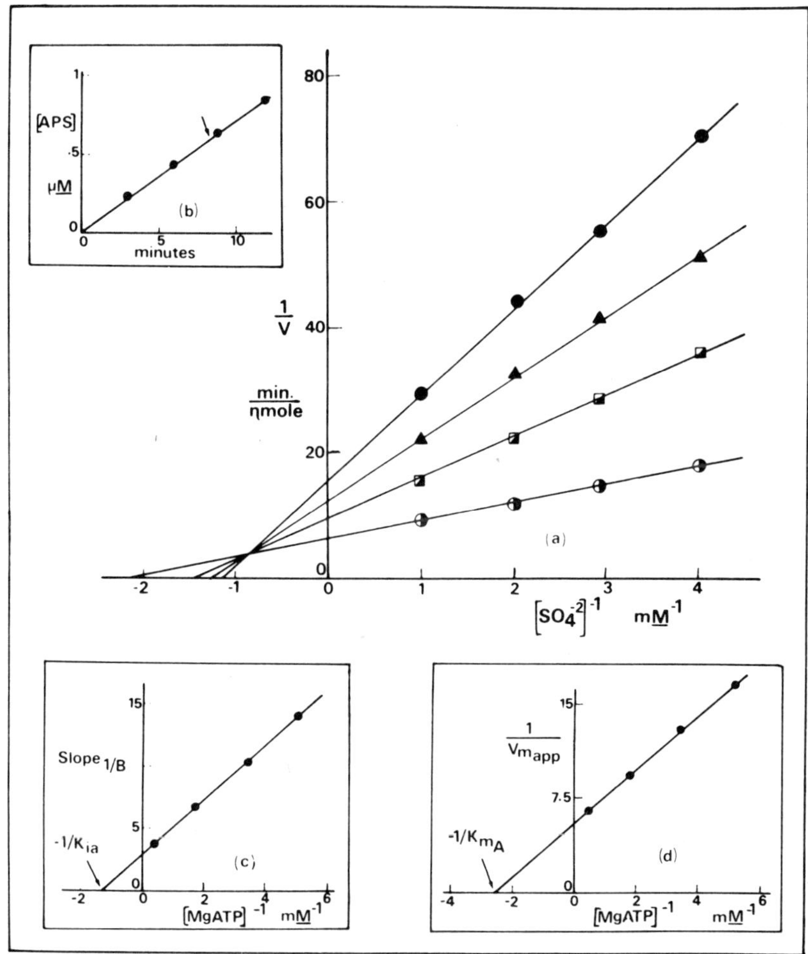

FIG. 1. Initial velocity in the absence of products. (a) Reciprocal of forward velocity versus $1/[SO_4^{2-}]$ at 8 mM total Mg^{2+} and the following fixed concentrations of ATP: 2.36 mM, ◑; 0.59 mM, ◪; 0.295 mM, ▲; 0.197 mM, ●. At the above concentrations essentially all the ATP was present as MgATP. (b) Linearity of reaction with time at maximum substrate concentrations used in (a). *Arrow* indicates the point used to calculate the velocity used in (a). (c) Slope replot. (d) $1/v$ axis intercept replot.

with the rapid equilibrium nature of the reverse reaction. Some of the kinetic constants obtained in these studies are summarized in Table 2.

The relationship between the kinetic constants of an enzyme and the equilibrium constant of the reaction catalysed is given by the Haldane equation. For an ordered Bi Bi reaction, the Haldane equation is:

$$K_{eq} = \frac{V_{max_f} K_{iq} K_{m_P}}{V_{max_r} K_{ia} K_{m_B}} \qquad (1)$$

We can now see how ATP sulphurylase has adapted to the extremely unfavourable equilibrium constant of the physiological reaction. The enzyme has sacrificed affinity characteristics in order to maximize the catalytic velocity in the forward direction. That is, the substrates in the forward direction have K_m's and K_i's in the millimolar range, while the products bind 1000-fold tighter (K_m's and K_i's are in the micromolar range). As a result, the ratio V_{max_f}/V_{max_r}, while only about 0.01, is about a million times greater than K_{eq}. If all ligands had the same affinity for the enzyme, the V_{max_f} would be only one hundred millionth of V_{max_r}.

TABLE 2

Kinetic constants of ATP sulphurylase[a]

Constant	*Value*[b]
K_{m_A} (K_m for MgATP)	0.39 mM
K_{m_B} (K_m for SO_4^{2-})	0.50 mM
K_{m_P} (K_m for $MgPP_i$)	0.65 μM
K_{iq} (K_i for APS)	0.3-1.0 μM
K_{IQ} (K_i for APS as a substrate inhibitor)	0.3 mM
K_{ia} (K_i for MgATP)	0.60-0.71 mM
K_{IA} (K_i for free ATP)	0.50-0.80 mM
$K_{i_{AMP}}$	0.55-0.80 mM
$K_{i_{MgADP}}$	0.65 mM
$K_{i_{NAD^+}}$	0.25 mM
$K_{i_{NADP^+}}$	0.18 mM
$K_{i_{NO_3^-}}$	0.18-0.24 mM
$K_{i_{ClO_3^-}}$	0.13-0.21 mM
$K_{i_{formate}}$	1.2 mM
$K_{i_{P_i}}$	8.7 mM
$K_{i_{MgP_i}}$ (in reverse reaction)	1.1-1.53 mM
V_{max_f}/V_{max_r} (physiological substrates)	0.011
$V_{max_f\,(MoO_4^{2-})}/V_{max_f\,(SO_4^{2-})}$	142
K_{eq} (pH 8.0, 30°C, experimental)	2.5×10^{-9}
K_{eq} (from Haldane equation)	6.0×10^{-9}-3.3×10^{-8}

[a]All kinetic constants were determined from appropriate replots of initial velocity data.
[b]Values for several of the constants were obtained by more than one method. The range reported indicates the maximum and minimum values obtained.

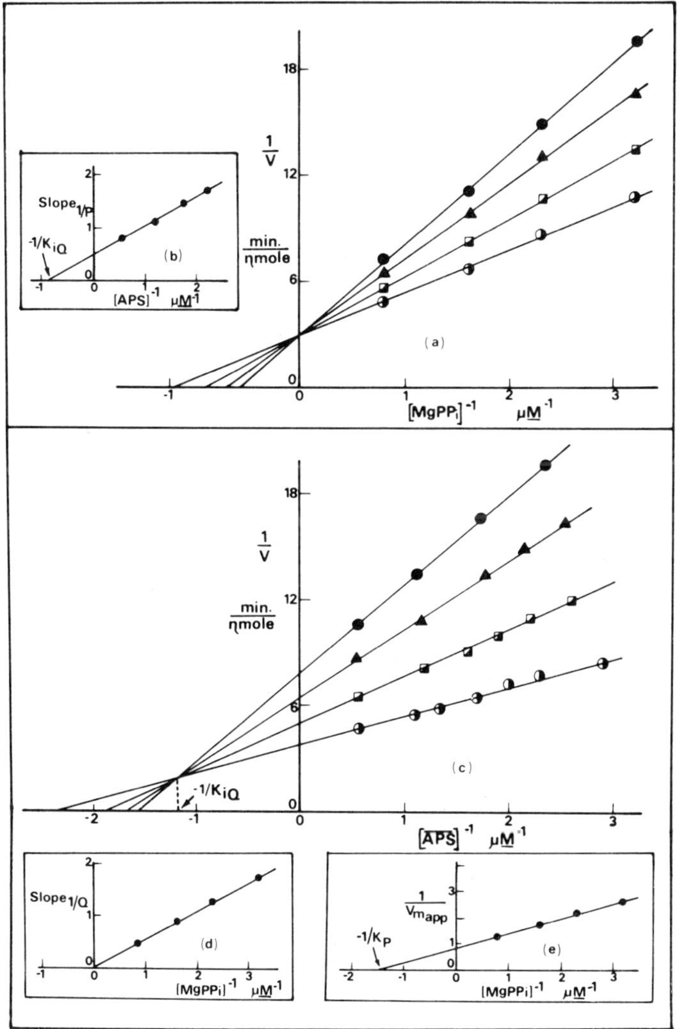

FIG. 2. Initial velocity of reverse reaction in the absence of products. (a) Reciprocal reverse velocity versus $1/[1MgPP_i]$ at 10 mM total Mg^{2+} and the following fixed concentrations of APS: 2 μM, ◑; 1 μM, ◪; 0.67 μM, ▲; 0.5 μM, ●. (b) Slope replot. (c) Reciprocal reverse velocity versus 1/average [APS] (averaged over the course of the reaction). Total Mg^{2+}, 10 mM was used with the following fixed concentrations of total PP_i: 8.0 μM, ◑; 4.0 μM, ◪; 2.8 μM, ▲; 2.0 μM, ●. (Corresponding to $MgPP_i$ concentrations of 1.25, 0.624, 0.437 and 0.312 μM.) Rationale for using average [APS] is given in Segel 1975, p 57-59. (d) Slope replot. (e) $1/v$ axis intercept replot. At the concentrations of APS used, inhibition by contaminating AMP is negligible.

Given that the kinetic mechanism is obligately ordered,

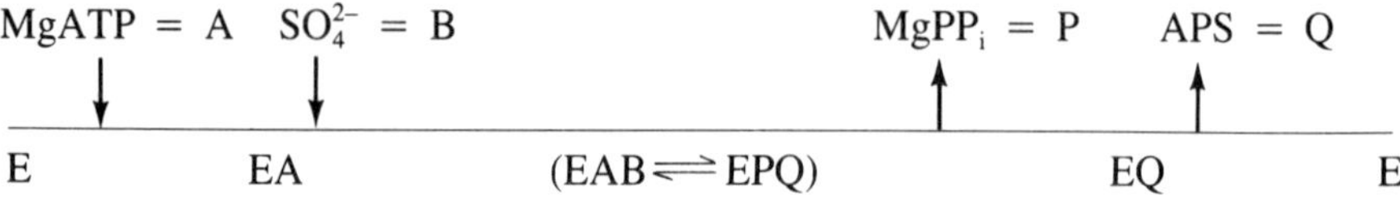

one may wonder *why* MgATP or APS must bind first. There are three possible explanations: (a) the binding of MgATP or APS induces a conformational change in the enzyme which exposes or forms the sulphate (or $MgPP_i$) binding site. (b) Sulphate utilizes enzyme-bound MgATP as part of its binding site; $MgPP_i$ utilizes APS as part of its binding site. (c) The binding site for the 'second' substrate does not appear until after catalytic cleavage of the α-β pyrophosphate or phosphosulphate bond of the 'first' substrate.

If either (a) or (b) is the reason for the ordered sequence, it seems reasonable that a compound that resembles substrate A (or Q) should induce the required conformational change (or provide the groups) to bind B (or P). Thus, we might expect free ATP, CaATP, CrATP, or Mg-α,β-methylene ATP to induce sulphate binding. (These inhibitors bind as well as MgATP to the free enzyme.) If this occurred, a dead-end EIB complex would form (where I = a non-cleavable MgATP analogue). The versatile tool of steady-state kinetics can be used to disclose the existence of such a complex. If an EIB complex formed, substrate inhibition by sulphate should be observed. The velocity equation for this situation is given on p 803 of Segel (1975). No such substrate inhibition was observed experimentally.

As an added technique, the inhibition patterns of nitrate or chlorate (= X) in the presence of a MgATP analogue (= I) were determined. The equations describing the velocity behaviour when both types of inhibitors are present simultaneously are:

$$\frac{v}{V_{max}} = \frac{[A]}{K_{m_A}\left(1 + \frac{K_{ia}K_{m_B}}{K_{m_A}[B]}\right)\left(1 + \frac{[I]}{K_i} + \frac{[I]\,[X]}{K_i K_x'}\right) + [A]\left[1 + \frac{K_{m_B}}{[B]}\left(1 + \frac{[X]}{K_X}\right)\right]} \quad (2)$$

$$\frac{v}{V_{max}} = \frac{[B]}{K_{m_B}\left[\left(1 + \frac{[X]}{K_X}\right) + \frac{K_{ia}}{[A]}\left(1 + \frac{[I]}{K_i} + \frac{[I]\,[X]}{K_i K_X'}\right)\right] + [B]\left[1 + \frac{K_{m_A}}{[A]}\left(1 + \frac{[I]}{K_i} + \frac{[I][X]}{K_i K_X'}\right)\right]} \quad (3)$$

Thus, if the MgATP analogue induces the binding of nitrate to the sulphate site (as MgATP does), nitrate will act as a mixed-type inhibitor with respect to MgATP. If

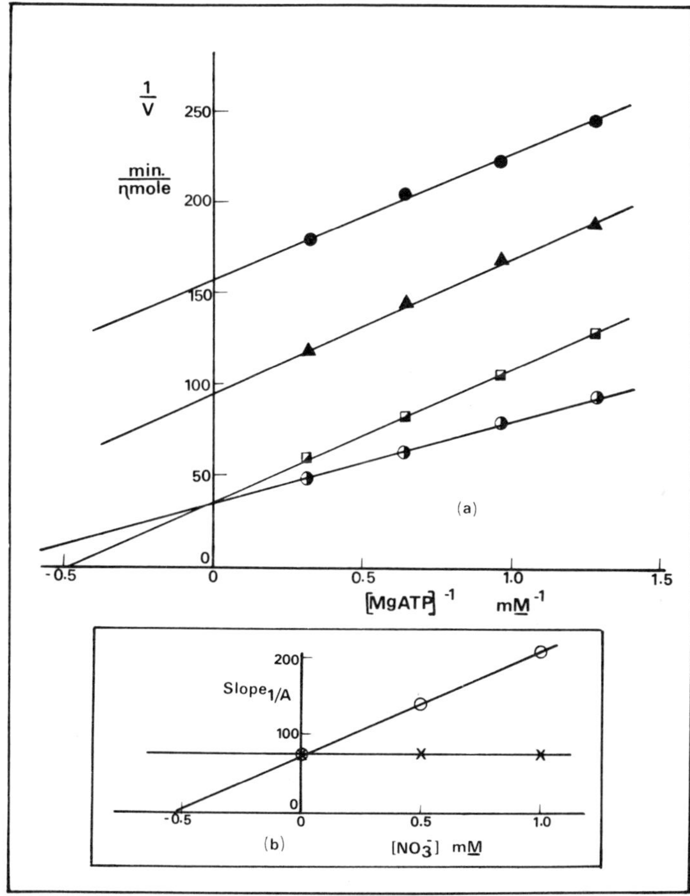

FIG. 3. Multiple dead-end inhibition by NO_3^- and Mg-α,β-methylene ATP. (a)Reciprocal forward velocity versus 1/[MgATP] with 8 mM total Mg^{2+} and the concentration of SO_4^{2-} fixed at 0.5 mM. The following fixed concentrations of inhibitors were used: no inhibitor (◑), 1.05 mM α, β-methylene ATP (◪), 1.05 mM α, β-methylene ATP plus 0.5 M NO_3^- (▲), and 1.05 mM α,β-methylene ATP plus 1 mM NO_3^- (●), (b) Replot of slope versus [NO_3^-] for the lines in (a) that contain α,β-methylene ATP (×). The theoretical slope (calculated from Equation 1) is also plotted as a function of [NO_3^-] assuming $K'_X = K_x$ = 0.2 mM(○).

no EIX complex forms, the $[I][X]/K_iK'_X$ term will be absent from the factor multiplying K_{mA} in equation (2) and nitrate will remain an uncompetitive inhibitor. Similarly, if EIX forms, nitrate will become a mixed-type inhibitor with respect to sulphate. If EIX does not form, the $[I][X]/K_iK'_X$ terms in both parts of equation (3) will be absent; nitrate remains a competitive inhibitor with respect to sulphate.

Experimentally, no kinetic evidence for the formation of an EIX complex was obtained (Fig. 3). The obvious conclusions are that *either* (a) the structural requirement for the metal-nucleotide is so stringent that only MgATP can induce the conformational change that forms the sulphate site, *or* (b) the mere binding of a metal-nucleotide is not sufficient to promote the binding of sulphate. The latter implies that the 'first' substrate must be catalytically cleaved before the 'second' site appears. (CaATP, CrATP and Mg-α,β-methylene ATP do not serve as alternative substrates – they are true dead-end inhibitors.)

EQUILIBRIUM BINDING STUDIES

Direct equilibrium binding measurements (Farley et al 1978) support the ordered reaction sequence deduced from the steady-state kinetics. The enzyme will bind [^{14}C]ATP in the presence or absence of Mg^{2+}, but no binding of $^{35}SO_4^{2-}$ could be detected in the presence or absence of Mg^{2+}, or free ATP. Concentrations of $^{35}SO_4^{2-}$ up to 10 times K_{m_B} were tested. Analyses of the binding data by reciprocal and Scatchard plots revealed that at saturation, 7.6–8.4 moles of [^{14}C]ATP were bound per 440 000 g enzyme. The *n* value of approximately 8 is consistent with the SDS and native gel electrophoresis studies which suggest that the enzyme is composed of eight identical subunits of mol.wt. 55 000. The experimental K_S for [^{14}C]ATP was 0.50–0.53 mM. This value is in good agreement with the values for K_{ia} and K_{IA} determined from kinetic measurements (Table 2). It should be noted that the ^{14}C-labelled material bound in the presence of Mg^{2+} was probably a mixture of ^{14}C-labelled MgATP and [^{14}C]AMP. (The latter derived from hydrolysis of the former.) The Scatchard plots were linear because both compounds have essentially the same K_S. If [^{14}C]AMP did not bind to the enzyme, the plots would be non-linear (Builder & Segel 1978).

The binding of AP^{35}S to the enzyme could be demonstrated qualitatively. But, in prolonged incubations, the AP^{35}S was hydrolysed and the amount of enzyme-bound ^{35}S decreased. No binding of $^{32}PP_i$ could be demonstrated in the presence or absence of Mg^{2+}. Concentrations of $^{32}PP_i$ up to 20-times K_{m_P} were tested.

SPECULATIONS ON THE MECHANISM OF THE REACTION

Several lines of evidence suggest that the unireactant cleavage of MgATP and APS are part of the normal reaction sequence: (a) the multiple inhibition studies clearly established that the sulphate subsite can be occupied only if the first substrate to bind is a reactive nucleotide. (b) The homogeneous enzyme catalyses the hydrolysis of MgATP and of APS; APS hydrolysis in the absence of $MgPP_i$ is as rapid as the normal reverse reaction. (c) The homogeneous enzyme catalyses an $Mg^{32}PP_i$–MgATP exchange in the absence of sulphate. Finally, (d) the large difference between the values of K_{ia} (ca. 1 mM) and K_{iq} (ca. 1 μM) is consistent with the 'EQ' complex being predominantly E~AMP·SO_4. The reasoning is as follows. If an E~AMP-activated complex is involved in the reaction mechanism it means that the binding of the nucleotides is a two-step phenomenon, involving both association and cleavage. The two-stage MgATP binding reaction can be written as

$$E + A \underset{}{\overset{K_A}{\rightleftharpoons}} EA \underset{}{\overset{K_A'}{\rightleftharpoons}} EA'$$

where EA′ represents the enzyme~AMP·$MgPP_i$ complex. K_{ia} may then be defined as $[E][A]/[EA]_t = [E][A]/([EA] + [EA'])$, so that $K_{ia} = K_A/(1 + 1/K_A')$, where $K_A = [E][A]/[EA]$ and $K_A' = [EA]/[EA']$. If (a) $K_A \simeq 10^{-3}$ M (this is the observed dissociation constant for all non-reactive MgATP analogues tested), (b) ΔG′ for the hydrolysis of the PP_i-AMP bond of the E·MgATP complex is approximately −10 000 cal/mol, and (c) ΔG′ for hydrolysis of the E~AMP bond of the E~AMP·$MgPP_i$ complex is approximately −13 000 cal/mol, then ΔG′ for the formation of E~AMP·$MgPP_i$ is about +3000 cal/mol. Therefore, K_A' is about $10^{2.2}$, and K_{ia} is about 10^{-3} M. The calculation implies that EA_t is mostly in the EA form (Michaelis complex). In considering the reverse reaction, we can assume that $K_Q \simeq 10^{-3}$ M (i.e., that APS associates with the enzyme as any other nucleotide). The ΔG′ for hydrolysis of the phophosulphate anhydride bond of APS is approximately −18 000 cal/mol. If ΔG′ for hydrolysis of the E~AMP bond in the activated complex (in this case, E~AMP·SO_4^{2-}) is about −13 000 cal/mol, then ΔG′ for the formation of E~AMP·SO_4^{2-} is about −5000 cal/mol, and $K_Q' = [EQ]/[EQ']$ is $10^{-3.66}$. Therefore, $K_{iq} = K_Q/(1 + 1/K_Q') \simeq 0.22$ μM, and EQ_t is composed mainly of the E~AMP·SO_4^{2-} complex.

Further evidence for the isomerization of EA is provided by the inconsistent value of $k_{-1_{calc}}$ (the unimolecular rate constant for EA dissociation calculated from $V_{max_f}K_{ia}/K_{m_A}$) (Segel 1975, p 588). The calculated value (0.25 unit/mg of protein) is significantly less than (and therefore inconsistent with) V_{max_r}. V_{max_r} can be equal to or less than k_{-1}, but not greater. Thus, $k_{-1_{calc}} \neq k_{-1}$ and the rate constants composing K_{ia}, K_{m_A} and V_{max_f} are not those of a simple ordered Bi Bi reaction.

Fig. 4 illustrates our current ideas about the number and nature of the subsites on ATP sulphurylase. Fig. 5 summarizes the suggested reaction sequence and kinetic mechanism.

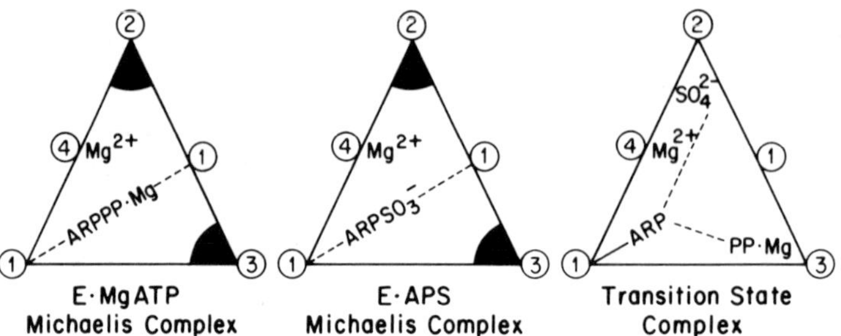

FIG. 4. Proposed schematic model of ATP sulphurylase. The nucleotide substrate is bound initially at subsite 1 which recognizes the AMP moiety. Catalytic cleavage of the α-β pyrophosphate or phosphosulphate anhydride bond results in the formation of E~AMP at part of subsite 1 and a conformational change in the enzyme that exposes subsites 2 and 3. Subsite 2 accepts SO_4^{2-} (or the sulphate portion of APS). Subsite 3 accepts $MgPP_i$ (or the $MgPP_i$ portion of MgATP). Water may be able to occupy subsites 2 and 3 or, possibly, water has no actual site but occupancy of subsites 2 or 3 sterically hinders the attack of water on E~AMP. Subsite 4 is highly specific for Mg^{2+} (required for APS cleavage/synthesis). Ca^{2+} can bind to subsite 4, but with much less affinity than Mg^{2+}.

CHEMICAL MODIFICATION STUDIES DESIGNED TO DISCLOSE ESSENTIAL AMINO ACID RESIDUES

The *P. chrysogenum* enzyme was subjected to a number of specific chemical modifications designed to identify amino acid residues that are responsible for its functional integrity, including residues that may be at the active site (Means & Feeney 1971, Glazer et al 1975). The results are summarized below.

Effect of thiol reagents

ATP sulphurylase of *P. chrysogenum* is insensitive to *p*-chloromercuribenzoate (PCMB), iodoacetate, iodoacetamide and 5,5′-dithiobis(2-nitrobenzoic acid) (DTNB). In order to assess the extent of actual sulphydryl modification, the enzyme was incubated with DTNB and the increase in absorbance at 412 nm was monitored. Although the reaction proceeded to completion until eight sulphydryls per 440 000 daltons were modified, no change in activity was detectable. The addition of Mg^{2+}, ATP, or MgATP to the incubation mixture had no effect on the rate or extent of the reaction.

Effect of 2-hydroxy-5-nitrobenzyl bromide (a 'tryptophan-specific' reagent)

No effect on ATP sulphurylase activity was observed after incubating the enzyme with the water-soluble sulphonium salt of Koshland's reagent. A spectrophotomet-

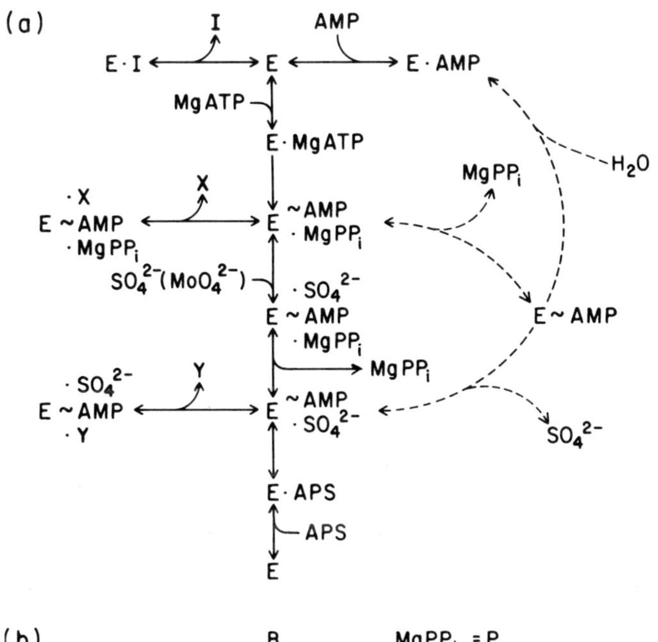

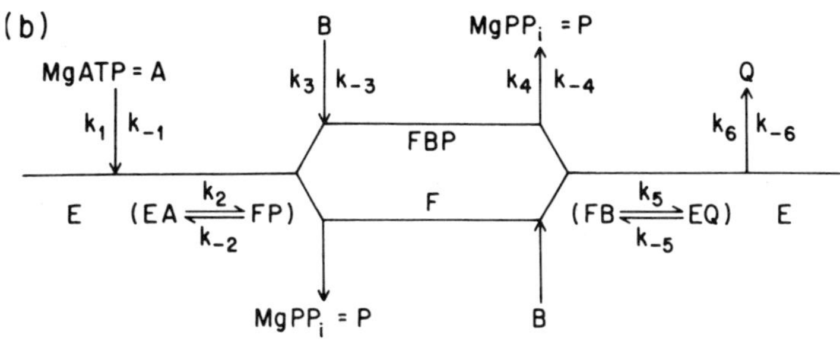

FIG. 5. (a) Reactions catalysed by ATP sulphurylase. I represents a non-reactive MgATP analogue (e.g., free ATP, CrATP, Mg-α,β-methylene ATP). X represents a non-reactive sulphate analogue (e.g., chlorate, nitrate, formate, phosphate). The E~AMP·$MgPP_i$·X complex is inactive in the hydrolysis and exchange reactions. Y represents a non-reactive $MgPP_i$ analogue (e.g., MgP_i). The rate-limiting step of the normal forward reaction must be the synthesis or release of APS. (The maximum $Mg^{32}PP_i$-MgATP exchange rate in the presence of SO_4^{2-} is over 100 times faster than the V_{max} of APS synthesis. This places the rate-limiting step for the overall forward reaction somewhere beyond $MgPP_i$ release.) The value of k_4 calculated from $V_{max_r}K_{iq}/K_{m_Q}$ (122 units × mg enzyme^{-1}) is considerably larger than V_{max_f}. This suggests that the isomerization reaction E~AMP·SO_4^{2-} → E·APS, rather than APS dissociation, limits the rate of the normal forward reaction. The rate-limiting step of the normal reverse reaction (MgATP synthesis) may be sulphate release from E~AMP·$MgPP_i$·SO_4^{2-} or isomerization of this central complex. (The reverse reaction appears to be *rapid equilibrium ordered*.) The hydrolysis of APS and of MgATP proceed via the same E~AMP complex. Yet, the V_{max} of APS hydrolysis is 100 times greater than the V_{max} of MgATP hydrolysis. This suggests that the rate-limiting step in MgATP hydrolysis (and in the $Mg^{32}PP_i$-MgATP exchange reaction in the absence of SO_4^{2-}) is the release of $MgPP_i$ from the E~AMP·$MgPP_i$ complex. (b) Proposed schematic representation of the reactions catalysed by ATP sulphurylase. If B = SO_4^{2-}, the upper route is much faster, yielding Q = APS. In the absence of sulphate, the lower route with B = H_2O may predominate with Q = AMP.

ric analysis after overnight dialysis revealed that 3.9 tryptophan residues per subunit were actually modified. In the presence of 7 M-urea, 4.1 tryptophan residues were modified per subunit. Our earlier amino acid analysis showed that the enzyme contains only four or five tryptophan residues per subunit (Tweedie & Segel 1971a).

Effect of 'lysine-specific' reagents

There was no discernible effect on ATP sulphurylase activity after incubating the enzyme with various chemical reagents regarded as specific for the modification of free amino groups, such as pyridoxal, pyridoxal-phosphate, trinitrobenzene sulphonic acid (TNBS), or formaldehyde, with or without subsequent $NaBH_4$ reduction.

Effect of 'arginine-specific' reagents

Incubation of ATP sulphurylase with diacetyl (2,3-butanedione) in the presence of sodium borate resulted in a first-order inactivation. APS (10 μM) and MgATP plus NO_3^- (5 mM each) protected the enzyme from modification. Nitrate alone did not protect; MgATP alone was much less effective than MgATP plus NO_3^-. A replot of the log of the reciprocal of the half-life for inactivation versus the log of the diacetyl concentration had a slope of 0.94, indicating that the modification of about one arginine residue per active centre is sufficient to account for the observed decrease in activity. The modification of the arginine residue reduces the $V_{max_{app}}$ of the reaction but has no effect on the apparent K_m for MgATP. Chemical modification with diacetyl plus borate inhibits all the catalytic activities of ATP sulphurylase at the same rate and to the same extent. However, ATP *binding* was unaffected (n = 8; K_S = 0.5 mM).

Effect of 'histidine-specific' reagents

ATP sulphurylase activity was unaffected by incubation with diethylpyrocarbonate (ethyoxyformic anhydride, DEP). The titration curve for modification, as reflected by the change in absorbance at 240 nm, levelled off at four histidines per subunit, implying that the other 10 residues per subunit were inaccessible to this reagent. However, photo-oxidation with methylene blue inactivated all activities except ATP binding. The inactivation reaction followed pseudo-first-order kinetics, was light dependent, and could be significantly retarded by 10 μM-APS. The kinetics of the inactivation indicate that the modification of only 0.85 residues per active centre could account for the observed inhibition. The histidine modification reduced the $V_{max_{app}}$ of the molybdolysis but had no effect on the K_m for MgATP. Other studies revealed no change in the apparent value of K_i for AMP (consistent with the lack of effect on binding). The inactivation reaction is strongly pH-dependent with an apparent pK of about 6.8 to 6.9, a value that is typical of histidine residues. Other possible photo-oxidizable residues (such as methionine) are not expected to have a pK_a of 6.8–6.9. Tyrosine residues were unaffected (see below).

To assess the difference between modification by DEP and by methylene blue (both of which are relatively specific for histidine residues) ATP sulphurylase that had been treated with methylene blue was dialysed and then treated with DEP. The change in absorbance at 240 nm indicated that only 0.3 reactive histidine residues per subunit remained out of the four that would normally have been accessible to DEP. ATP sulphurylase that had been first modified with DEP (with no apparent loss of activity) could be inactivated by subsequent treatment with methylene blue. The results suggest that the essential histidine is buried in a hydrophobic pocket, inaccessible to DEP, but accessible to methylene blue.

Effect of tetranitromethane

Treatment of ATP sulphurylase with tetranitromethane (TNM) resulted in *partial inhibition* when assayed with non-saturating substrates. Inactivation followed pseudo-first-order kinetics up to about 20% remaining activity and levelled off at about 10% remaining activity (Fig. 6). The enzyme could be protected from modification by APS (10 μM) or MgATP plus NO_3^- (5 mM each) but not by free ATP or MgATP alone. The kinetics of the inactivation (Fig. 7a) revealed that only 0.94 tyrosyl residue per active centre need be modified to account for the observed decrease in activity. In contrast to the arginine and histidine-specific reagents, TNM modification increased the K_m for MgATP from 0.5 mM to 4.0 mM, but had little effect on V_{max} of the molybdolysis reaction (Fig. 7b) or K_S for [^{14}C]ATP binding.

The actual extent of modification of tyrosine residues with TNM was measured by the change in the absorbance at 428 nm after overnight dialysis to remove nitroformate from the reaction mixture. (TNM nitrates the ring of tyrosine, not the phenolic-OH group whose ionization can be monitored.) Prolonged incubation of the enzyme with excess TNM modified only three tyrosine residues out of a total of 15 present (Tweedie & Segel 1971a). Incubation of the enzyme for shorter times with lower levels of TNM resulted in the nitration of only one tyrosine per residue. The resulting nitrotyrosine had a pK_a of about 6.0. When all three accessible tyrosines were modified, the nitrotyrosyl–enzyme titration curve exhibited two pK_a's, one at about 6.0, which was observed before, and another at pH 7.7, corresponding to the ionization of the two more slowly reacting residues.

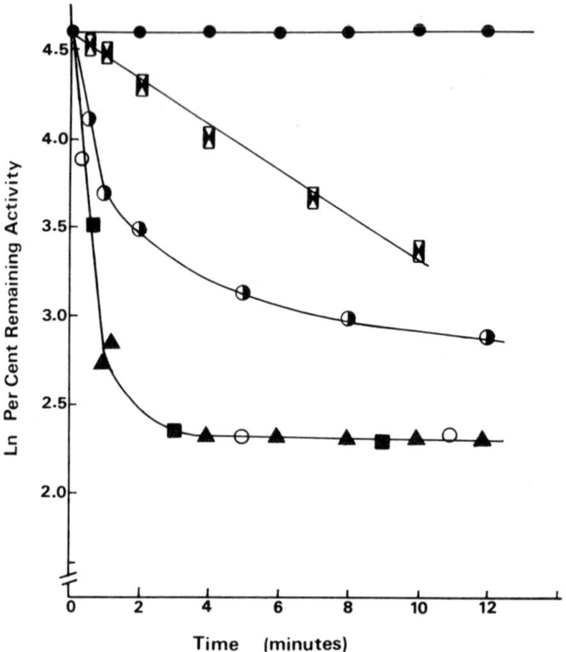

FIG. 6. Modification of ATP sulphurylase with tetranitromethane: Ln percentage remaining activity (SO_4^{2-}-dependent $Mg^{32}PP_i$-MgATP exchange) versus time. ATP sulphurylase was incubated at 0.5 mg/ml in 0.25 ml 80 mM-Tris buffer, pH 8.0, containing: no additions, ●; 64 μM-TNM, ▲; 64 μM-TNM plus 8 mM Mg^{2+}, ■; 64 μM-TNM plus 6.5 mM ATP, ○; 64 μM-TNM plus 6.5 mM-MgATP plus 5 mM-NO_3^-,◑; and 64 μM-TNM plus 9.6 μM-APS,⊠. The reactions were quenched by dilution and assayed for remaining SO_4^{2-}-dependent $Mg^{32}PP_i$-MgATP exchange activity. MgATP behaved identically to free ATP or free Mg^{2+} (no protection). The plateau at about 2.3 on the ordinate represents about 10% remaining activity. The same plateau was observed with 128 μM-TNM.

The pH profiles for the rates of inactivation by TNM (Fig. 8) showed that the pK_a of the essential tyrosine in the native enzyme is <7.0 even before modification. Other experiments established that photo-oxidation with methylene blue did not destroy the three TNM-reactive tyrosines. Thus, it is quite certain that both a histidine and a tyrosine (in addition to an arginine) are important to the functional integrity of the enzyme. (TNM oxidizes the free SH groups, but, as described earlier, these residues do not seem to play any role in the catalytic reaction.)

Effect of N-acetylimidazole

In order to confirm the presence of an essential tyrosine we reacted the enzyme with *N*-acetylimidazole (NAI). This reagent inhibited both the molybdolysis and

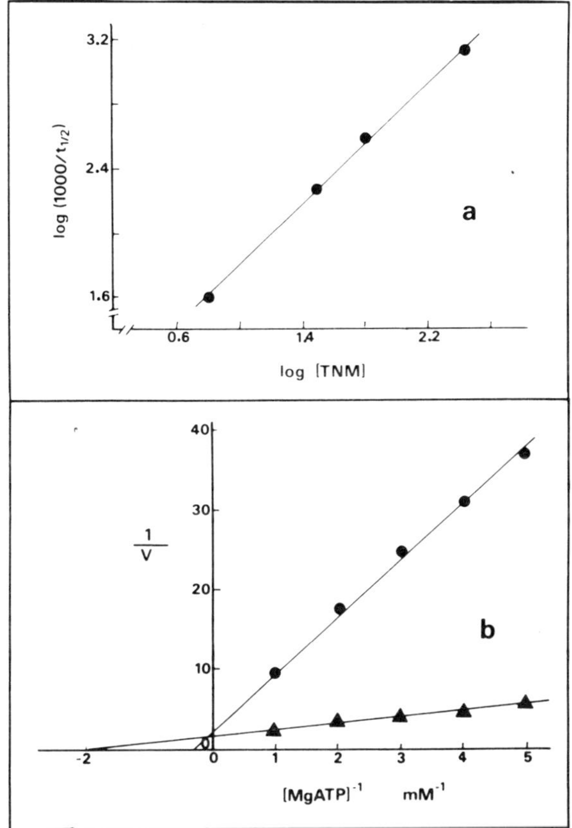

FIG. 7. Reaction kinetics of tetranitromethane (TNM) modification. (a) Log ($10^3/t_{1/2}$) versus log [TNM], where $t_{1/2}$ is the half-life for SO_4^{2-}-dependent $Mg^{32}PP_i$-MgATP exchange activity. ATP sulphurylase was incubated at 0.49 mg/ml in 80 mM-Tris buffer, pH 8.0 (total volume of 0.25 ml) with various concentrations of TNM (6.4 to 320 μM). The reactions were quenched by dilution at various times and assayed for remaining SO_4^{2-}-dependent $Mg^{32}PP_i$-MgATP exchange activity. (b) Reciprocal initial velocity (molybdolysis, relative units) versus reciprocal MgATP concentration for the TNM-modified enzyme (●) and the control (▲). The modification reaction mixtures contained 4.8 mg/ml ATP sulphurylase, 50 mM-Tris buffer, pH 8.0, and 128 μM-TNM in a total volume of 0.405 ml. The reactions were allowed to proceed for 20 min. The reactions were quenched by dialysis and assayed for remaining molybdolysis activity.

the sulphate-dependent $Mg^{32}PP_i$-MgATP exchange reactions. In both cases, the effect was on $V_{max_{app}}$, in contrast to the effect of TNM. The result is consistent with the derivatization of an essential tyrosyl-OH. A subsequent treatment with hydroxylamine under appropriate conditions restored the molybdolysis activity to 76% of a control that had also been incubated with hydroxylamine, but not with

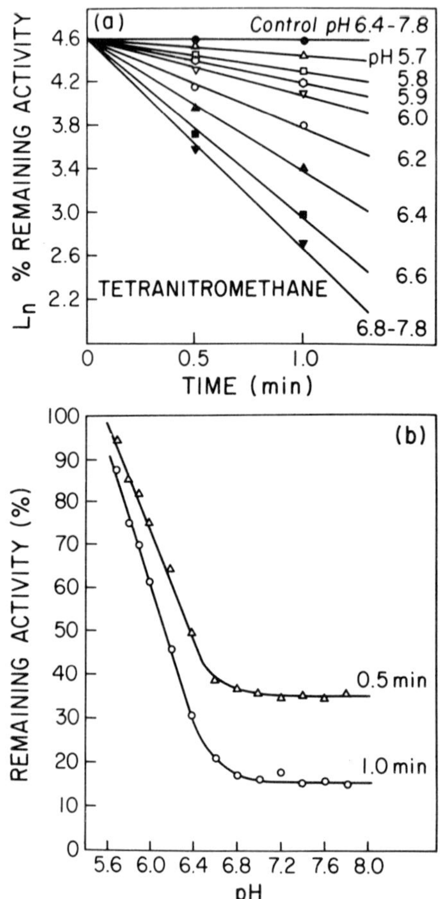

FIG. 8. Effect of pH on the rate of inactivation of ATP sulphurylase by tetranitromethane (TNM). The enzyme was incubated at about 0.5 mg/ml with 64 μM-TNM at the indicated pH values. The buffers used were 0.05 M-MES (pH 5.7-6.2) and 0.05 M-MOPS (pH 6.0-7.8). (a) Inactivation curves. (b) Molybdolysis activity remaining after 0.5 and 1.0 minute.

NAI. (Hydroxylamine itself inactivates the enzyme.) The results confirm the identity of the acetylated residue as tyrosine. The pH profile for the NAI-promoted inactivation is shown in Fig. 9.

Effect of water-soluble carbodiimides

Treatment of the enzyme with 1-cyclohexyl-3-(2-morpholinyl)-4-ethyl-carbodiimide (CMC) resulted in a pH-dependent, first-order inactivation of the

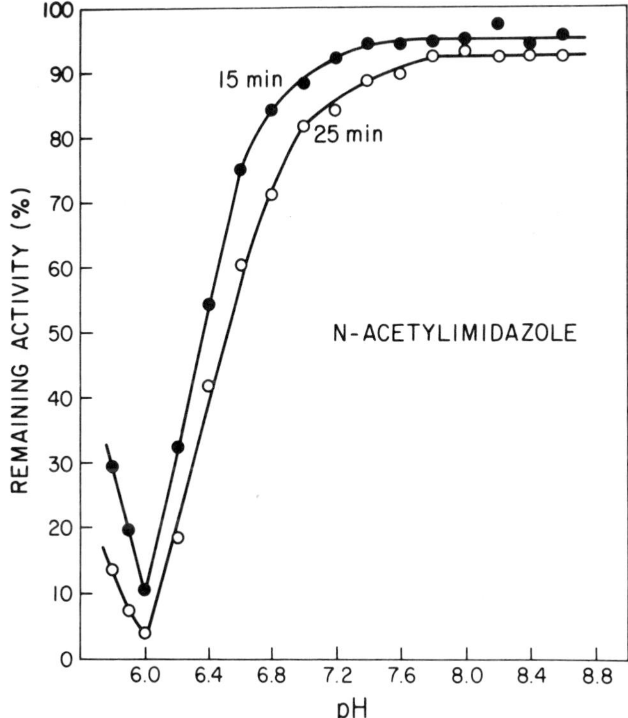

FIG. 9. Effect of pH on the extent of enzyme inactivation by 60 mM-*N*-acetylimidazole after 15 and 25 minutes of preincubation with the reagent. The conditions were the same as described in Fig. 8. Tris buffers (0.05 M) were used between pH 7.8 and 8.6. Activity was measured by the molybdolysis reaction. The values at pH 6.0 and 6.2 were corrected for enzyme inactivation at the low pH.

enzyme. The pK_a of the reactive residue is about 6.5. Diimides are reasonably specific for carboxyl groups, but phenolic, sulphydryl, and amino group residues also react with the reagents to some extent (Means & Feeney 1971). The pH profile (Fig. 10) eliminates amino groups as the site of attack. (The reagent attacks uncharged groups.) The modification of cysteine would not affect enzymic activity. The inhibition by CMC could not be reversed by hydroxylamine, suggesting that the reactive residue is not a tyrosine. Thus, it is possible that in addition to histidine, arginine and tyrosine, a carboxyl group may be essential for the functional integrity of the enzyme.

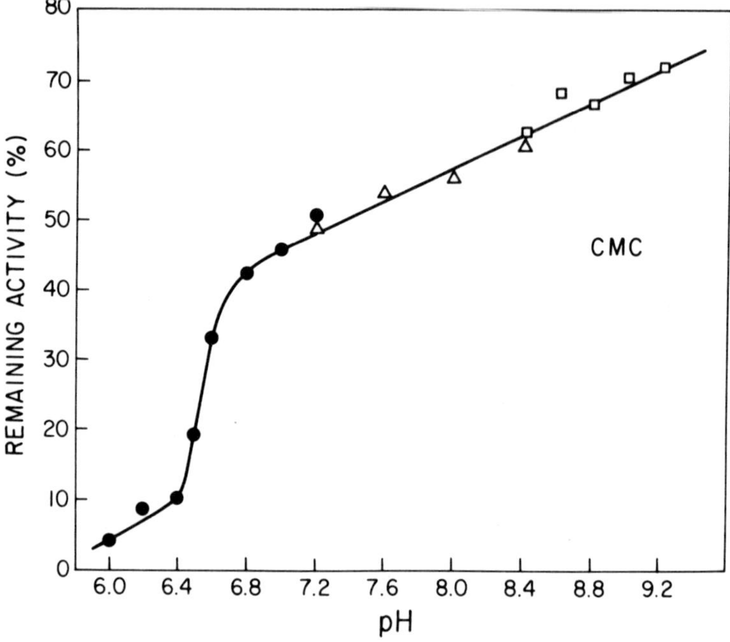

FIG. 10. Effect of pH on the extent of enzyme inactivation by 40 mM-cyclohexyl-3-(2-morpholinyl)-4-ethyl-carbodiimide after 12 minutes. MOPS buffers were used between pH 6.0 and 7.2. Tris buffers were used between pH 7.2 and 9.2. Activity was measured by the molybdolysis reaction. The values at pH 6.0 and 6.2 were corrected for the inactivation by low pH.

SPECULATION ON THE ROLE OF THE ESSENTIAL TYROSINE

We know two effects of nitrating the essential tyrosine: (a) the pK_a of the OH group decreases, i.e., the tyrosine becomes a stronger acid, and as a result (b) the K_m for MgATP increases. In contrast, all the other modification reagents simply decrease V_{max}; that is, they irreversibly inactivate the enzyme. The unique effect of TNM suggested that it might be possible to identify the specific step(s) in which the tyrosine participates. The question then is: which rate constants, if altered, would yield an increase in K_{m_A} without significantly changing K_{ia} and V_{max_f}? The rate constant composition of these kinetic constants in an ordered Bi Bi reaction in which both EA and EQ isomerize (Fig. 5b, p 33) is shown below.

$$K_{ia} = \frac{k_{-1}}{k_1\left(1 + \dfrac{k_2}{k_{-2}}\right)} \qquad (4)$$

$$V_{\max_f} = \frac{[E]_t}{\dfrac{1}{k_2} + \dfrac{1}{k_4} + \dfrac{1}{k_5} + \dfrac{1}{k_6} + \dfrac{k_{-5}}{k_5 k_6}} \qquad (5)$$

$$K_{m_A} = \left(\frac{k_{-1} + k_2}{k_1}\right) \frac{1}{k_2\left(\dfrac{1}{k_4} + \dfrac{1}{k_5} + \dfrac{1}{k_6} + \dfrac{k_{-5}}{k_5 k_6}\right) + 1} \qquad (6)$$

The k_2 seems to be a likely candidate. If k_2 is of the same magnitude, or somewhat less than k_{-1} (a reasonable assumption), then a decrease in k_2 would markedly increase K_{m_A} (because k_2 *multiplies* a number of other terms in the denominator of Equation 6). Furthermore, if nitration decreases k_2, then it is likely that the analogous k_{-5} would also decrease. (The reaction is symmetrical in the forward and reverse directions; both k_2 and k_{-5} are rate constants for the isomerization of an enzyme nucleotide Michaelis complex and would certainly involve the same reactive residue.) A decrease in k_{-5} would augment the increase in K_{m_A}. $V_{\max_f}$ would be relatively unaffected by a decrease in k_2 because $1/k_2$ is only one of several *added* constants in the denominator of equation 5. But, even if $1/k_2$ makes a significant contribution to the total denominator value of equation 5, the effect of a decrease in k_2 would be compensated for by the decrease in the analogous k_{-5}. K_{ia} would not change significantly if k_2/k_{-2} (i.e., $1/K'_A$) is $<<1$, as is suggested by the nearly identical K_i values for MgATP and a variety of non-reactive nucleotides.

Thus, there is a good possibility that the essential tyrosine plays a specific role in the isomerization of the E·MgATP and E·APS complexes. Perhaps the tyrosine serves as an adenylyl group acceptor (Fig. 11). The chemical consequences of nitration are consistent with this suggestion. Decreasing the pK_a of the phenolic-OH by nitration of the ring would make a tyrosyl-OAMP less stable (more 'energy rich'); that is, the ratios k_2/k_{-2} and k_{-3}/k_3 would decrease. This could very well include a decrease in k_2 and k_{-5}. The protection afforded by APS or MgATP plus nitrate, but not by MgATP alone, is also consistent with the essential tyrosine playing a role in nucleotide isomerization. In the free enzyme or E·MgATP complex, the tyrosyl-OH would be free to ionize, making the ring susceptible to nitration by TNM. (Only a minute fraction of the total EA complex would exist in the E~AMP·MgPP$_i$ form.) But, in the presence of MgATP and excess nitrate, the enzyme would be driven to the dead-end E~AMP·MgPP$_i$·NO$_3$ complex in which the tyrosine cannot ionize to the TNM-reactive species.

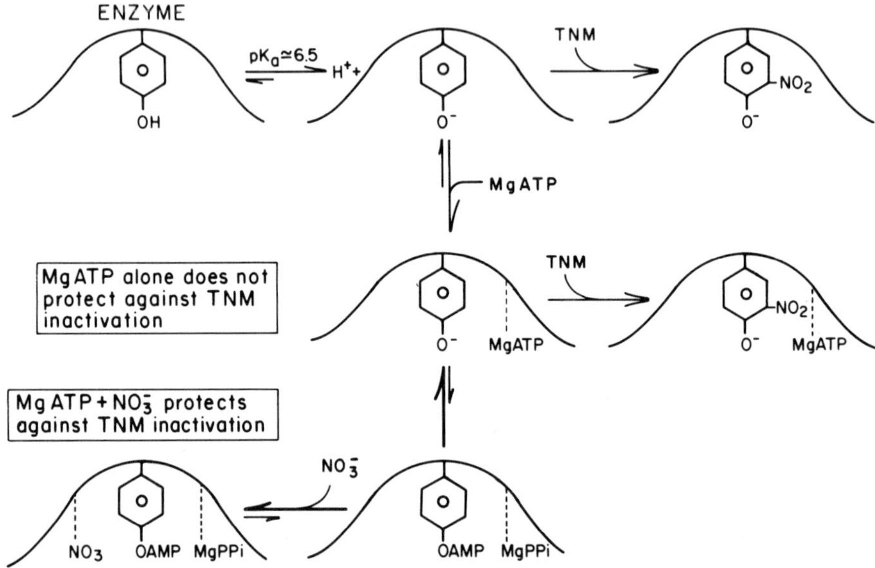

FIG. 11. Proposed mechanism of protection of ATP sulphurylase against inactivation by tetranitromethane (TNM). MgATP alone does not protect because the EA complex exists predominantly in the E·MgATP form in which the tyrosine is still susceptible to attack by TNM. The addition of excess nitrate drives the EA complex to the dead-end E~AMP·MPP$_i$·NO_3^- complex in which the tyrosine can no longer ionize to the TNM-reactive species.

ACKNOWLEDGEMENTS

All the research on the *P. chrysogenum* ATP sulphurylase described in this paper and the cited references was supported by the National Science Foundation. Our current research is supported by NSF Grant PCM77-12193.

References

Adams CA, Nicholas DJD 1972 Adenosine 5′-pyrophosphate sulphurylase in baker's yeast. Biochem J 128:647-654

Builder S, Segel IH 1978 Equilibrium ligand binding assays using labeled substrates: nature of the errors introduced by radiochemical impurities. Anal Biochem 85:413-424

Burnell JN, Anderson JW 1973 Adenosine diphosphate sulphurylase activity in leaf tissue. Biochem J 133:417-428

Chandler CJ, Segel IH 1978 Mechanism of the antimicrobial action of pyrithione: effects on membrane transport, ATP levels, and protein synthesis. Antimicrob Agents Chemother 14:60-68

Farley JR, Cryns DF, Yang YHJ, Segel IH 1976 Adenosine triphosphate sulfurylase from *Penicillium chrysogenum:* steady state kinetics of the forward and reverse reactions. J Biol Chem 251:4389-4397

Farley JR, Nakayama G, Cryns D, Segel IH 1978 Adenosine triphosphate sulfurylase from *Penicillium chrysogenum:* equilibrium binding, substrate hydrolysis, and isotope exchange studies. Arch Biochem Biophys 185:376-390

Farley JR, Mayer S, Chandler CJ, Segel IH 1979 ATP sulfurylase from *Penicillium chrysogenum:* is the internal level of the enzyme sufficient to account for the rate of sulfate utilization? J Bacteriol 137:350-356

Glazer AN, Delange RJ, Sigman DS 1975 Chemical modification of proteins. North-Holland/American Elsevier, New York

Means GE, Feeney RE 1971 Chemical modification of proteins. Holden-Day, San Francisco

Reuveny Z, Filner P 1976 A new assay for ATP sulfurylase based on differential solubility of the sodium salts of adenosine-5′-phosphosulfate and sulfate. Anal Biochem 75:410-428

Robbins PW, Lipmann F 1958 Separation of the two enzymatic phases in active sulfate synthesis. J Biol Chem 233:681-685

Segel IH 1975 Enzyme kinetics. Chapters 6 and 9, Wiley-Interscience, New York

Shaw WH, Anderson JW 1972 Purification, properties and substrate specificity of adenosine triphosphate sulphurylase from spinach leaf tissue. Biochem J 127:237-247

Shaw WH, Anderson JW 1974 The enzymology of adenosine triphosphate sulphurylase from spinach leaf tissue. Biochem J 139:27-35

Tweedie JW, Segel IH 1971a ATP sulfurylase from *Penicillium chrysogenum*. I. Purification and characterization. Prep Biochem 1:91-117

Tweedie JW, Segel IH 1971b Adenosine triphosphate sulfurylase from *Penicillium chrysogenum*. II. Physical, kinetic, and regulatory properties. J Biol Chem 246:2438-2446

Wilson LG, Bandurski RS 1958 Enzymatic reactions involving sulfate, sulfite, selenate and molybdate. J Biol Chem 233:975-981

Discussion

Bright: Where is ATP sulphurylase localized in *Penicillium chrysogenum?*

Segel: I suspect that the enzyme is in the cytoplasm, because the internal level is rather high — about 0.1–0.5% of the total extractable soluble protein.

Swaisgood: Carbodiimide also reacts with tyrosyl residues. Is it possible that it is reacting with tyrosyl residues in the enzyme rather than with carboxyl groups?

Segel: Yes. One could in theory distinguish this by using hydroxylamine to regenerate the modified tyrosine; it wouldn't regenerate a modified carboxyl group. However, our enzyme is inactivated to a significant degree by the hydroxylamine that is meant to reactivate it, so this is a difficult distinction to make. If we use *N*-acetylimidazole to modify the tyrosine, we can partially regenerate activity with hydroxylamine, but hydroxylamine will not reactivate the enzyme at all after it has been treated with a carbodiimide.

Swaisgood: With carbodiimide, does the modification of one residue per mole cause inactivation?

Segel: We don't know yet how many carboxyl groups are essential for the 'functional integrity' of the enzyme. For that matter, we really don't know how many 'essential' histidines, arginines, and tyrosines are present. The kinetics of inactivation only suggest that one modification per active subunit is sufficient to affect enzymic activity.

Kägi: Is there any evidence from chemical modification studies that tyrosine is involved in adenylate transfer?

Segel: We have only the indirect evidence reported in our paper.

Meister: What concentration of hydroxylamine inactivates the enzyme?

Segel: The enzyme is inactivated by prolonged incubation with 0.5–1.0 M-hydroxylamine. (These are the concentrations usually used to regenerate carbodiimide-modified and *N*-acetylimidazole-modified proteins.)

Meister: It has been our experience that many enzymes are insensitive to low concentrations of hydroxylamine; if you find one that is very sensitive, that is interesting. But 1M is a rather high concentration.

Roy: Hydroxylamine is a powerful inhibitor of sulphatase A (EC 3.1.6.1), provided that traces of metal are present. There is no inhibition under stringent metal-free conditions or if EDTA is present. EDTA can reverse the inhibition, which we believe to be due to Cu^{+} formed by the reduction of Cu^{2+} (Roy 1970).

Whatley: The ATP-coupling factor appears to have a tyrosine at its 'active centre' (e.g. Ferguson et al 1974). This may be relevant to Dr Segel's ideas.

Roy: You say that the rate of inactivation with tetranitromethane shows that the tyrosyl residue involved in the reaction is very acidic. How do you distinguish between ionization of the hydroxyl group and changes in the conformation of the protein with changing pH?

Segel: The tetranitromethane (TNM)-reactive species is the tyrosyl *anion*, and the ratio of Tyr-O^{-}/Tyr-OH varies with pH. So it is reasonable to assume that the pH profile for the TNM-induced inactivation reflects, to a substantial degree, the effect of pH on the Tyr-O^{-}/Tyr-OH ratio. If changing pH also changes the accessibility of the reactive tyrosine residue, then the observed pK_a would be a composite pK_a of two processes.

Roy: Have you tried reducing the nitrotyrosine?

Segel: Yes. The reducing agent (sodium hydrosulphite) killed the enzyme.

Kredich: We obtained an apparent K_m for the ATP sulphurylase from *Salmonella typhimurium* of less than 50 μM at saturating ATP concentrations. This is different by a factor of ten from the K_m value (0.5 mM) that you obtain, Dr Segel. On the same point, you mentioned a K_m of 10^{-5} M for sulphate uptake. There is a curious feature of the growth of *Salmonella* on sulphate which could be explained by a good concentrating device, and may have nothing to do with the sulphurylase. This bacterium grows at a normal rate until sulphate is completely gone, and then stops growing. How does *P. chrysogenum* grow in those conditions? Does it slow down gradually?

Segel: The growth of *P. chrysogenum* gradually slows as the culture density increases. However, a direct comparison between the nutrient-limited growth patterns of *Salmonella* and *Penicillium* can't be made easily. *Penicillium* is a filamentous fungus that grows mainly at the hyphal tip, so growth is more linear than exponential. We see exponential growth only in very young (dilute) cultures.

Kredich: If you limit the available sulphate by using only half the amount needed for saturating growth, do you see any difference in the initial phase of growth from when you have saturating levels of sulphate?

Segel: I don't think so.

Whatley: With *Salmonella,* Dr Kredich has said that with a given concentration of sulphate the cells just 'run to the end' of the available sulphate and then stop growing; are there concentrations at which the initial growth rate is lower but which then also run to the end and stop? This happens for some other substrates in *E. coli.*

Kredich: I haven't looked at that, but if you start with a sulphate concentration that gives half maximal final growth, *S. typhimurium* grows with a normal initial growth rate and then flattens out once the sulphate is completely utilized.

Hamilton: Dr Segel, you did kinetic work in the past on sulphate transport in *Penicillium* showing that sulphate is cotransported inwards with calcium and protons, and that the carrier returns to the outside with some other anion. Have you any information on how the carrier functions in bacterial systems, specifically in dissimilatory sulphate reducers, where the sulphide produced is also required to leave the cell?

Segel: I haven't seen anything on that — nor, in fact, much new work on sulphate transport in any bacteria. The isolation of the sulphate-binding protein put physiologists off doing *in vivo* studies (too old-fashioned!). There is still a lot to be done on how sulphate goes into cells.

Ziegler: Since soil contains nitrate, would the binding of nitrate to ATP sulphurylase act as a negative effector, and would this have any physiological significance?

Segel: I don't know. It might be a way of coordinating nitrate and sulphate reduction, both of which must go on in some constant ratio to provide soil microorganisms with amino acids for protein synthesis. The K_i of the enzyme for nitrate is about 0.5 mM—close to the K_m for sulphate.

Postgate: I have always been bothered by the fact that nitrate is a competitive inhibitor of sulphate for these enzymes, because it is not a structural analogue of sulphate. One would expect selenate or monofluorophosphate to be more active as a competitive inhibitor (see Postgate 1979). Yet you seem to imply that nitrate functions at the sulphate site on the enzyme. The implication is that it is actually binding to the sulphate-binding site.

Segel: The kinetic evidence does indeed suggest that nitrate (and chlorate) bind to the sulphate subsite.

Postgate: A compound like monofluorophosphate is a fairly exact analogue of sulphate and should be a spectacular inhibitor.

Segel: We haven't tried monofluorophosphate as an inhibitor of ATP sulphurylase yet. The compound does competitively inhibit sulphate transport in *P. notatum*, but its K_i is about ten times the K_m for sulphate at pH 5.5 (T. Formosa, C.J. Chandler & I.H. Segel, unpublished work).

Jones-Mortimer: What are the intracellular concentrations of nitrate and sulphate?

Segel: In nature, the internal levels of nitrate and sulphate would probably be dictated in part, at least, by the concentrations of these and other ions in the environment. *Penicillium* is capable of accumulating sulphate internally to a concentration of 10–20 mM. External nitrate has no effect on the activity of the sulphate permease. The nitrate permease and nitrate reductase are repressed in the presence of high NH_4^+ or amino acids, and induced by NO_3^- in the absence of NH_4^+ and amino acids (Goldsmith et al 1973). So, the only time the cells would be faced with high internal nitrate concentrations is when nitrate is the sole nitrogen source.

Roy: Are the subunits of the octomer active, Dr Segel? Do you know anything about the association constants?

Segel: I don't know if the monomers are active. So far, we have induced dissociation only with sodium dodecyl sulphate.

Dodgson: Turning again to the essential amino acid residues in the enzyme, arginine is increasingly being implicated as an ionic binding site in many enzyme systems that deal with acidic substrates such as phosphate esters. We have also recently made a similar finding for testicular hyaluronidase and its polyanionic substrate, hyaluronic acid (Gacesa et al 1979). Is arginine playing a similar role in ATP sulphurylase?

Segel: The enzyme appears to bind ATP even after modification of the 'essential' arginine with diacetyl plus borate. Perhaps there is another arginine or lysine involved in nucleotide *binding* that is not accessible to modifying reagents.

Dodgson: And arginine is not binding sulphate or nitrate?

Segel: The free enzyme does not bind sulphate. Presumably, sulphate binds only to the E·MgATP complex. We haven't yet looked for sulphate binding to residue-modified enzyme in the presence of MgATP.

Dodgson: What is known about the hydrolysis of APS by ATP sulphurylase?

Segel: It might be due to a contaminant, but I don't think so, because the ratio of V_{max} forward to V_{max} reverse for the natural substrates is the same as the V_{max} ratio for the hydrolysis of ATP and APS. APS hydrolysis is inhibited by ATP, and ATP hydrolysis is inhibited by nitrate. So it looks as if this one enzyme, ATP sulphurylase, hydrolyses these nucleotide substrates if the natural second substrate (sulphate or Mg PP_i) is not present.

Mudd: Do you know the physiological concentration of APS in the cell?

Segel: No.

Meister: Can you say more about the mechanism of action of ATP sulphurylase and the molybdolysis?

Segel: The molybdolysis simply substitutes molybdate for sulphate; selenate works also. The advantage of using molybdate is that the reaction becomes irreversible because the 'APMo' that forms spontaneously breaks down to AMP and molybdate. Since sulphate is not used, there would be no product inhibition by APS. The PP_i produced is converted to P_i (with pyrophosphatase) and measured colorimetrically. The $^{32}PP_i$–ATP exchange in the absence of sulphate may proceed by the lower (minor) route shown in Fig. 5b (p 33).

Meister: Is the pyrophosphate exchange much slower, then, in the absence of sulphate?

Segel: Yes, about 1% of the exchange rate seen in the presence of saturating sulphate. I should point out that in the presence of sulphate, the PP_i–ATP exchange is also about 100 times faster than the overall forward reaction (APS synthesis). This is why I suggest that the overall forward reaction is rate-limited at the APS synthesis step.

Jones-Mortimer: What happens to the 3′-phosphoadenosine 5′-phosphate (PAP)?

Segel: Good question! I would assume that a 3′-phosphatase acts on PAP, returning 5′-AMP to the general adenylate pool.

References

Ferguson SJ, John P, Lloyd WJ, Radda GK, Whatley FR 1974 Selective and reversible inhibition of the ATPase of *Micrococcus denitrificans* by 7-chloro-4-nitrobenzo-2-oxa-1,3-diazole. Biochim Biophys Acta 357:457-461

Gacesa P, Dodgson KS, Olavesen AH 1979 Functional arginine residues in bovine testicular hyaluronidase. Biochem Soc Trans 7: 1058-1060

Goldsmith J, Livoni J, Norberg CL, Segel IH 1973 Regulation of nitrate uptake in *Penicillium chrysogenum* by ammonium ion. Plant Physiol 52:362-367

Postgate JR 1979 The sulphate-reducing bacteria. Cambridge University Press, Cambridge

Roy AB 1970 The sulphatase of ox liver. XIII. The action of carbonyl reagents on sulphatase A. Biochim Biophys Acta 198:76-81

Pathways of assimilatory sulphate reduction in plants and microorganisms

JEROME A. SCHIFF

Institute for Photobiology of Cells and Organelles, Brandeis University, Waltham, Massachusetts 02154, USA

Abstract Assimilatory sulphate reduction, largely restricted to plants and microorganisms where it provides reduced sulphur for the formation of amino acids and proteins, nucleic acids, and various sulphur-containing coenzymes, begins with the activation of sulphate through reaction with ATP to form adenosine 5′-phosphosulphate (APS) and adenosine 3′-phosphate 5′-phosphosulphate (PAPS). Two pathways of assimilatory sulphate reduction are known. One, found in some blue-green algae (cyanobacteria) and in all oxygen-envolving eukaryotes, begins with APS where the sulpho group is transferred via APS sulphotransferase to a thiol acceptor (glutathione (G-S^-) in *Chlorella)* to form the organic thiosulphate (G-S-SO_3^-). The organic thiosulphate appears to be reduced further by an organic thiosulphate reductase employing reduced ferredoxin to form G-S-S^-. The terminal sulphur is then thought to be reductively transferred to *O*-acetylserine via *O*-acetylserine sulphydrase to form cysteine. A second pathway, found in bacteria and fungi, begins with PAPS where the sulpho group is transferred via PAPS sulphotransferase to an acceptor thiol to form an organic thiosulphate. Since thioredoxin is indispensable, this molecule may be the carrier or may serve to reduce the carrier. NADPH via thioredoxin reductase or glutathione and glutathione reductase reduces thioredoxin. These reactions release sulphite which is further reduced to sulphide by sulphite reductase, employing NADPH. Sulphide is then thought to react with *O*-acetylserine to form cysteine via *O*-acetylserine sulphydrase. The cellular location and evolution of these pathways is discussed.

THE PLACE OF SULPHATE REDUCTION IN THE SULPHUR CYCLE

Unlike most of the other elements required by living systems, carbon, nitrogen and sulphur undergo extensive metabolic transformations. Sulphur, unlike carbon and nitrogen, can be utilized in its most highly oxidized naturally occurring form, sulphate; sulphate reduction is necessary for the formation of sulphur-containing amino acids and proteins (Schiff & Hodson 1973, Roy & Trudinger 1970, Siegel, in Greenberg 1975).

Fig. 1 shows the relation of sulphate to the many reactions which sulphur

Sulphur in biology
(Ciba Foundation Symposium 72) p 49-69

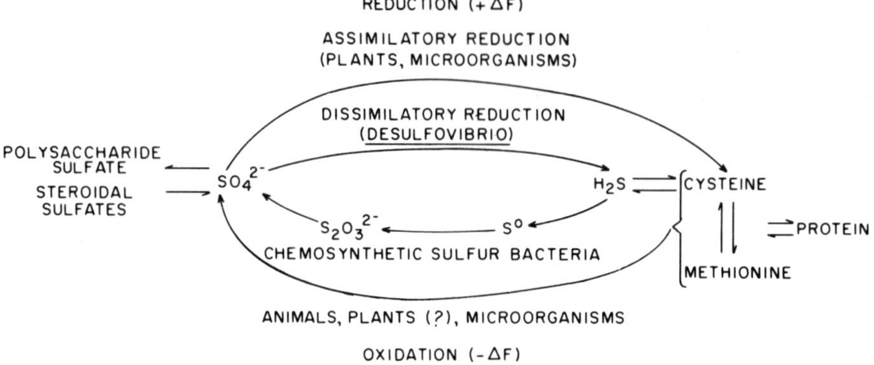

FIG. 1. Reactions of sulphur in the biosphere.

undergoes in the biosphere. Sulphate is used by living systems to form sulphate esters of polysaccharides, phenols, steroids and other organic compounds through a series of activation and transfer reactions. Sulphate is reduced by certain anaerobic organisms such as the bacterium *Desulfovibrio* in their energy metabolism where, during anaerobic respiration, sulphate is used in place of oxygen to oxidize substrates, releasing energy and hydrogen sulphide. Many organisms reduce sulphate to the thiol level found in many coenzymes, and the amino acids cysteine and methionine which serve as building blocks for peptides and proteins; this process is called assimilatory sulphate reduction. Most organisms (including higher animals and higher plants) oxidize reduced sulphur to sulphate although the chemosynthetic bacteria are the only organisms which have been observed to couple the energy released (some 180 kcal per mole) to the reduction of carbon dioxide. The anoxygenic photosynthetic bacteria use reduced sulphur compounds as photosynthetic electron donors, thereby oxidizing them to sulphate.

PHYLOGENETIC DISTRIBUTION OF REACTIONS INVOLVING SULPHATE TRANSFER AND REDUCTION

Reactions resulting in the esterification of organic compounds by sulphate are widely distributed (Schiff & Hodson 1973, DeMeio, in Greenberg 1975). Sulphate esters of organic compounds are formed by higher animals in the form of steroid sulphates, phenol sulphates, and polysaccharide sulphates such as chondroitin sulphate. The algae produce sulphated polysaccharides as wall constituents, frequently in copious amounts. The better-known sulphated polysaccharides of commercial importance such as carrageenan and agar are produced by marine members of the red algae or Rhodophyta. There seem to be no reliable reports of sulphated polysaccharides among the prokaryotes (bacteria and cyanobacteria or blue-green algae) or higher plants. These gelatinous materials may have evolved in the algae as adaptations to periodic wetting and drying as water-conservation mechanisms, a particular problem for benthic and terrestrial algae. In the higher animals, the sulphated polysaccharides are used to provide a matrix for cellular structures such as cartilage and skin. From our present knowledge the occurrence of sulphated polysaccharides seems to represent the adaptation of particular groups of organisms to special needs.

The distribution of reduction reactions, however, seems to follow a more regular evolutionary pattern (Fig. 2) (Schiff & Hodson 1973). Reduction reactions are expensive, energetically, and require the investment of hundreds of kcal per mole in order to take place (Fig. 2). The reduction of nitrate, carbon dioxide and sulphate seem to have been lost in the evolution of the primitive animals, the protozoa, and

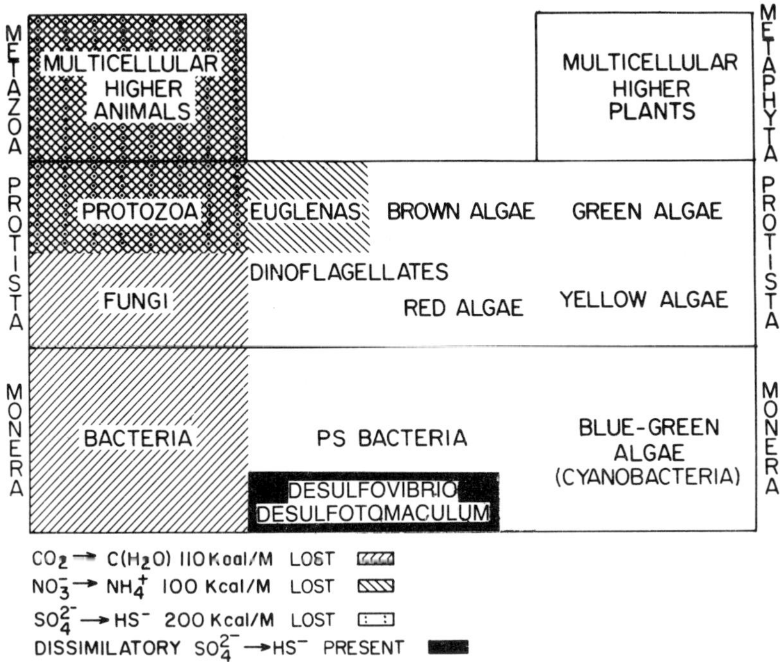

FIG. 2. Loss of energetically expensive reduction reactions in various contemporary groups of organisms. Dissimilatory sulphate reduction is restricted to two small genera of anaerobic bacteria (*Desulfovibrio* and *Desulfotomaculum*). The clear areas indicate organisms that can reduce sulphate, nitrate and carbon dioxide (a small group of aerobic chemosynthetic bacteria (not shown) are capable of reducing carbon dioxide, given oxidizable inorganic substrates as energy sources). The various patterns indicate the loss of these reduction reactions in various groups. All three reduction pathways have been lost in the evolution of the protozoa and higher animals.

this deficiency persists in the evolution of the multicellular animals. These reactions may have been lost when photosynthetic reactions trapping light energy were lost. Animals require energy for movement and having lost photosynthetic energy found themselves with a tight energy budget. Being predators they could fulfil their architectural and energy requirements for reduced sulphur, nitrogen and carbon by eating organisms which could perform these reductions, such as microorganisms or plants, thereby originating the food chains we know today. Contemporary animals obtain their reduced sulphur compounds by eating or harbouring sulphate-reducing plants or microorganisms, or by eating other animals which do.

SULPHATE UPTAKE, ACTIVATION AND TRANSFER

Sulphate uptake and its control has been well studied in many systems (Schiff & Hodson 1973, Siegel, in Greenberg 1975). Uptake seems to be accomplished through active transport mediated by a carrier enzyme system.

As far as is known, the cellular metabolism of sulphate begins with a series of activation reactions with ATP to form the nucleoside phosphosulphates adenosine 5′-phosphosulphate (APS) and adenosine 3′-phosphate 5′-phosphosulphate (PAPS) (Fig. 3) (Schiff & Hodson 1973, De Meio, in Greenberg 1975). APS is formed against an unfavourable equilibrium since the free energy of hydrolysis of the phosphosulphate bond is higher than that of the pyrophosphate linkage (Roy &

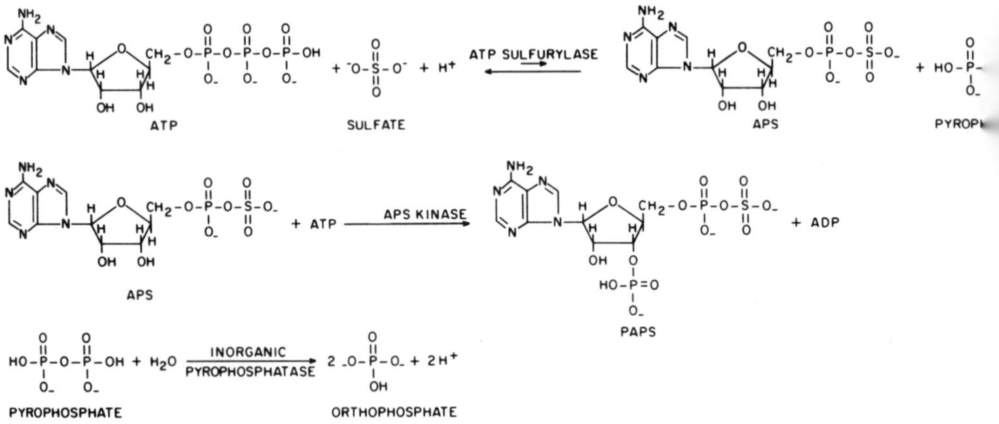

FIG. 3. Enzymic reactions involved in sulphate activation. The enzymes shown are ATP sulphurylase (ATP: sulphate adenylyltransferase, EC 2.7.7.4), APS kinase (ATP: adenylylsulphate 3′-phosphotransferase, EC 2.7.1.25), and inorganic pyrophosphatase (pyrophosphate phosphohydrolase, EC 3.6.1.1).

Trudinger 1970). To offset this unfavourable equilibrium, pyrophosphate removal through inorganic pyrophosphatase activity and/or reaction of APS with another molecule of ATP to form PAPS are used to allow an accumulation of nucleoside phosphosulphates. A 3′(2′),5′-diphosphonucleoside 3′(2′)-phosphohydrolase (DPNPase) is widely distributed which converts PAPS to APS and may facilitate APS formation despite the unfavourable equilibrium of the first reaction forming APS (Fig. 4) (Tsang & Schiff 1976c). This enzyme may also provide control of APS and PAPS levels in organisms where both are required for different purposes.

All sulphate transfer reactions resulting in sulphate esterification appear to use PAPS as the sulphate donor, so far as is known (DeMeio, in Greenberg 1975, Schiff & Hodson 1973). APS is a substrate for hydrolases that release sulphate (Tsang & Schiff 1976c) and for a novel enzyme reaction (adenylylsulphate:NH_3 adenylyltransferase (APSAT)) which converts APS and ammonia to adenosine 5′-phosphoramidate (Fankhauser et al 1979) (this activity was previously thought to be a cyclase forming cyclic AMP; cyclic AMP shares many properties with adenosine 5′-phosphoramidate: Tsang & Schiff 1976c). Sulphonic acids, which contain sulphur at the redox level of sulphite, occur in many living systems, the plant sulpholipid of chloroplast thylakoids being an important example (Schiff & Hodson 1973). Although the formation of the sulphonic acid group of compounds such as taurine can arise through oxidation of reduced sulphur in animals (Siegel, in Greenberg 1975), the formation of the sulphonic acid group of the plant sulpholipid appears to come more readily from sulphate via as yet unknown reactions. The formation of sulphonic acids is an area requiring more study.

SULPHATE REDUCTION

Dissimilatory sulphate reduction occurs in certain anaerobic bacteria such as *Desulfovibrio* which use sulphate as an oxidant in place of molecular oxygen, and in this process APS is the nucleotide sulphate donor for reduction, which accumulates hydrogen sulphide and results in oxidative phosphorylation (Roy & Trudinger 1970, Siegel, in Greenberg 1975, Peck 1970).

Assimilatory sulphate reduction, where sulphate is reduced to form the thiol groups of sulphur-containing amino acids, coenzymes and other organic compounds, now appears to exhibit two distinct patterns (Tsang & Schiff 1975, Schiff & Hodson 1973):

(1) In most oxygen-evolving photosynthesizers (including all eukaryotic algae and some prokaryotic blue-green algae (Tsang & Schiff 1975, Schmidt 1977), all higher plants tested (Schmidt 1975a) and spinach chloroplasts (Schmidt & Schwenn 1971, Schmidt 1975b), the nucleotide sulphate donor for reduction is APS via a highly specific APS sulphotransferase (adenylylsulphate: thiol sulphotransferase) which

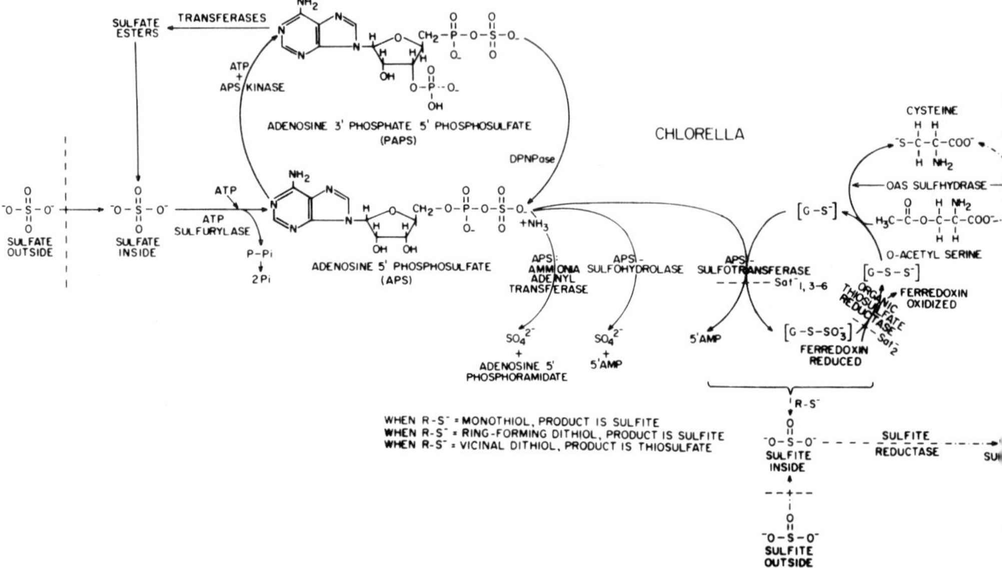

FIG. 4. APS pathway of sulphate reduction in *Chlorella*. Reactions thought to be on the main path of sulphate reduction *in vivo* are shown as solid lines. Reactions blocked in various mutants that cannot grow on sulphate (*Sat*$^-$ mutants) are shown by dashes through the reactions deleted. Side-reactions are shown by alternate dashes and dots. Sulphite (or thiosulphate) is only produced through chemical reactions requiring thiols and probably only occurs *in vivo* when sulphite (or thiosulphate) is available to the cells from outside. G-S$^-$ designates reduced glutathione.

will not use PAPS (Fig. 4) (Tsang & Schiff 1976a). Where this form of reduction has been studied in detail; e.g. *Chlorella, Euglena* and spinach chloroplasts (Schmidt & Schwenn 1971, Brunold & Schiff 1976, Schmidt et al 1974, Abrams & Schiff 1973) a low molecular weight thiol carrier (which appears to be glutathione in *Chlorella:* Tsang & Schiff 1976a) is also involved but the APS sulphotransferase will transfer promiscuously to other added thiols in the presence or absence of the carrier (Tsang & Schiff 1976a). Ferredoxin (in the *Chlorella* and higher plant system) or NADPH (in *Euglena* extracts) must be added to achieve reduction of the sulpho group of APS to the thiol level of amino acids.

(2) In organisms which lack oxygen-evolving photosynthesis (besides animals, which do not reduce sulphate to the thiol level: Schiff & Hodson 1973), such as yeast and *Escherichia coli* and other bacteria (Tsang & Schiff 1975, Torii & Bandurski 1967, Tsang & Schiff 1978b, Dreyfuss & Monty 1963, Fujimoto & Ishimoto 1961, Tsang & Schiff 1976b, Tsang & Schiff 1978b, Wilson & Bierer 1976) and a few oxygen-evolving blue-green algae (Schmidt 1977), PAPS is the nucleoside phosphosulphate donor for reduction via a specific PAPS sulphotransferase (3′-

phosphoadenylylsulphate: thiol sulphotransferase) which will not use APS (Tsang & Schiff 1974, Peck 1961) (Fig. 5).

The transfer of the sulpho group of PAPS and further reduction requires thioredoxin in yeast and *E. coli* (Porqué et al 1970, Tsang & Schiff 1978b, Tsang & Schiff 1976b). Only reduced pyridine nucleotide need be added to obtain reduction, but the PAPS sulphotransferase will only function in the presence of thioredoxin (Tsang & Schiff 1976b) or an analogous molecule (Tsang & Schiff 1978b); other added thiols in the absence of thioredoxin are inactive.

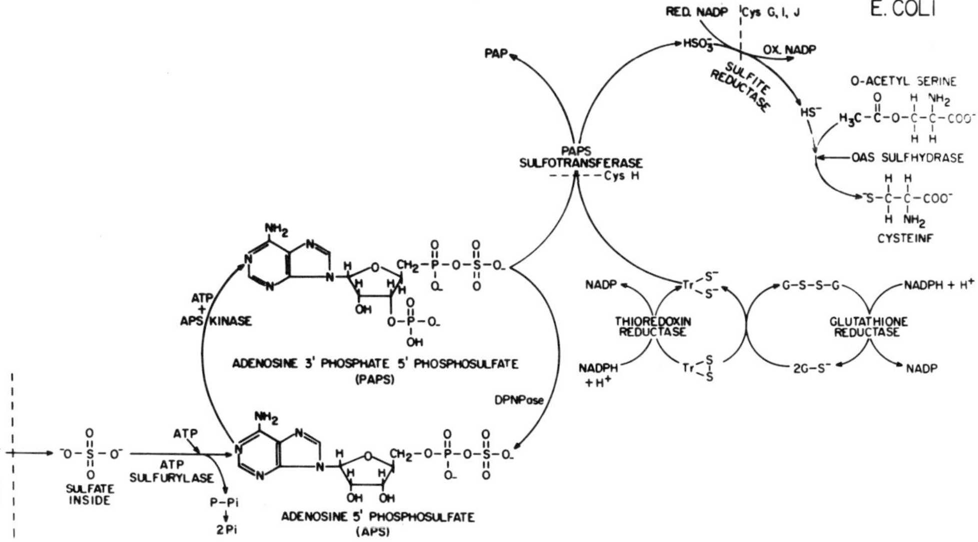

FIG. 5. PAPS pathway of sulphate reduction in *E. coli*. Thioredoxin is required in the PAPS sulphotransferase reaction and formation of sulphite, but the reactions it undergoes have not been elucidated fully. Mutants *(Cys)* blocked in various enzyme reactions are shown by dashes through the reactions deleted. G-S^- designates reduced glutathione. $Tr\langle^{S^-}_{S^-}$ designates reduced thioredoxin.

DETAILED REACTIONS OF THE TWO ASSIMILATORY PATHWAYS

The APS pathway

Our most detailed information comes from work on the enzymes from *Chlorella, Euglena,* spinach chloroplasts (Schmidt & Schwenn 1971, Schmidt 1975b, Brunold & Schiff 1976, Schmidt et al 1974, Abrams & Schiff 1973, Tsang & Schiff 1976a, Tsang & Schiff 1978a) and duckweed *(Lemna)* (Brunold & Schmidt 1978) (Fig. 4). It has taken some time to sort out the reactions of APS sulphotransferase because the enzyme transfers the sulpho group of APS not only to the normal carrier (glutathione) (Tsang & Schiff 1978a) in *Chlorella,* to a somewhat larger molecule in spinach chloroplasts (Schmidt & Schwenn 1971) but to practically any thiol added (Tsang & Schiff 1976a). Thiols forming rings on oxidation (such as dithiothreitol) form sulphite in this reaction, monothiols (such as glutathione and thioethanolamine) form the Bunte salt (R-S-SO_3^-) of the thiol and in the presence of excess thiol, sulphite. Vicinal dithiols (such as 2,3-dithiopropanol, BAL) form thiosulphate ($^-$S-SO_3^-) where the oxidized sulphur comes from APS and the reduced sulphur from the thiol. Of all the thiols, glutathione stands out because it has a much higher activity with the enzyme than any other monothiol and shows kinetics suggestive of a regulatory interaction. Glutathione seems to be the only small molecule in *Chlorella* extracts which is active in the APS sulphotransferase reaction (Tsang & Schiff 1978a). If thiols are omitted from the reaction, it is possible to demonstrate that the enzyme transfers the sulpho group of APS to an enzyme-bound acceptor in the crude extracts (Abrams & Schiff 1973). This labelled material shows the properties of a Bunte salt or organic thiosulphate (R-S-SO_3^-) and is thought to be the sulpho form of the thiol carrier. Reconstruction experiments using glutathione-S-SO_3^- show that sulphite and thiosulphate can be formed chemically by a non-enzymic reaction with the appropriate thiol (Tsang & Schiff 1976a). While the transfer of the sulphate group from APS to the enzyme-bound carrier is heat labile, the subsequent reaction of the enzyme-bound Bunte salt with added thiols is non-enzymic and the same products are formed with each thiol as in the model experiments (Tsang & Schiff 1976a, Abrams & Schiff 1973).

Thus we view the reaction catalysed by APS sulphotransferase as the transfer of the sulpho group of APS to glutathione (G-S^-) to form the Bunte salt (G-S-SO_3^-). As will be discussed shortly, further physiological reduction appears to involve this bound sulpho group. The reductive release of sulphite or thiosulphate by addition of thiols results from chemical side-reactions of this carrier-bound sulpho group.

There are two candidates for enzymes catalysing reduction to the thiol level. One is sulphite reductase (hydrogen-sulphide:$NADP^+$ oxidoreductase, EC 1.8.1.2) which is present in these systems and will bring about the reduction of free sulphite to free sulphide with reduced pyridine nucleotides (Siegel, in Greenberg 1975, Schmidt et al 1974, Tsang & Schiff 1976b). The other enzyme is an organic thiosulphate reductase (ferredoxin: sulphoglutathione oxidoreductase) (formerly called 'thiosulphonate reductase') which reduces glutathione-S-SO_3^- to bound sulphide (probably glutathione-S-S^-), or dithionite to sulphide, with reduced ferredoxin (Schmidt 1973, Schmidt et al 1974). Studies of a mutant of *Chlorella* (*Sat_2^-*) which cannot grow on or reduce sulphate have shown that organic thiosulphate reductase activity is extremely low or absent while normal levels of sulphite reductase activity are present (Schmidt et al 1974). Thus the presence of sulphite reductase alone does not allow sulphate reduction to proceed *in vivo* and shows that organic thiosulphate reductase is the preferred enzyme for sulphate reduction in this system. We view the physiological reaction as the reduction of G-S-SO_3^- to G-S-S^-, which seems to be the primary reduction reaction *in vivo* in the APS system. Sulphite reductase would probably only act on free sulphite if it arose through side-reactions *in vivo* or if it was taken up from outside by the cell.

These systems contain a very active *O*-acetylserine sulphydrase (*O*-acetyl-L-serine acetate-lyase [adding H_2S], EC 4.2.99.8) which forms cysteine from free sulphide or from carrier-bound sulphide (Brunold & Schiff 1976, Schmidt et al 1974). Two electrons would be required to convert G-S-S^- to the thiol group of cysteine in this reaction.

An important advance in studies of this system was finding conditions for the reduction of APS to form cysteine in cell-free extracts without the addition of thiols (Schmidt et al 1974) which form free sulphite and sulphide through chemical side-reactions (Tsang & Schiff 1976a). Studies of this cell-free system from wild-type *Chlorella* and mutants blocked for organic thiosulphate reductase and APS sulphotransferase have proved to be consistent with the interpretations offered above and have led to a scheme in which what are thought to be the normal reactions *in vivo* are shown in solid lines and the side-reactions in broken lines (Fig. 4, p 54).

The cellular location and synthesis of the various enzymes of this pathway have been studied. The activating enzymes for sulphate, APS sulphotransferase, organic thiosulphate reductase (reductant, ferredoxin) and *O*-acetylserine sulphydrase, have been found in spinach chloroplasts (Schmidt & Schwenn 1971, Schwenn & Trebst 1976, Fankhauser & Brunold 1979). It is likely that this is also the case in *Chlorella,* judging from the close similarity of green algal and higher plant systems. In *Euglena,* however, which is an animal-like cell resembling the protozoa but containing chloroplasts, APS sulphotransferase is found in the mitochondria and microbodies and the organic thiosulphate reductase (reductant, NADPH) and *O*-acetylserine sulphydrase are found in the mitochondria (Brunold & Schiff 1976); thus the mitochondria of *Euglena* contain all the known enzymes of the bound pathway of APS reduction. It is interesting that green algal and higher plant cells cannot be induced to lose their chloroplasts, perhaps because certain enzyme systems (such as sulphate reduction) which are essential for viability are localized in the plastids. In *Euglena,* however, chloroplasts and plastid DNA can be lost (Schiff 1973). *Euglena* does not seem to localize essential reactions (other than those essential for photosynthesis) in the plastids.

In the course of studies on the APS sulphotransferase two other activities were found which act on APS (Tsang & Schiff 1978c). One is an APS sulphohydrolase (adenylylsulphate sulphohydrolase, EC 3.6.2.1) which forms sulphate and AMP from APS. PAPS is much more stable than APS in extracts; perhaps this is why the

DPNPase reaction forming APS from PAPS is advantageous as an APS source. The second activity forms adenosine 5′-phosphoramidate (Fankhauser et al 1979) previously thought to be cyclic AMP (Tsang & Schiff 1976c), from APS and ammonia.

The PAPS pathway

This system has been best studied in yeast and *Escherichia coli* (Tsang & Schiff 1975, Torri & Bandurski 1967, Fujimoto & Ishimoto 1961, Dreyfuss & Monty 1963, Tsang & Schiff 1976b, Wilson & Bierer 1976) (Fig. 5). PAPS sulphotransferase (also called 'PAPS reductase' before thioredoxin was identified as a component) is specific for PAPS as the sulphate donor and is also completely specific for thioredoxin; other thiols will not replace thioredoxin (Tsang & Schiff 1976b) or an analogous molecule (Tsang & Schiff 1978b). As in *Chlorella*, an enzyme-bound intermediate with the properties of a Bunte salt (R-S-SO_3^-) can be detected as a product of the PAPS sulphotransferase reaction (Tsang & Schiff 1976b, Torii & Bandurski 1967). The formation of free sulphite requires thioredoxin reductase (NADPH: oxidized-thioredoxin oxidoreductase, EC 1.6.4.5) and reduced pyridine nucleotide (Torii & Bandurski 1967, Porqué et al 1970) but these can be replaced by appropriate thiols (Tsang & Schiff 1976b) to yield sulphite or thiosulphate as in the APS system. In certain conditions this system can accept electrons from the glutathione-glutathione reductase system ((NAD(P)H:oxidized-glutathione oxidoreductase, EC 1.6.4.2)) (Tsang & Schiff 1978b, Holmgren et al 1978). The product of this PAPS sulphotransferase-thioredoxin system is sulphite when NADPH is used as the reductant. (Free sulphite is reduced to sulphide with reduced pyridine nucleotides via sulphite reductase: Siegel, in Greenberg 1975). In *Chlorella*, two reductase activities are present, one that acts on free sulphite and one that acts on a bound intermediate in the form of a Bunte salt or organic thiosulphate, the latter being the physiological reaction in sulphate reduction (Schmidt et al 1974). However, *E. coli* mutants blocked for sulphate reduction lack sulphite reductase activity, which suggests that this enzyme system is the physiological reaction in these organisms (Tsang & Schiff 1978b). *O*-Acetylserine sulphydrase is also present (Kredich & Tompkins 1966) which can form cysteine from *O*-acetylserine and sulphide. Taking together the information presented here, and various suggestions made previously by various workers, it is possible to write a scheme consistent with current knowledge (Fig. 5).

The role of thioredoxin in this system is still being studied. Reduced thioredoxin ($\mathrm{Tr}{<}^{S^-}_{S^-}$) could act as the thiol carrier acceptor for the sulpho group of PAPS in the PAPS sulphotransferase reaction, forming $\mathrm{Tr}{<}^{S-SO_3^-}_{S^-}$ (sulphothioredoxin). Elimination of sulphite would yield oxidized thioredoxin ($\mathrm{Tr}{<}^{S}_{S}|$) which could

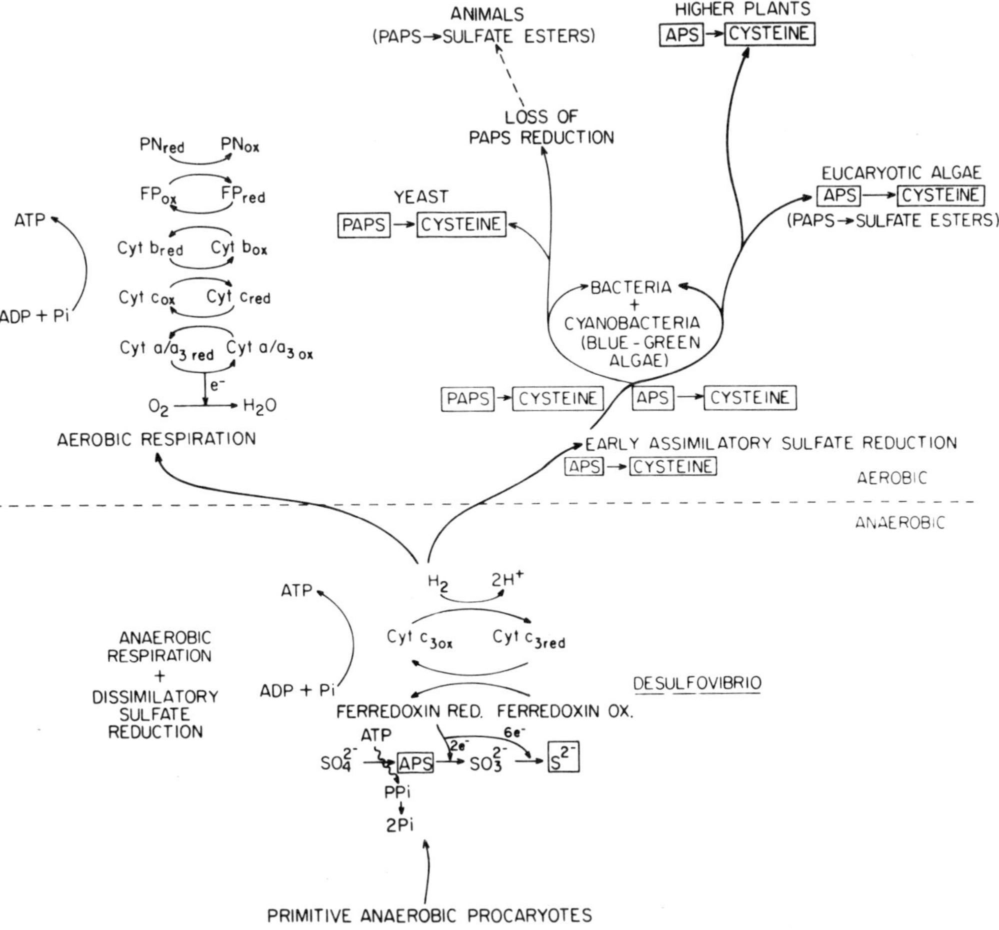

FIG. 6. Suggested evolutionary relationships of sulphate-reducing pathways in various groups. Dissimilatory sulphate reduction, as part of anaerobic respiration, probably arose in the anaerobic phase of the origin of life and has persisted in the modern prokaryotic anaerobes *Desulfovibrio* and *Desulfotomaculum* which use sulphate as an oxidant in respiration, forming ATP and sulphide. This pathway begins with APS. With the release of oxygen into the atmosphere, oxygen became the electron acceptor in respiration and a separation of respiration and sulphate reduction occurred, leading to the establishment of assimilatory sulphate reduction, still using APS as the activated sulphate for reduction. For as yet unknown reasons the PAPS pathway for reduction evolved from the APS pathway and modern prokaryotes (bacteria and blue-green algae or cyanobacteria) have one or the other. The APS pathway has persisted in the evolution of the eukaryotic photosynthetic oxygen evolvers including the eukaryotic algae and higher plants. The PAPS pathway leads to yeast, among the fungi. Animals have lost the ability to reduce PAPS to the thiol level of the amino acids but retain PAPS as a donor of the sulpho group in esterification reactions via specific sulphotransferases. Eukaryotic algae also use PAPS as the donor of the sulpho group in esterification reactions although they use APS as the substrate for reduction.

then be reduced as shown in Fig. 5. In view of the ready elimination of sulphite from Bunte salts of non-vicinal dithiols (Tsang & Schiff 1976a), some mechanism would be necessary to prevent the immediate release of sulphite from sulphothioredoxin, since the bound intermediate formed from PAPS in *E. coli* is rather stable (Tsang & Schiff 1976b). Perhaps linkage of one of the thiol groups of thioredoxin to a thiol group of the transferase via an S—S bond accomplishes this. Another possibility is suggested by recent work which shows that a large proportion of the thioredoxin has a thiol group substituted with phosphate (Pigiet & Conley 1978). Perhaps this substitution of one thiol group by phosphate prevents spontaneous elimination of sulphite from sulphothioredoxin until timely enzymic removal of the phosphate. It is possible, of course, that thioredoxin is not the carrier and that the bound intermediate contains the sulpho group bound to a different acceptor. In this case, thioredoxin would act as the immediate reductant.

It is worth noting that although the bound intermediates in both the APS and PAPS systems show many reactions of Bunte salts, frequently not all of the activity is exchangeable with sulphite (Tsang & Schiff 1976b, Torii & Bandurski 1967) and further work is needed to prove that these are definitely organic thiosulphates and that they account for all of the activity in both types of systems.

SPECULATIONS ON THE ORIGIN AND EVOLUTION OF PATHWAYS OF SULPHATE REDUCTION

It is likely that the most primitive form of sulphate reduction is the dissimilatory pathway of the anaerobes (Roy & Trudinger 1970, Siegel, in Greenberg 1975, Peck 1970) (Fig. 6) which may have survived with little modification from the anaerobic phase of the origin of life and evolution. In the absence of molecular oxygen, other oxidants were pressed into service to oxidize substrates. The result was anaerobic respiration employing sulphate or nitrate as oxidants with the consequent formation of their reduced forms, sulphide and nitrogen gas. In organisms such as *Desulfovibrio* and *Desulfotomaculum* the formation of ATP from the oxidation of substrates is coupled to dissimilatory sulphate reduction which serves as a sink for the electrons removed in the oxidation of the substrate (hydrogen in this example). This dissimilatory pathway uses APS as the activated form of sulphate for reduction and the pathway proceeds through the formation of free sulphite and sulphide. In an anaerobic environment there is no danger from chemical oxidation of these intermediates by molecular oxygen. Anaerobic conditions also allow the undisturbed participation of autoxidizable low potential reductants such as ferredoxin in the sulphate reduction pathway. Multi-electron reductions utilizing low potential reductants such as those involved here, in sulphite reduction, in nitrite reduction and in nitrogen fixation very likely evolved under the early anaerobic conditions in which life originated.

When oxygen came into the atmosphere through the activities of the first oxygenic photosynthesizers (very likely early prokaryotes related to the cyanobacteria or blue-green algae) (see Schiff 1973) aerobic respiration using oxygen as the electron acceptor evolved and organisms respiring anaerobically were forced into anaerobic niches (such as deep mud) to survive. With the availability of oxygen, I suppose, came the separation of respiration and sulphate reduction in the aerobic organisms. Respiration now employed the reduction of oxygen to water (Fig. 6) in the formation of ATP. Sulphate reduction, although no longer required for respiration, was still necessary for the formation of reduced sulphur compounds including cysteine and methionine of protein. This separation of sulphate reduction from respiration led to the establishment of assimilatory sulphate reduction as a separate process to provide these reduced sulphur compounds.

This assimilatory sulphate reduction process was, at first I suppose, similar to the process of dissimilatory sulphate reduction in the anaerobes in that it used APS as the activated intermediate and reduced ferredoxin as the reductant. Free sulphite and sulphide react with oxygen, however, and this may have been the point at which carrier-bound intermediates such as glutathione-S-sulphonate ($G\text{-}S\text{-}SO_3^-$) and glutathione persulphide ($G\text{-}S\text{-}S^-$), which are not autoxidizable, were introduced into the APS pathway. This process would have been present in the early aerobic prokaryotes and is found among contemporary prokaryotes such as the cyanobacteria or blue-green algae (Tsang & Schiff 1975, Schmidt 1977). It is the conviction of many modern biologists, including myself, that the chloroplasts and mitochondria of modern eukaryotes originated from the invasion of non-photosynthetic, non-respiratory eukaryotes by appropriate prokaryotic endosymbionts such as primitive blue-green algae and bacteria (Schiff 1973). In view of this, it is not surprising that the APS pathway of assimilatory sulphate reduction with ferredoxin as reductant is found in eukaryotic algae such as *Chlorella* (Tsang & Schiff 1975, Schmidt et al 1974) and higher plants (Schmidt 1975a, Brunold & Schmidt 1978), and in spinach has been shown to be localized in the chloroplasts (Schmidt & Schwenn 1971, Schwenn & Trebst 1976). This is probably one of the reasons why chloroplast function can never be eliminated completely from green algae and higher plants even when reduced carbon is available; essential functions such as sulphate reduction (and nitrite reduction) are localized in the plastids. Organisms more closely related to the animals, such as *Euglena*, do not put essential functions other than carbon dioxide fixation in their plastids and when provided with a reduced carbon source, they can eliminate plastid functions completely. In *Euglena*, sulphate reduction is localized in the mitochondria and microbodies (Brunold & Schiff 1976). In this organism, the APS pathway characteristic of oxygenic eukaryotes is present but the reductant has become NADPH rather than reduced ferredoxin which is available only in the chloroplasts of eukaryotes, so far as we know.

Although the APS pathway is pervasive among eukaryotes having oxygenic photosynthesis, another pathway appears to have evolved at the same time. While some prokaryotes use the APS pathway of assimilatory sulphate reduction, others use a PAPS pathway (Tsang & Schiff 1975, Schmidt 1977), suggesting that the PAPS pathway evolved from the APS pathway at or after the establishment of assimilatory sulphate reduction (Fig. 6). This pathway is found in bacteria such as *E. coli* (Tsang & Schiff 1975,1976b,1978b, Dreyfuss & Monty 1963, Fujimoto & Ishimoto 1961) and in fungi such as yeast (Torii & Bandurski 1967, Wilson & Bierer 1976) where the reductant is NADPH. Since PAPS is formed from APS it presumably arose later in evolution than APS, but why PAPS was selected for assimilatory sulphate reduction in these organisms is not readily explained. As we have already noted (Fig. 2, p 51), sulphate reduction was lost in the evolution of the protozoa and higher animals; the pathway from sulphate to PAPS formation has been conserved, however, and PAPS serves as the source of sulphate for esterification reactions via PAPS sulphotransferases of various specificities (De Meio, in Greenberg 1975). PAPS also serves as the donor for esterification reactions in organisms that use APS for sulphate reduction, suggesting that PAPS formation was fairly common among primitive prokaryotes but was selected for use in assimilatory sulphate reduction only in certain prokaryotes and in non-photosynthetic eukaryotes such as the fungi.

ACKNOWLEDGEMENTS

The support of a series of grants from the National Science Foundation (PCM 76-21486) is gratefully acknowledged. The technical assistance of Mrs Linda Corrado is appreciated, as is the secretarial help of Mrs Margaret King. Nancy O'Donoghue prepared the illustrations.

References

Abrams WR, Schiff JA 1973 Studies of sulfate utilization by algae. 11. An enzyme-bound intermediate in the reduction of adenosine-5′-phosphosulphate (APS) by cell-free extracts of wild type *Chlorella* and mutants blocked for sulfate reduction. Arch Microbiol 94:1-10

Brunold C, Schiff JA 1976 Studies of sulfate utilization by algae. 15. Enzymes of assimilatory sulfate reduction in *Euglena* and their cellular localization. Plant Physiol (Bethesda) 57:430-436

Brunold C, Schmidt A 1978 Regulation of sulfate assimilation in plants. 7. Cysteine inactivation of adenosine 5′-phosphosulfate sulfotransferase in *Lemna minor* L. Plant Physiol (Bethesda) 61:342-347

Dreyfuss J, Monty KJ 1963 Coincident repression of the reduction of 3′-phosphoadenosine 5′-phosphosulfate, sulfite and thiosulfate in the cysteine pathway of *Salmonella typhimurium*. J Biol Chem 238:3781-3783

Fankhauser H, Garber L, Schiff JA 1979 Adenylyl sulphate (APS): ammonia adenylyl transferase (APSAT) forming adenosine 5′-phosphoramidate (APA) from APS and ammonia. Plant Physiol (Bethesda) 63 : S-162

Fankhauser H, Brunold C 1979 Localization of *O*-acetyl-L-serine sulfhydrylase in *Spinacia oleracea*. Plant Sci Lett 14:185-192

Fujimoto D, Ishimoto M 1961 Sulfate reduction in *E coli*. J Biochem (Tokyo) 50:533-537

Greenberg DM (ed) 1975 Metabolic pathways. Academic Press, New York, vol VII, Metabolism of sulfur compounds

Holmgren A, Ohlsson I, Crankvist ML 1978 Thioredoxin from *Escherichia coli*. Radioimmunological and enzymatic determinations in wild type cells and mutants defective in phage T_7 DNA replication. J Biol Chem 253:430-436

Kredich N, Tomkins G 1966 The enzymic synthesis of L-cysteine in *Escherichia coli* and *Salmonella typhimurium*. J Biol Chem 241:4955-4965

Peck HD Jr 1961 Enzymatic basis for assimilatory and dissimilatory sulfate reduction. J Bacteriol 82:933-939

Peck HD Jr 1970 Sulfur requirements and metabolism of microorganisms. In: Muth OH, Oldfield JE (eds) Symp sulfur in nutrition. Avi, Westport, Conn, p 61-79

Pigiet V, Conley RR 1978 Isolation and characterization of phosphothioredoxin from *Escherichia coli*. J Biol Chem 253:1910-1920

Porqué PG, Baldesten A, Reichard P 1970 The involvement of the thioredoxin system in the reduction of methionine sulfoxide and sulfate. J Biol Chem 245:2371-2374

Roy AB, Trudinger PA 1970 The biochemistry of inorganic compounds of sulphur. Cambridge University Press, Cambridge

Schiff JA 1973 The development, inheritance and origin of the plastid in *Euglena*. Adv Morphog 10:265-309

Schiff JA, Hodson RC 1973 The metabolism of sulfate. Annu Rev Plant Physiol 24:381-414

Schmidt A 1973 Sulfate reduction in a cell-free system of *Chlorella*. The ferredoxin-dependent reduction of a protein-bound intermediate by a thiosulfonate reductase. Arch Microbiol 93:29-52

Schmidt A 1975a Distribution of APS sulfotransferase activity among higher plants. Plant Sci Lett 5:407-415

Schmidt A 1975b A sulfotransferase from spinach leaves using adenosine 5′-phosphosulfate. Planta (Berl) 124:267-275

Schmidt A 1977 Assimilatory sulfate reduction via 3′-phosphoadenosine 5′-phosphosulfate (PAPS) and adenosine 5′-phosphosulfate (APS) in blue green algae. FEMS (Fed Eur Microbiol Soc) Microbiol Lett 1:137-140

Schmidt A, Schwenn JD 1971 On the mechanism of photosynthetic sulfate reduction. Proc second int congr photosynthesis. Junk, The Hague, p 507-514

Schmidt A, Abrams WR, Schiff JA 1974 Studies of sulfate utilization by algae. 12. Reduction of adenosine 5′-phosphosulfate to cysteine in extracts from *Chlorella* and mutants blocked for sulfate reduction. Eur J Biochem 47:423-434

Schwenn JD, Trebst A 1976 Photosynthetic sulfate reduction by chloroplasts. In: Barber J (ed) The intact chloroplast. Elsevier, Amsterdam, p 315-334

Torii K, Bandurski RS 1967 Yeast sulfate-reducing system. III. An intermediate in the reduction of 3′-phosphoryl-5′-adenosine phosphosulfate to sulfite. Biochim Biophys Acta 136:286-295

Tsang MLS, Schiff JA 1975 Studies of sulfate utilization by algae. 14. Distribution of adenosine 5′-phosphosulfate (APS) and adenosine 3′-phosphate 5′-phosphosulfate (PAPS) sulfotransferases in assimilatory sulfate reducers. Plant Sci Lett 4:301-307

Tsang MLS, Schiff JA 1976a Studies of sulfate utilization by algae. 17. Reactions of the adenosine 5′-phosphosulfate (APS) sulfotransferase from *Chlorella* and studies of model reactions which explain the diversity of side products with thiols. Plant Cell Physiol 17:1209-1220

Tsang MLS, Schiff JA 1976b Studies of a sulfate-reducing pathway in *Escherichia coli* involving bound intermediates. J Bacteriol 125:923-933

Tsang MLS, Schiff JA 1976c Properties of enzyme fraction A from *Chlorella* and copurification of 3′(2′)5′-bisphosphonucleoside 3′(2′)-phosphohydrolase, adenosine 5′-phosphosulfate sulfohydrolase and adenosine-5′-phosphosulfate cyclase activities. Eur J Biochem 65:113-121

Tsang MLS, Schiff JA 1978a Studies of sulfate utilization by algae. 18. Identification of glutathione as a physiological carrier in assimilatory sulfate reduction by *Chlorella* extracts. Plant Sci Lett 11:177-183
Tsang MLS, Schiff JA 1978b Assimilatory sulfate reduction in a mutant of *Escherichia coli* lacking thioredoxin activity. J Bacteriol 134:131-138
Wilson LG, Bierer D 1976 The formation of exchangeable sulphite from adenosine 3′-phosphate 5′-sulphatophosphate in yeast. Biochem J 158:255-270

Discussion

Ziegler: The intermediate organic thiosulphates, of which sulphoglutathione is the most active, pose an interesting problem. Free sulphoglutathione (G–S–SO_3^-; Bunte salt) would be rapidly reduced by glutathione (Reaction 1), since chloroplasts undoubtedly contain thioltransferases.

$$GSSO_3^- + GSH \longrightarrow GSSG + HSO_3^- \qquad (1)$$

Glutathione reductase-catalysed reduction of glutathione disulphide by NADPH would drive the reaction far to the right. Since the *in vitro* work indicates that sulphoglutathione is the intermediate *in vivo,* some mechanism must prevent its rapid reduction.

Dodgson: The answer is probably compartmentation, since Dr Schiff was discussing a chloroplast system. Where are the mammalian transferases localized?

Ziegler: In a recent review Freedman (1979) suggested that mammalian tissues contain a number of different thioltransferases, some of which are membrane bound. I suspect that you are correct that some special form of compartmentation must protect the intermediate sulphoglutathione. The chloroplasts must contain thioltransferases; otherwise, enzymes with essential thiols would be inactivated.

Schiff:* I agree that compartmentation is one likely explanation, although these enzymes could be regulated to be inactive unless they are needed for specific purposes. One indication that they may be inactive in the normal cell is derived from studies of mutants blocked for sulphate reduction in *Chlorella*. Mutant Sat_2^- is blocked in the organic thiosulphate reductase step and cannot reduce sulphate to form cysteine. If enzymes which could form free sulphite from G–S–SO_3^- were present and active, free sulphite would form and be reduced by sulphite reductase (which is present) to form sulphide and cysteine. Since this doesn't happen, I conclude that the release of free sulphite from G–S–SO_3^- doesn't occur in the intact cells.

Segel: Several years ago we attempted to purify a cysteine-*S*-sulphate (*S*-sulphocysteine) reductase, but were unsuccessful. We concluded that the reduction of the compound *in vivo* could be completely accounted for by the non-

* Comments added in writing after the symposium.

enzymic reaction with reduced glutathione, coupled with the NADPH-linked glutathione reductase (Woodin & Segel 1968).

Ziegler: Above pH 7.6 that reaction is non-enzymic and extremely fast. However, below pH 7.2 the reaction is slow in the absence of thioltransferase.

Mudd: In certain rare conditions human beings accumulate S-sulphocysteine; in the rare genetic deficiency of sulphite oxidase there is secondary accumulation of sulphite. In the three such cases studied the hallmark abnormality has been the excretion of *S*-sulphocysteine. Certain mammals, civets, also excrete large amounts of *S*-sulphocysteine. *S*-Sulphocysteine in these circumstances is not completely reduced by reduced glutathione. The explanation for the civets is unknown. They are not sulphite oxidase-deficient!

Schiff: There are reports in the literature that animals can reduce sulphate to the level of sulphite, but of course one must be sure in each case that the microbial flora in the intestinal tract of the animal is not doing the reduction.

Segel: Professor Schiff suggests that the hydrolysis of PAPS to APS by a 3′-phosphatase helps to overcome the terribly unfavourable K_{eq} of the ATP sulphurylase reaction. However, the *simultaneous* sequential operation of ATP sulphurylase, APS kinase and PAPS-3′-phosphatase ($SO_4^{2-} \rightleftharpoons APS \rightleftharpoons PAPS \rightleftharpoons APS$) does not constitute an effective coupled reaction sequence. The ATP sulphurylase reaction cannot be pulled to the right by a reaction downstream that produces APS. On the other hand, PAPS *could* serve as a reservoir of potential APS if the 3′-phosphatase is regulated in a temporal manner; that is, if the 3′-phosphatase is inactive during the accumulation of PAPS. Then, *at some later time,* the hydrolysis of the 3′-phosphate of PAPS could produce a pool of APS larger than that which could be accumulated by the ATP sulphurylase reaction. This would require that (a) the cells have some way of switching the 3′-phosphatase off and on (a fascinating possibility), and (b) ATP sulphurylase be inhibited (when the 3′-phosphatase is activated) so that the APS formed is not converted back to ATP or hydrolysed to AMP plus sulphate.

Schiff: Dr Segel's point is a good one. The way I think of it is that since inorganic pyrophosphatase is widely distributed, hydrolysis of the pyrophosphate formed in the ATP sulphurylase reaction to phosphate prevents the back reaction of ATP sulphurylase when APS is formed from PAPS by the 3′-phosphohydrolase. Of course it is also possible, as Dr Segel suggests, that the phosphohydrolase is subject to regulation. Since we have purified the phosphohydrolase to homogeneity it should be possible to study its regulation by various molecules of interest.

Dodgson: There is an analogous 3′-phosphatase or 3′-nucleotidase in the cytosol of mammalian liver cells which is activated by cobalt (Denner et al 1973). Unfortunately, we did not investigate whether this enzyme, which converts PAPS to APS, was inhibited by any of the components of the system.

Meister: Dr Segel, haven't you told us that APS is enzyme bound?

Segel: The internal level of ATP sulphurylase in *Penicillium chrysogenum* is rather high — about 5 μm in active sites. The affinity of the enzyme for APS is also very high; K_{iq} is about 1 μM. Consequently, at any time during the formation of PAPS, a substantial fraction of the total intracellular APS must be bound to ATP sulphurylase. In *P. chrysogenum,* APS has no known function except to serve as the substrate for APS kinase. Perhaps ATP sulphurylase and APS kinase form a complex such that the APS produced by the former is transferred directly to the latter — a form of biochemical channelling. It makes a nice story, but we have no proof yet.

Roy: There have been reports that APS remains bound to the ATP sulphurylases of sheep liver (Pannikar & Bachhawat 1968) and of mouse mastocytoma (Shoyab & Marx 1972). It was also reported that in the latter case the subsequent action of APS kinase produced a free PAPS.

Dodgson: Dr Segel, can you offer any explanation for the fact that the cell apparently wastes an ATP molecule to make PAPS from APS when the latter compound is already in an 'activated' form?

Segel: The extra ATP used to convert APS to PAPS isn't really wasted. The APS kinase reaction permits the cell to accumulate a reasonable pool of 'active' sulphate as PAPS — far more than could accumulate as APS from the ATP sulphurylase and inorganic pyrophosphatase reactions alone. The overall $\Delta G^{\circ\prime}$ for the reactions

$$SO_4^{2-} + MgATP + H_2O \rightleftharpoons APS + 2P_i + Mg^{2+}$$

is about +6000 cal/mole; that is, the overall reaction lies far to the left. Even under non-standard state conditions at high levels of internal sulphate and MgATP, only a minute amount of APS could accumulate. On the other hand, the overall sequence

$$SO_4^{2-} + 2MgATP + H_2O \rightleftharpoons PAPS + MgADP + 2P_i + Mg^{2+}$$

has a $\Delta G^{\circ\prime}$ of about zero.

Kelly: It is a wasteful process in the sense that APS would surely be drawn away by the subsequent biosynthetic reactions in any case; so why does the cell need to make PAPS and then regenerate APS?

Postgate: I would like some explanation of that. What is the direct evidence for accumulation of a pool of APS?

Segel: I have no evidence for the accumulation of a measurable internal pool of APS in *P. chrysogenum*. Nor do we have any evidence for a 3′-phosphatase that converts PAPS to APS such as Professor Schiff has demonstrated in green plants. There would be no need for such an enzyme in filamentous fungi where, presumably, PAPS rather than APS is the substrate for sulphate reduction.

Ziegler: What is the concentration of PAPS in the cell? Is there a pool of accumulated PAPS or is it used almost immediately?

Whatley: PAPS accumulates to a small extent in *Paracoccus denitrificans* and in spinach, *inter alia;* it is visible on an electrophoretogram.

Segel: PAPS can be demonstrated in extracts of *P. chrysogenum* after sulphur-starved mycelia have been pulsed with very high specific activity $^{35}SO_4^{2-}$ in 'Calvin type' experiments (Segel & Johnson 1973). I don't know what the steady-state level of PAPS is in mycelia grown in batch culture on inorganic sulphate as sole sulphur source. However, the K_m of choline sulphokinase for PAPS is about 1.2×10^{-5} M (Renosto & Segel 1977), so I would guess that the usual internal pool of PAPS is somewhere between 0.1 K_m and K_m. (Choline-*O*-sulphate is accumulated to relatively high levels in the mycelium.)

Mudd: Surely the idea that the enzyme–APS complex would serve as a reservoir of APS would be negated by the hydrolytic capacity of APS sulphurylase?

Segel: I don't think anyone is seriously suggesting that E·APS serves as a 'reservoir' of APS — just that APS may be transferred directly to APS kinase without equilibrating as free APS in the cytoplasm (where it may be degraded by APS hydrolases). There have been reports of APS hydrolases. I wonder whether some of these reports are describing the action of ATP sulphurylase in the absence of $MgPP_i$.

Dodgson: I shall discuss this later in our paper (p 163-176). My only comment at this point is that by converting APS to PAPS, the cell has presumably introduced a new specificity determinant. In APS there is a potential for hydrogen-bonding between the ester sulphate group and the 3′-hydroxyl group of ribose — a potential that is lacking in PAPS where the ester sulphate group would tend to be repulsed by the 3′-phosphate group. A new dimension, in terms of specificity for biological systems, has therefore been introduced with the acquisition of the 3′-phosphate.

Segel: If APS were used universally as the sulphate donor for sulphate ester formation the sulphokinases would have to have an extremely low K_m for APS (because the internal level of APS is so low). It may have been less difficult to devise a sulphotransferase with a K_m of about 10^{-5} M for the sulphate donor. This may explain why APS kinase evolved — to produce an 'active' sulphate that could be accumulated by cells to a level of about 10^{-5} M.

Postgate: This may be getting us towards an answer to Professor Kelly's question, that it is useful to the organism to expend some ATP to accumulate a pool of PAPS, since it cannot accumulate a pool of APS.

Meister: This is a wasteful process anyway, because you are splitting ATP to PP_i, which has then to be split. If one were designing this for maximum efficiency one would have split ATP to ADP in the first place.

Whatley: But then the $\Delta G^{\circ\prime}$ would be -18, not -12 kcal!

Meister: That may be the answer!

Whatley: If you add the first three equations and get an equilibrium constant of zero, you could drive the synthetic reaction very simply by altering the concentrations of ATP and ADP.

Segel: Exactly. The $\Delta G^{\circ\prime}$ values simply tell us which way the reaction would proceed under a standard state condition which is 1 M in all reactants and products, or equimolar concentrations of reactants and products for reactions containing equal numbers of reactants and products (e.g., $A + B \rightleftharpoons P + Q$). The internal steady-state levels of ATP and sulphate in cells growing on sulphate as sole sulphur source are about 10^{-3} M and 10^{-2} M, respectively. The internal levels of APS and inorganic pyrophosphate are unknown. But if they are about 10^{-7} M each, the product concentration/substrate concentration ratio would be about 10^{-9}, which is 0.1 times the K_{eq} of 10^{-8}, and so the net flux would be in the direction of APS + PP_i formation.

By the way, the questions of the energy expense (i.e. the use of two ATP molecules) in activating sulphate may be irrelevant. Aerobic cells probably have ATP to spare.

Kredich: Speed is important, however.

Segel: Yes. As any well-educated biochemist knows, the kinetics of an enzyme-catalysed reaction are far more informative than the thermodynamics! It's not a question of which way a reaction will go that is important but, rather, how fast it goes.

Dodgson: Dr Segel, have you found PAPS sulphatase in *Penicillium chrysogenum?*

Segel: No, but we didn't really look. We studied choline sulphokinase with a partially purified preparation and obtained no indication that PAPS in the assay mixture was being rapidly destroyed (Renosto & Segel 1977).

Postgate: Professor Schiff says in his Abstract (p 49) that assimilatory sulphate reduction is *largely* restricted to plants and microorganisms. I would have said that it was exclusively confined to them.

Schiff: As I noted earlier (p 65), there are reports of the reduction of sulphate to sulphite in animals where the gut flora does not seem to be involved. Also, the protozoan *Leishmania tarentolae* has been shown to grow on sulphate as the sole source of sulphur (see Schiff & Hodson 1973).

Kredich: It has been said that tissues like bovine corneal epithelium can reduce sulphate to sulphide (Wortman 1963).

Rose: Dr Schiff has stated that we are not certain about the precise mechanism by which the sulphonic acid group of plant sulpholipids is derived from sulphate. In the light of Professor Kelly's paper on the sulphur cycle (p 3-18), and also the recent paper by Dr Harwood from our department (Harwood & Nicholls 1979) on the importance of plant sulpholipid in sulphur recycling, this is something we should note as worth particular attention for the future.

References

Denner HB, Stokes AM, Rose FA, Dodgson KS 1973 Separation and properties of the soluble 3′-phosphoadenosine 5′-phosphosulphate-degrading enzymes of bovine liver. Biochim Biophys Acta 315:394-401

Freedman B 1979 How many distinct enzymes are responsible for the several cellular processes involving thiol: protein-disulphide interchange? FEBS (Fed Eur Biochem Soc) Lett 97:201-210

Harwood JL, Nicholls R 1979 The plant sulpholipids — a major component of the sulphur cycle. Biochem Soc Trans 7:440-447

Pannikar KR, Bachhawat BK 1968 Purification and properties of ATP-sulphate adenylyl transferase from liver. Biochim Biophys Acta 151:725-729

Renosto F, Segel IH 1977 Choline sulfokinase of *Penicillium chrysogenum:* partial purification and kinetic mechanism. Arch Biochem Biophys 180:416-428

Schiff JA, Hodson RC 1973 The metabolism of sulfate. Annu Rev Plant Physiol 24:381-414

Segel IH, Johnson MJ 1973 Intermediates in inorganic sulphate utilization by *Penicillium chrysogenum*. Arch Biochem Biophys 103:216-226

Shoyab M, Marx W 1972 Enzyme–substrate complexes of ATP-sulphurylase from mouse mastocytoma. Biochim Biophys Acta 258;125-132

Woodin TS, Segel IH 1968 Gluthathione reductase-dependent metabolism of cysteine-*S*-sulfate by *Penicillium chrysogenum*. Biochim Biophys Acta 167:78-88

Wortman B 1963 Pyridine nucleotide-stimulated production of 3′-phosphoadenosine 5′-phosphosulphate by beef cornea epithelia extract. Biochim Biophys Acta 77:65-72

Oxidative phosphorylation linked to the dissimilatory reduction of elemental sulphur by *Desulfovibrio*

G.D. FAUQUE, L.L. BARTON* and J. LE GALL

Laboratoire de Chimie Bactérienne, C.N.R.S., 13274 Marseille Cedex 2, France

Abstract Hydrogenase and cytochrome c_3 purified from *Desulfovibrio gigas* and *D. desulfuricans* strain Norway form a soluble complex which is capable of transferring electrons from molecular hydrogen to colloidal sulphur (S^0). In this reaction, sulphur is reduced to hydrogen sulphide. Since both strains are capable of growth using elemental sulphur as terminal electron acceptor, it was of interest to check for oxidative phosphorylation in this sulphur reduction system. Membranes isolated from *D. gigas* or *D. desulfuricans* strain Norway contain hydrogenase and *c*-type cytochromes and catalyse the $H_2 \rightarrow S^0$ reaction. With *D. gigas*, esterification of orthophosphate is coupled to the membrane-mediated transfer of electrons from H_2 to S^0. A P/2e ratio of 0.1 was observed and this value could be reduced by the addition of methyl viologen or cytochrome c_3. These results indicate that the reaction of colloidal sulphur with c_3 may be more than a purely chemical reaction. Since whole cells can use sulphur flower while cell-free extracts react only with colloidal sulphur, it is evident that cells handle sulphur in a way which is not yet fully understood.

An important contribution to the knowledge of the sulphur cycle has been the discovery that some bacteria are able to utilize elemental sulphur as terminal electron acceptor (Pfennig & Biebl 1976, Biebl & Pfennig 1977).

These bacteria belong to the newly described genus *Desulfuromonas* bacteria that are unable to use sulphate as terminal electron acceptor, or to some species of *Desulfovibrio*, the well-known sulphate-reducing bacteria.

Since hydrogen sulphide, which is formed during the reduction of sulphur, can be reoxidized by green sulphur bacteria, a syntrophic relationship can be established between the two types of organisms, thus providing a simplified sulphur cycle (Fig. 1) Of course, such a 'consortium' can only operate with phototrophic bacteria such as *Chlorobium* which are able to export colloidal sulphur into the culture medium.

**On leave from the Department of Biology, University of New Mexico, Albuquerque, New Mexico 87131, USA.*

Sulphur in biology
(Ciba Foundation Symposium 72) p 71-86

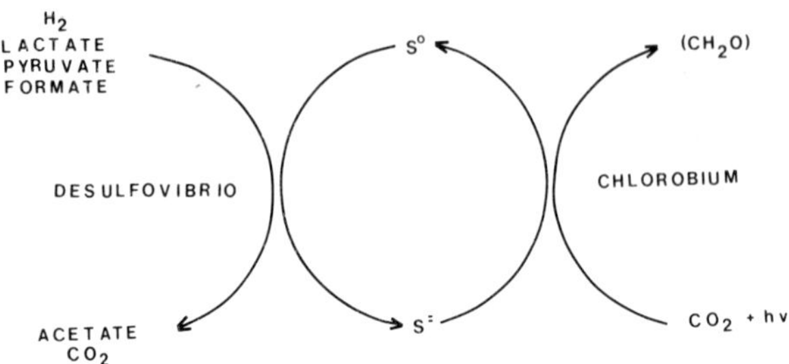

FIG. 1. Anaerobic sulphur cycle.

In fact, the discovery that elemental sulphur constitutes a respiratory substrate was due to a careful analysis of the so-called '*Chloropseudomonas etylica* N_2', which was proved to be a mixed culture containing *Chlorobium limicola, Desulfuromonas acetoxidans* and the strain of sulphate-reducer 9974.

It is interesting to note that this discovery has been delayed by the fact that the species of *Desulfovibrio* which has been so far the most intensively studied, namely *D. vulgaris* Hildenborough, is unable to grow using elemental sulphur. Furthermore, when cytochrome c_3 from this species was found to be able to reduce colloidal sulphur (Ishimoto et al 1958), it was concluded that this reaction had no physiological significance.

Recently (Fauque et al 1979b) the problem of the involvement of cytochrome c_3 in the reduction of elemental sulphur was reconsidered. It was found that cytochrome c_3 from *Desulfovibrio desulfuricans* strain Norway 4 (NCIB 8310) purified together with the sulphur reductase activity, thus establishing the involvement of this haemoprotein as a terminal reductase for sulphur.

Cytochromes c_3 from *Desulfovibrio gigas* (NCIB 9332) and the strain of sulphate-reducer 9974 were also able to perform the same reaction. In contrast, strong inhibition by the product of the reaction, hydrogen sulphide, was noted with cytochrome c_3 from *D. vulgaris* strain Hildenborough (NCIB 8303). This fact is sufficient to explain the inability of this strain to grow using elemental sulphur instead of sulphate.

That *Desulfovibrio* species can perform electron transfer-linked biosynthesis of ATP was shown by Peck (1966) in the reaction leading from molecular hydrogen to the reduction of sulphite. Later, oxidative phosphorylations were also shown to occur during the reduction of fumarate by hydrogen in particulate fractions from *D. gigas* (Barton et al 1970).

Since elemental sulphur has been proved to be a respiratory substrate in *Desulfovibrio,* it was tempting to check for the occurrence of ATP synthesis during its reduction by molecular hydrogen.

MATERIALS AND METHODS

The colloidal suspension of 'hydrophilic' sulphur was obtained from thiosulphate acidification with concentrated sulphuric acid by the method of Roy & Trudinger (1970).

Purified hydrogenase from *D. desulfuricans* Norway 4 was prepared as described by Fauque (1979). Pure cytochrome c_3 from the same strain was obtained by the method of Bruschi et al (1977).

Pure hydrogenase from *D. gigas* was a gift from C.E. Hatchikian.

Cultures of *D. gigas* and *D. desulfuricans* Norway were grown as previously described (Le Gall et al 1965). The cells were disrupted by a single passage through a French pressure cell under a pressure of 3 to 4 thousand pounds per square inch. Membranes were isolated from the extracts using differential sucrose gradient centrifugation (Barton et al 1970).

Sulphur reduction was followed by the Warburg respiratory method used by Fauque et al (1979b).

Esterification of orthophosphate was tested according to the method described by Barton et al (1970).

RESULTS

As already described (Fauque et al 1979b), the reduction of S^0 to H_2S was quite rapid when hydrogenase was mixed with the cytochrome c_3 in a hydrogen gas phase. Values of 1.96 μmoles H_2 oxidized per minute for *D. gigas* cytochrome c_3 and of 2.57 μmoles H_2 oxidized per minute for *D. desulfuricans* Norway cytochrome c_3 were obtained when 14 nanomoles of cytochrome c_3 were added to the 3 ml reaction vessel containing 11 μmoles S^0, excess hydrogenase and 100 μmoles phosphate buffer, pH 7.0.

During the reduction of S^0 a significant increase of turbidity was observed (Fauque et al 1979). Since centrifugation of this precipitate does not decrease the amount of cytochrome c_3 (from strain 9974) in solution, this turbidity is probably due to the appearance of a transitory form of sulphur rather than a complex between the colloidal substrate and the protein molecules.

Membranes isolated from both *D. gigas* and *D. desulfuricans* Norway 4 were rich in hydrogenase and cytochrome c_3. The addition of these membranes to S^0 resulted in hydrogen oxidation (Fig. 2).

The cytochrome content of these membrane fractions was determined to be 6.2×10^{-10} moles for 10.2 mg protein-membrane of *D. gigas* and 1.97×10^{-9} moles for 12.4 mg protein-membrane of *D. desulfuricans*. Molar extinction coefficients of 1.08×10^5 and 1.3×10^5 respectively were used (at 553 nm for the ferrous forms of

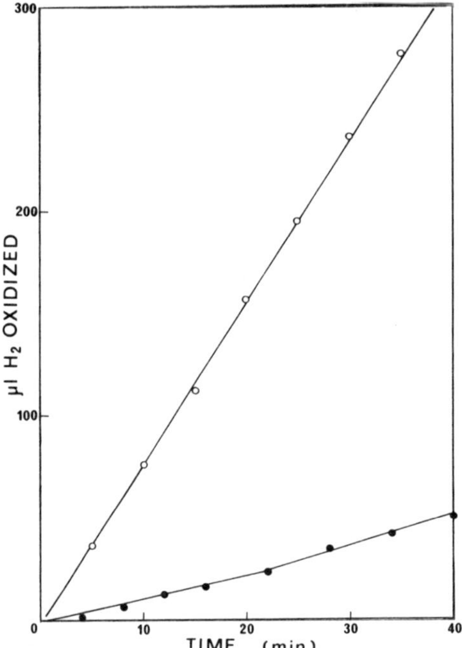

FIG. 2. Reduction of colloidal sulphur by membranes from *Desulfovibrio*. Reaction contained per 3 ml : 30 μmoles Tris-HCl buffer (pH 7.4) with 10.2 mg of membranes from *D. gigas* (●) or 12.4 mg of membranes from *D. desulfuricans* Norway (○) and with 5 μmoles of colloidal sulphur.

the cytochromes). No correction for the eventual presence of cytochrome c_3 (Mol.wt. 26 000) in *D. gigas* or of cytochrome $c_{553(550)}$ in *D. desulfuricans* Norway 4 (Fauque et al 1979a) was made.

The faster reaction rate with *D. desulfuricans* Norway membranes could be attributed to greater amounts of cytochrome.

As shown in Fig. 3, the ratio of S^0 to membrane is of particular importance, since maximal hydrogen oxidation rates occur at a certain concentration of sulphur for a given amount of membrane. This suggests the existence of a specific interaction between membranes and sulphur micelles.

If the ΔG′ value of – 6.5 kcal/reaction is correct for the reaction:

$$H_2 + S^0 \rightarrow H_2S$$

sufficient energy is released to account for the production of one ATP for two electrons.

The coupling of ATP production to electron flow from hydrogen to S^0 was examined with the *D. gigas* particulate fraction. The hydrogen fumarate system

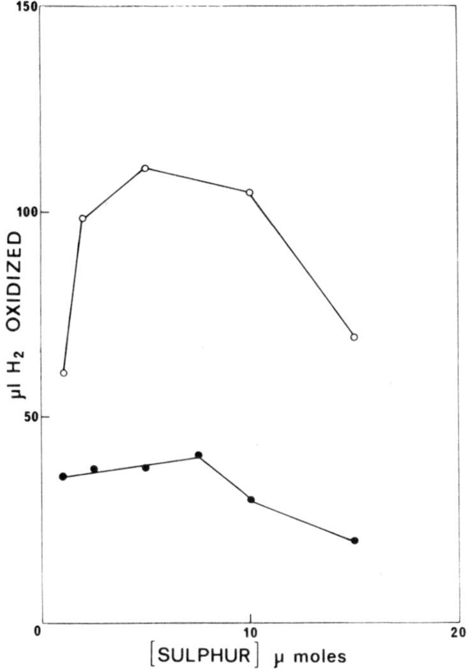

FIG. 3. Hydrogen oxidation at different concentrations of colloidal sulphur. Reaction contained per 3 ml : 30 μmoles Tris-HCl buffer (pH 7.4) with 10.2 mg of membranes from *D. gigas* (●) or 12.4 mg of membranes from *D. desulfuricans* Norway (○). Time of incubation for *D. gigas* was 28 min and for *D. desulfuricans* Norway, 15 min.

was used as a reference. Esterification values are given in Table 1. A P/2e ratio of 0.1 was obtained, a value similar to the one obtained with the hydrogen–sulphite system (Peck 1966). This suggests some similarity to the mechanism for oxidative phosphorylation coupled to the flow of electrons from hydrogen to sulphur compounds. However, an important difference is that in ATP production coupled to S^0 reduction, only the membrane fraction is required and no soluble components are needed. It is also to be noted that the electron transfer chain from hydrogen to S^0 can be reconstituted using two soluble components, namely hydrogenase and cytochrome c_3, while similar attempts in reconstituting either sulphite or fumarate reduction have failed so far.

Uncoupling of phosphorylation was observed with pentachlorophenol (10^{-4} M) and the addition of methyl viologen (0.3×10^{-4} M) or 1.3 nmoles of *D. gigas* cytochrome c_3. The effect of adding these two soluble electron transfer components may be explained by a short-circuiting of the membranous electron transfer mechanism.

TABLE 1

Oxidative phosphorylation coupled to sulphur reduction in *D. gigas*

Reaction mixture[a]	*H_2oxidized (μmol)*	*P_i esterified*[b] *(μmol)*	*P/2e⁻*
Complete H_2–fumarate system	4.1	1.32	0.3
Complete H_2–S^0 system	2.7	0.26	0.1
plus methyl viologen	15.1	0.04	0.0

[a]Reaction time was 30 min and 12.5 mg of membranes from *D. gigas* were used per reaction.
[b]P_i esterified was calculated by subtracting the control value, which had no electron acceptor, from each reaction.

DISCUSSION

That some bacteria can use elemental sulphur as terminal electron acceptor is of particular importance, since four times as much hydrogen sulphide is formed when S^0 replaces SO_4^{2-} as electron acceptor:

$$8[H] + 4S^0 \rightarrow 4H_2S$$
$$8[H] + H_2SO_4 \rightarrow H_2S + 4H_2O$$

The concern for H_2S toxicity is therefore increased when *Desulfovibrio* and *Desulfuromonas* grow in natural environments containing S^0 (Biebl & Pfennig 1977). A study done on various tropical soils has shown the ecological importance of this process (Traoré 1978).

As with phosphorylating systems from hydrogen to either sulphite or fumarate, only *D. gigas* was able to give active particulate fractions. The reason for this is unknown. A possible explanation is that the phospholipid composition of *D. gigas* membranes allows a better stabilization of the vesicles. *D. gigas* contains 70% of phosphatidylglycerol and 30% of phosphatidylethanolamine. These percentages are almost exactly reversed in *D. desulfuricans* Norway 4 and *D. vulgaris* (Makula & Finnerty 1974). In view of its anomalous phospholipid composition, these authors classify *D. gigas* as an atypical Gram-negative bacterium. Since another particularity of this bacterium is its size (Le Gall 1963), it is tempting to speculate that a more stable membrane is necessary to maintain the structure of this 'gigantic' organism.

The very speculative model for S^0 reduction by particles of *D. gigas*, presented in Fig. 4, deserves comment. The rupture of bacteria by French pressure cells is recognized as producing inverted (inside-out membranes) vesicles in many bacteria. Direct evidence for such a configuration is given by a study of ATPase

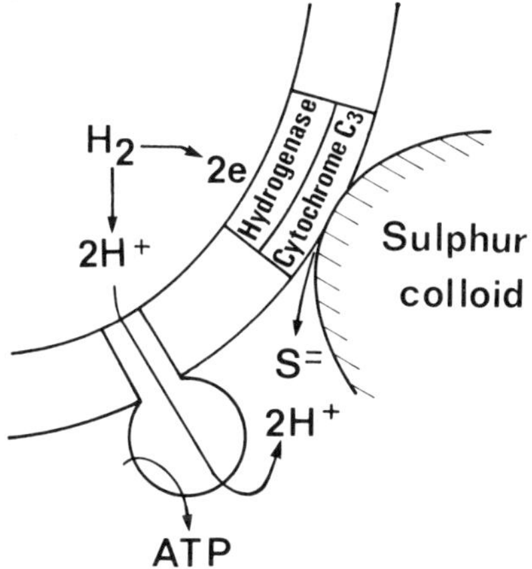

FIG. 4. Model for reduction of colloidal sulphur in *D. gigas*.

activity by Guarraia & Peck (1971). Since *D. gigas* hydrogenase is in the periplasmic space (Bell et al 1974), in inverted membranes the hydrogenase would be on the inside, as depicted in the model. The movement of protons across the membrane and the coupled ATP production is according to the Mitchell hypothesis, since the reaction seems to occur in closed vesicles. Of course, the more general model presented by Williams (1978) can also be used to describe this phosphorylation. Low P/2e ratios could be attributed to the adsorption of some hydrogenase onto the surface of the inverted vesicles. Hydrogen oxidation at the outer surface of these vesicles would then result in the reduction of coupled phosphorylation, due to the reduction of the proton gradient across the membrane.

The uncoupling effect of the addition of soluble cytochrome c_3 would be due to direct reduction of these molecules by membrane-bound cytochrome or hydrogenase.

Since the reduced soluble molecules of cytochrome c_3 are active, they can then react directly with colloidal sulphur.

Since the establishment of the tertiary structure of cytochrome c_3 is now in progress (Frey et al 1976) there is hope that a detailed mechanism for the reduction of S^0 will be available soon. Unfortunately, little is known about the structure of colloidal sulphur itself and it is not clear whether cytochrome c_3 reduces S_8 rings or long chains of sulphur atoms, which are both thought to be constituents of the sulphur micelles. An interesting fact concerning the reduction of colloidal sulphur by the species *Desulfuromonas acetoxidans* has been found by Bache (1978). This organism contains a cytochrome c_7 (Probst et al 1977) which is related to cytochrome c_3 but has only three haem groups instead of four. This cytochrome is unable to reduce S^0 itself; the complete reaction requires another cytochrome fraction. The loss of sulphur reduction activity as a result of the absence of one haem and the related peptides may be due either to changes in redox properties of the molecule or to the absence of a specific sulphur attachment site on the cytochrome.

The electron transfer chain coupled to oxidative phosphorylation may appear surprisingly short, since it can be reconstituted in solution with only two proteins, namely hydrogenase and cytochrome c_3. However, hydrogenase from *D. gigas* has been shown to contain three clusters each having four iron and four labile sulphur atoms (Hatchikian et al 1978), while cytochrome c_3 is a tetrahaemoprotein in which each haem group possesses a distinct redox potential (Xavier et al 1979). Thus a total of seven redox centres can be counted, a number comparable to the centres found in mitochondrial phosphorylating site II (Chance 1972).

It is noteworthy that all elements are present that have either been purified or can be purified, that theoretically allow a complete reconstitution of phosphorylating vesicles: ATPase, phospholipids and the two electron transfer proteins.

Cytochrome c_3 is the sulphur reductase in some strains of sulphate-reducing bacteria, particularly strains Norway 4 and 9974. However, cytochrome c_3 from *D. vulgaris* Hildenborough is inhibited by the product of the reaction, hydrogen sulphide. This fact explains why this strain is unable to grow by reducing elemental sulphur. This shows that a drastic change in the function of cytochrome c_3 took place in *D. vulgaris* Hildenborough, where it has only the function of an electron carrier in the electron transfer chain, instead of being a terminal reductase.

References

Bache R 1978 Untersuchungen von c-Typ Cytochromen und eines Eisenswefelproteins aus der Cytoplasmafraktion von *Desulfuromonas acetoxidans*. Thesis, University of Göttingen

Barton LL, Le Gall J, Peck HD Jr 1970 Phosphorylation coupled to oxidation of hydrogen with fumarate in extracts of the sulfate-reducing bacterium *Desulfovibrio gigas*. Biochem Biophys Res Commun 41:1036-1042

Bell GR, Le Gall J, Peck HD Jr 1974 Evidence for the periplasmic location of hydrogenase in *D. gigas*. J Bacteriol 120:994-997

Biebl H, Pfennig N 1977 Growth of sulfate-reducing bacteria with sulfur as electron acceptor. Arch Microbiol 112:115-117

Bruschi M, Hatchikian CE, Golovleva LA, Le Gall J 1977 Purification and characterization of cytochrome c_3, ferredoxin, and rubredoxin isolated from *D. desulfuricans* Norway. J Bacteriol 129:30-38

Chance B 1972 The nature of electron transfer and energy coupling reactions. FEBS (Fed Eur Biochem Soc) Lett 23:3-20

Fauque G 1979 Réduction physiologique du soufre colloidal chez les bactéries sulfato-réductrices. Thesis, University of Aix-Marseille I

Fauque G, Bruschi M, Le Gall J 1979a Purification and some properties of cytochrome $c_{553(550)}$ isolated from *D. desulfuricans* Norway. Biochem Biophys Res Commun 86:1020-1029

Fauque G, Herve D, Le Gall J 1979b Structure-function relationship in hemoproteins: the role of cytochrome c_3 in the reduction of colloidal sulfur by sulfate-reducing bacteria. Arch Microbiol 121:261-264

Frey M, Haser R, Pierrot M, Bruschi M, Le Gall J 1976 Preliminary crystallographic study on cytochrome c_3 of *D. desulfuricans* (strain Norway). J Mol Biol 104:741-743

Guarraia LJ, Peck HD Jr 1971 Dinitrophenol-stimulated adenosine triphosphatase activity in extracts of *D. gigas*. J Bacteriol 106:890-895

Hatchikian EC, Bruschi M, Le Gall J 1978 Characterization of the periplasmic hydrogenase from *D. gigas*. Biochem Biophys Res Commun 82:451-461

Ishimoto M, Kondo Y, Kameyama T, Yagi T, Shiraki M 1958 The role of cytochrome in the enzyme system of sulfate-reducing bacteria. In: Proc int symp enzyme chemistry, Tokyo and Kyoto. Maruzen, Tokyo, p 229-234

Le Gall J 1963 A new species of *Desulfovibrio*. J Bacteriol 86:1120

Le Gall J, Mazza G, Dragoni N 1965 Le cytochrome c_3 de *D. gigas*. Biochim Biophys Acta 99:385-387

Makula RA, Finnerty WR 1974 Phospholipid composition of *Desulfovibrio* species. J Bacteriol 120:1279-1283

Mitchell P 1967 Proton-translocation phosphorylation in mitochondria, chloroplasts and bacteria: natural fuel cells and solar cells. Fed Proc 26:1370-1379

Peck HD Jr 1966 Phosphorylation coupled with electron transfer in extracts of the sulfate reducing bacterium, *D. gigas*. Biochem Biophys Res Commun 22:112-118

Pfennig N, Biebl H 1976 *Desulfuromonas acetoxidans* gen. nov. and sp. nov., a new anaerobic, sulfur-reducing, acetate-oxidizing bacterium. Arch Microbiol 110:3-12

Probst I, Bruschi M, Pfennig N, Le Gall J 1977 Cytochrome $c_{551.5}$ (c_7) from *Desulfuromonas acetoxidans*. Biochim Biophys Acta 460:58-64

Roy AB, Trudinger PA 1970 The biochemistry of inorganic compounds of sulphur. Cambridge University Press, Cambridge

Traoré SA 1978 Contribution à l'étude *in situ* des bactéries sulfo-réductrice dans quelques sols tropicaux. Cahiers de l'Office de la Recherche Scientifique et Technique Outre-Mer. Centre O.R.S.T.O.M. de Dakar. Sér Biol

Xavier AV, Moura JJG, Le Gall J, Dervartanian DV 1979 Oxidation-reduction potentials of the hemes in cytochrome c_3 from *D. gigas* in the presence and absence of ferredoxin by EPR spectroscopy. Biochimie (Paris), 61:689-695

Williams RJP 1978 The history and the hypothesis concerning ATP formation by energized protons. FEBS (Fed Eur Biochem Soc) Lett 85:9-19

Discussion

Hamilton: I was interested in your inverted-vesicle model for the reduction of colloidal sulphur. But how do you think colloidal sulphur, or flowers of sulphur, get into the intact cell?

Le Gall: Pfennig & Biebl (1976) were able to grow *Chlorobium* separately from *Desulfotomaculum* (the sulphur-reducing organism). Colonies of the latter organism can grow on a Petri dish at some distance from *Chlorobium* colonies. Since colloidal sulphur is insoluble, how can it diffuse fast enough to allow the two organisms to grow through interspecies sulphur transfer? This question, which is similar to yours, has received no answer yet.

Hamilton: Sulphate reduction has a very unfavourable mid-point potential

(–207mV) and that is overcome by making adenosine 5′-phosphosulphate (APS) the form reduced. The sulphur mid-point potential is even worse (–276mV) than the sulphate one, so how does the cell overcome this problem and get any energy out of the reaction at all?

Le Gall: How much can one rely on this figure for sulphur reduction, however? On the one hand you have something which is not soluble; on the other hand you have something which *is* soluble. It is supposed to be a reversible reaction, but how can you reverse it and go from sulphide to the same form of sulphur again?

Postgate: You say that sulphur is not soluble, but the saturation solubility of sulphur in water could be as much as 5 μg-atom/ml, according to *Gmelin's Handbuch der anorganischen Chemie,* which is not really low. With a large surface, such as you have with a suspension of colloidal sulphur, the kinetic situation is very different from what it is with flowers of sulphur hydrophobically floating on water. So I don't regard this as a very serious problem. Sulphur evaporates; its vapour pressure is around 1.7×10^{-6} Torr at 23 °C, so the partial pressure of sulphur vapour in the aqueous environment could also be appreciable. Many microbes grow, albeit slowly, with 5 μM substrate concentrations or in dilute solutions of relatively insoluble gases.

This leads me to a different point. In my original experiments we took a colloidal sulphur sol and an ultrafiltrate thereof and saw that sulphate-reducing bacteria grew equally well on both of them. That implies that colloidal sulphur solutions are always accompanied by some truly soluble oxidized sulphur that passes very fine filters. I concluded that it isn't colloidal sulphur that is the substrate of the organism, but something like sulphane monosulphonic acid or polysulphide anions, which inevitably form whenever you manipulate colloidal sulphur solutions in air. So, going back to the point about the enzyme for which you don't know the structure of the substrate, would it be unfair to suggest that you don't even know what the substrate is?

Le Gall: No, it isn't, since this is exactly what we wrote ourselves (Fauque et al 1979)! Of course c_3 doesn't reduce tetra- or trithionate. Nobody knows what trithionate reductase is because it has not been isolated, although it must be important in the dissimilatory pathway. Tetrathionate has been partially purified. We haven't tried the highest polythionate yet. It could be that c_3 doesn't see the smaller polythionates, only the larger ones. Dr Cammack (personal communication) has done an electron-paramagnetic-resonance spectroscopy (e.p.r.) experiment with hydrogenase, c_3 and sulphur. The *g* values of the haems were not changed, which suggests that sulphur doesn't go very close to the haem iron. It also shows that as soon as the reaction starts a radical is formed which doesn't exist in the suspension of sulphur.

Kelly: Have you looked at the ultraviolet spectra of either of the reactions, before they get going, or of the sulphur solutions? An absorbance at about 320 nm might indicate polysulphide chains; as the chain gets longer absorbance values shift towards the higher wavelengths, 320–340 nm.

Le Gall: No, we haven't done this.

Postgate: What is the current view on sulphur utilization in thiobacilli?

Kelly: It hasn't really changed; there hasn't been much progress in elucidating attack on elemental sulphur. Apart from the fact that lipids are no longer believed to be involved, we are back to the idea that the solubility of sulphur at 1.7 mg/l may be adequate for its uptake and oxidation.

Whatley: What was the radical, Dr Le Gall? Where was it on the *g*-value scale?

Le Gall: The *g* value is 2.04. The fact that S^0 does not seem to get close to the haem iron is puzzling, since we found evidence that the nitrogen of compounds like NO_2^- and NH_2OH, which are also reduced by c_3, does get close to the haem (Der Vartanian & Le Gall 1974).

Postgate: When you get turbidity on adding sulphur to your system, when you centrifuge it, you say that the proteins remain in solution. Does that mean that whatever complex is forming that turbidity does not include c_3? This is difficult to understand.

Le Gall: We simply centrifuged the reaction mixture and the c_3 in solution remained the same.

Postgate: But you had hydrogenase there as well?

Le Gall: Yes.

Segel: Is there a resuspendable pellet that you could look at?

Le Gall: No, because the turbidity is not stable in air. It seems to be switching from one form of sulphur to another.

Postgate: The colloidal sulphur must be interacting with the protein in some sense, which means that presumably protein is binding to colloidal particles and then aggregating as in an immunological reaction, as a cluster which becomes sedimentable. This process seems to be reversible. Surely there must be some other protein in the preparation which you haven't seen?

Le Gall: If so, it has to be less than 5% of the c_3, but $\pm 5\%$ means nothing in this kind of experiment. We know we need an interaction between hydrogenase and c_3 for the reaction to occur. And hydrogenase alone doesn't reduce colloidal sulphur.

Postgate: But does pure hydrogenase give turbidity with colloidal sulphur?

Le Gall: No, and furthermore, it is not active in the absence of c_3.

Hamilton: You can get oxidation with hydrogenase and c_3 as partially purified proteins more or less in solution, but for phosphorylation you have to add membranes to provide the ATPase. Are you also adding other components of the electron transport chain which are necessary for phosphorylation?

Le Gall: No. It is probably a coincidence that one can purify the two redox components which are able to perform the redox function, but there are many electron transfer proteins in whole cells (see Le Gall et al 1979). But it is possible that the two proteins are sufficient, since they contain a total of seven redox centres: four haems in cytochrome c_3 and three iron-sulphur centres in hydrogenase, which is as many as some phosphorylating complexes from mitochondria. However, it is probably an oversimplified picture, as are all pictures of electron transfer chains.

Hamilton: Particularly if the hydrogenase and c_3 are on the outer surface of the membrane, which will give chemiosmotic problems unless there are other components in the electron transport chain.

Le Gall: Cytochrome c_3 is never found entirely in the periplasmic space. The fact that the coupling between S^0 reduction and ATP synthesis is bad is probably due to a messy distribution of c_3 on the vesicles. This is illustrated by the fact that addition of soluble c_3 uncouples the phosphorylation.

Postgate: In nature, does *Desulfovibrio* often find itself in an environment where there is free sulphur?

Le Gall: I don't know, but artificial conditions can be created where sulphur becomes abundant: sulphur-coated urea is widely used as a slow-release form of nitrogen. If it is put in rice paddy fields there are problems, because in this anaerobic environment you start making four times more H_2S than with sulphates, which may kill the young rice plants.

Kelly: Could we just complete the sulphur cycle in the rice paddies by mentioning that *Beggiatoa* will oxidize the sulphide and precipitate iron and thereby relieve the inhibition of the rice plants (Joshi & Hollis 1977).

Le Gall: Not always. Hydrogen sulphide can be toxic in water-logged soil.

Postgate: An important cause of crop damage in rice cultivation is high sulphide contamination.

Dodgson: Is there any information on how *Desulfovibrio* makes cysteine? We know that 3′-phosphoadenosine 5′-phosphosulphate (PAPS) plays the central role in assimilatory sulphate reduction in bacteria such as *Escherichia coli*. However, *Desulfovibrio* apparently produces only APS and no one seems to have established how the organism produces its cysteine.

Postgate: There are at least six different sulphite reductases reported in various strains of *Desulfovibrio* and *Desulfotomaculum,* some of which are thought to be assimilatory rather than dissimilatory.

Le Gall: They have a lot of ways of making H_2S, in fact. In the same organism you can find two different sulphite reductases, for example in *D. vulgaris*. One enzyme is actually the bisulphite reductase (dissimilatory) and when pure it makes trithionate. The other makes H_2S directly from sulphite (assimilatory sulphite reductase).

Kelly: Is trithionate an accepted intermediate on the pathway from sulphite to sulphide?

Le Gall: No, not by everybody. The main objection is that trithionate reductase has never been purified, whereas thiosulphate reductase has.

Kägi: The interrelationship between the sequestration of sulphide and of iron in geological deposits, which Dr Kelly referred to earlier, is interesting. What are the causal relationships in these events? Is the precipitation an incidental accompaniment of the metabolic activities of the microorganisms involved in assimilatory sulphide reduction? Or is it possible that these deposits have arisen from iron-sulphur proteins formed biosynthetically in these organisms?

Postgate: This falls into two parts, as I understand it: first, what one might call the plain chemistry, that if you reduce sulphate and generate sulphide in an evironment where there is iron, as there is in all environments, you will get precipitation of iron sulphides, so that is coincidental. But secondly there is the question mentioned earlier (p 16) of how organisms obtain iron for their own metabolic uses in the presence of free sulphide and whether they make use of iron sulphide incidentally or by some deliberate metabolic process.

Le Gall: The iron protein of *Desulfovibrio* found by Hatchikian & Bruschi (1979) is loaded with iron sulphide. It also contains molybdenum, but no function is known for this protein. It doesn't seem to undergo redox cycles. It is just iron sulphide, hooked into a protein. Because of the large amounts that the cells produce, it could play a role in the synthesis of iron-sulphur proteins.

Hamilton: You said that there is a coincidental aspect in iron and sulphur precipitating together in the environment. But there is also organically and biologically a function of the association of iron and sulphur, in that so many iron-sulphur proteins are involved in redox reactions, while certain microorganisms have the option of oxidizing reduced sulphur or reduced iron.

Kelly: The sedimentary copper, zinc and silver sulphides are biogenic as well as iron sulphides. There is evidence that this happens away from the organism, so there is a possibility of precipitation on the organism *or* away from it, by a purely chemical process of diffusion. You can simulate the system that operates in, say, an anaerobic sediment where sulphate reduction is going on, with sulphide diffusion through a layer of silt and metal diffusing down, in a large test-tube with a layer of sand, sulphur coming from the bottom, and metal from the top (Bubela & MacDonald 1969).

Depending on the solubility products of the different sulphides banding develops, like the banding seen for example in the Australian sulphides of Mt. Isa. At one time it was reckoned that there was positive precipitation of metal sulphides on *Desulfovibrio*-type organisms and a two-stage process – accumulation of the metal and then sulphide formation. I don't see how one distinguishes whether it was really sulphide precipitation on the surface or active accumulation of metal and sulphide formation.

Brierley: Our discussion on metallic sulphide deposition seems to suggest that *all* metal sulphides are mediated by biological activity. However, many deposits are formed hydrothermally. Anaerobic bacteria which produce hydrogen sulphide from sulphate isotopically fractionate the sulphur, resulting in an enrichment of ^{32}S. Measurement of the ratio of ^{32}S to ^{34}S will distinguish whether a sulphide mineral deposit is of hydrothermal or biogenic origin (Jensen 1959).

Kelly: One can date the origin of sulphate reduction on the evidence of $^{34}S/^{32}S$ isotope ratios as about 3.2×10^9 years ago, when there was a discontinuity and the enrichment of ^{32}S in precipitated sedimentary sulphide, because of this biological preference for ^{32}S sulphides (Trudinger 1976).

Kredich: Is it conceivable that this disequilibrium which has been used to date the origin of life on earth could have been formed by a non-biotic diffusional process?

Postgate: It is possible, but unlikely. Sulphur isotope fractionation is a kinetic isotope effect which occurs while the reaction is progressing. But as the reaction approaches completion the isotope ratio returns to the original isotope distribution.

The use of the sulphur isotope distribution in geochemistry has relied on areas where oxidized and reduced sulphur occur simultaneously in the same place, so that at some stage the sulphur cycle was in process. But if the cycle goes to completion you ought to end up with the same isotope distribution as was there at the beginning.

Kelly: You see this in German copper deposits where copper sulphide was deposited over a period of about five thousand years in a closed basin, so that the sulphate was used up, the enrichment declined and the isotope ratio composition tended back towards the initial ratio (Trudinger 1976).

Postgate: We have discussed the question of the accessibility of a relatively insoluble substrate like sulphur; the same applies to an insoluble substrate like iron when it is present as iron sulphide.

Le Gall: Desulfovibrio can make much iron protein and iron-sulphur protein: the only evidence for some sort of iron semi-starvation is the fact that flavodoxin is synthesized together with ferredoxin in normal media. Knight & Hardy (1966) found that when you starve *Clostridium* of iron it starts making a flavoprotein, flavodoxin, instead of ferredoxin, and that flavodoxin replaces ferredoxin in all its functions. In *Desulfovibrio,* by contrast, flavodoxin is always made even when the growth medium contains a lot of iron. The regulatory mechanism perhaps works differently, or *Desulfovibrio* may always be slightly starved for iron.

Postgate: This doesn't take us much further, because you cannot be sure that there is a lot of iron available, even if you have added it, because it could mostly be inaccessible to the organism as FeS.

Le Gall: We are in a situation where two proteins are being synthesized at the same time, and apparently have the same role in the cell.

Whatley: The equivalent is found for oxygen evolution from the blue-green algae: if these algae are starved of iron they produce flavodoxin. There the functions of the two proteins must again be identical; the principal function is to reduce pyridine nucleotides.

Ferredoxin is an example of how to make an enzyme without actually trying! If you put the apoprotein of ferredoxin, which has no catalytic activity, in a reducing medium with H_2S and ferrous iron, an active enzyme is formed which has a redox potential the same as that of hydrogen gas.

Postgate: It may be that the iron-sulphur cluster is the one enzyme prosthetic group that does not need to be synthesized; perhaps it just comes together.

Kelly: Is there any biochemical reason why facultatively aerobic sulphate-reducing bacteria haven't been found? That is, organisms that could use either oxygen or sulphate.

Le Gall: Desulfovibrio can tolerate oxygen but not use it as a terminal electron acceptor.

Postgate: I have tried hard to make it do so, because it has an active and effective oxidative metabolism, and if you couple those oxidations to oxygen reduction you ought to generate ATP. There is no metabolic reason why it shouldn't be a facultative aerobe.

Kelly: It is very odd, since it has had the longest time to evolve this capacity.

Postgate: An important consideration is that its pyrophosphatase is endogenously inactivated by air, so not only sulphate reduction but most biosyntheses cannot then take place (Ware & Postgate 1971). If mutants able to use O_2 have appeared, they have lost the property of assimilatory sulphate reduction very quickly, since nobody has isolated this sort of organism.

References

Bubela B, MacDonald JA 1969 Formation of banded sulphides: metal ion separation and precipitation by inorganic and microbial sulphide sources. Nature (Lond) 221:465-466

Der Vartanian DV, Le Gall J 1974 Cytochrome c_3: a monomolecular electron transfer chain. Biochim Biophys Acta 449:275-284

Fauque GD, Herve D, Le Gall J 1979 Structure–function relationship in hemoproteins: the role of cytochrome c_3 in the reduction of colloidal sulfur by sulfate-reducing bacteria. Arch Microbiol 121:126-264

Hatchikian EC, Bruschi M 1979 Isolation and characterization of a molybdenum iron-sulfur protein from *Desulfovibrio africanus*. Biochem Biophys Res Commun 86:725-734

Knight E, Hardy BWF 1966 Isolation and characterization of flavodoxin from nitrogen fixing *Clostridium pasteurianum*. J Biol Chem 241:2752

Jensen ML 1959 Sulfur isotopes and hydrothermal mineral deposits. Econ Geol 54:374-394
Joshi MM, Hollis JP 1977 Interaction of *Beggiatoa* and rice plant: detoxification of hydrogen sulfide in the rice rhizosphere. Science (Wash DC) 195:179-180
Le Gall J, Der Vartanian DV, Peck HD, Jr 1979 Flavoproteins, iron proteins and hemoproteins as electron transfer components of the sulfate-reducing bacteria. Curr Top Bioenerg 9: in press
Pfennig N, Biebl M 1976 *Desulfuromonas acetoxidans* gen. nov. and sp. nov., a new anaerobic sulfur-reducing acetate-oxidizing bacterium. Arch Microbiol 110:3-12
Trudinger PA 1976 Microbiological processes in relation to ore genesis. In: Wolf KH (ed) Handbook of strata-bound and stratiform ore deposits. Elsevier, Amsterdam, p 135-190
Ware DA, Postgate JR 1971 Physiological and chemical properties of a reductant-activated inorganic pyrophosphatase from *Desulfovibrio desulfuricans*. J Gen Microbiol 67:145-160

Synthesis of L-cysteine in *Salmonella typhimurium*

NICHOLAS M. KREDICH, M. DANUTA HULANICKA* and SCOTT G. HALLQUIST

*Howard Hughes Medical Institute at Duke University Medical Center, Durham, North Carolina 27710, USA and *Institute of Biochemistry and Biophysics, Polish Academy of Sciences, Warsaw, Poland*

Abstract In *Salmonella typhimurium* and *Escherichia coli* the biosynthesis of L-cysteine from L-serine and inorganic sulphate proceeds along a branched convergent pathway along one arm of which sulphate is reduced to sulphide, while on the other L-serine is acetylated to *O*-acetyl-L-serine. This system is subject to positive genetic control in which growth on a poor sulphur source, *O*-acetyl-L-serine and the product of the *cysB* regulatory gene are all required for derepression. The final step consists of the formation of L-cysteine from *O*-acetyl-L-serine and sulphide. We find that in *S. typhimurium* this reaction is catalysed by two different enzymes, *O*-acetylserine sulphydrylase A and *O*-acetylserine sulphydrylase B, coded for by *cysK* and *cysM* respectively. Both enzymes are under the control of the cysteine regulon, and either alone is sufficient for cysteine prototrophy during aerobic growth. Although the advantage to the bacterium of having two separate enzymes to carry out the same reaction is unclear, preliminary data suggest that *O*-acetylserine sulphydrylase B is preferentially utilized for cysteine biosynthesis during anaerobic growth. We speculate that one enzyme may prefer free sulphide as a substrate while the other may use a bound form of sulphide.

PATHWAY AND GENETICS

Early studies on cysteine biosynthesis in *Escherichia coli* and *Salmonella typhimurium* were aided considerably by the availability of large numbers of mutant strains classified as 'cysteine auxotrophs' (Clowes 1958a,b, Mizobuchi et al 1962). Strictly speaking, most of these proved to be sulphide auxotrophs since by custom the definition of cysteine prototrophy refers to the ability of wild type to grow on minimal medium containing sulphate as the sole sulphur source. Therefore, from the microbial geneticist's viewpoint the cysteine biosynthetic pathway includes reactions involved in the transport and assimilatory reduction of inorganic sulphate as well as those which lead directly from L-serine to L-cysteine (Fig. 1). Subsequent studies on the regulation of cysteine biosynthesis have shown that the

Sulphur in biology
(Ciba Foundation Symposium 72) p 87-99

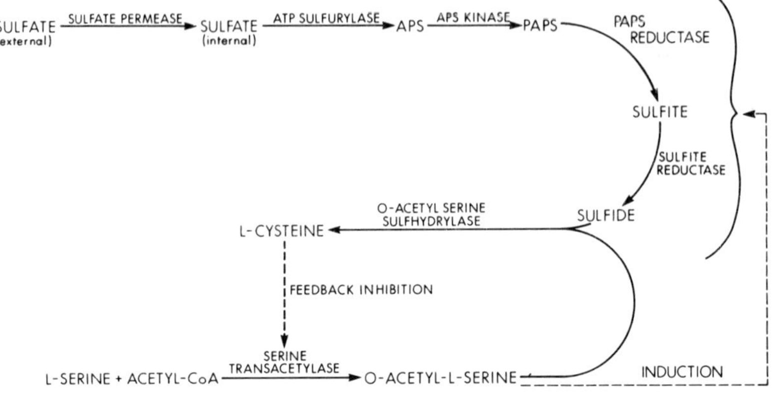

FIG. 1. Pathway of L-cysteine biosynthesis in *S. typhimurium* and *E. coli*. The regulatory features shown include feedback inhibition of serine transacetylase by L-cysteine and induction of the entire pathway, save serine transacetylase, by *O*-acetyl-L-serine.

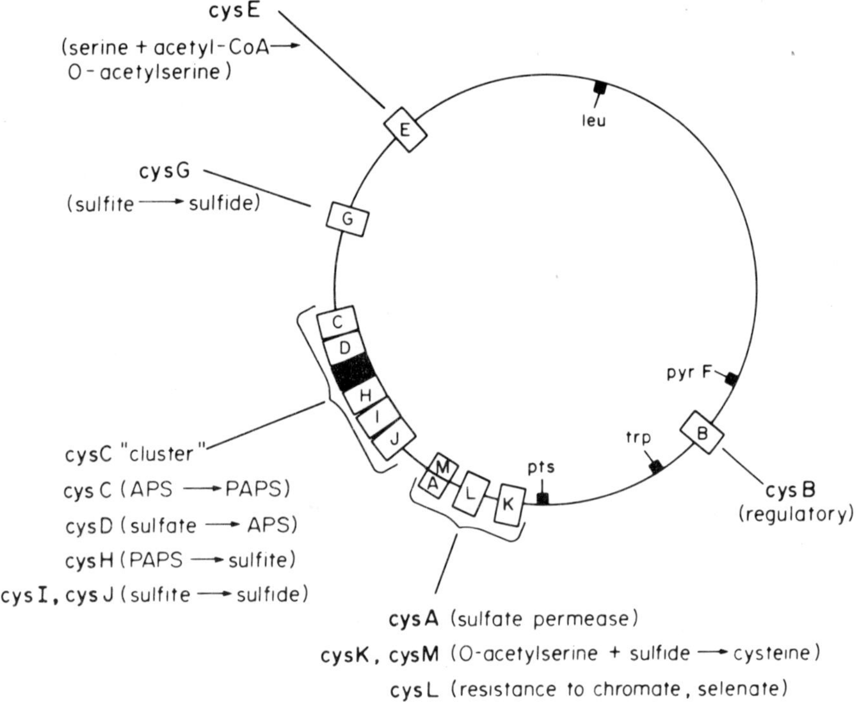

FIG. 2. Genetic loci for L-cysteine biosynthesis in *S. typhimurium*. *cysM* is located adjacent to *cysA*, but the order is not known.

bacterium shares our perspective in this regard, for both arms of this branched convergent pathway are under the same system of genetic control.

Genes involved with cysteine biosynthesis in *S. typhimurium* are scattered throughout the chromosome with clustering noted in the *cysCDHIJ* region and to a lesser extent around *cysA* (Fig. 2). Most have been defined by mutations which cause cysteine auxotrophy and are structural genes for enzymes in the biosynthetic pathway (Dreyfuss & Monty 1963). Thus, *cysA* codes for elements of the sulphate permease system (Dreyfuss 1964), *cysD* for ATP sulphurylase (EC 2.7.7.4), *cysC* for adenosine-5′-phosphosulphate kinase (EC 2.7.1.25), *cysH* for 3′-phosphoadenosine-5′-phosphosulphate reductase (now known as PAPS sulphotransferase: Tsang & Schiff 1978), *cysI* and *cysJ* for sulphite reductase apoenzyme (EC 1.8.1.2) (Siegel et al 1971), *cysG* probably for an enzyme which synthesizes the sulphite reductase co-enzyme sirohaem (Murphy & Siegel 1973), *cysE* for serine transacetylase (EC 2.3.1.30) (Kredich & Tomkins 1966), *cysK* for *O*-acetylserine sulphydrylase A (*O*-acetyl-L-serine acetate-lyase [adding H_2S]; EC 4.2.99.8) (Hulanicka et al 1974), and as I shall describe below *cysM* appears to be the structural gene for *O*-acetylserine sulphydrylase B. Other loci include *cysB*, which codes for an element of positive genetic control (Jones-Mortimer 1968, Kredich 1971, Jagura et al 1978), and *cysL*, mutations in which give resistance to selenate but have no known effect on cysteine biosynthesis (Hulanicka 1972).

REGULATION OF CYSTEINE BIOSYNTHESIS

Regulation of the cysteine biosynthetic pathway is accomplished by an interacting, dual system of gene repression-derepression and direct modulation of enzyme activity by metabolites. This latter form of control is most striking in the case of serine transacetylase, which is feedback-inhibited by L-cysteine with an apparent K_i of 1×10^{-6}M (Kredich et al 1969). *O*-Acetyl-L-serine, the product of this enzyme, is not only a direct precursor of L-cysteine but also serves as an internal inducer, which is required for derepression of all the enzymes in the pathway except for serine transacetylase itself, whose expression is not subject to this form of regulation (Jones-Mortimer et al 1968, Kredich 1971). Thus, growth on a poor sulphur source such as glutathione or L-djenkolate causes *O*-acetyl-L-serine levels to rise with resultant derepression of *O*-acetylserine sulphydrylase A and the enzymes of sulphate reduction. Conversely, growth on L-cysteine decreases *O*-acetyl-L-serine levels through end-product inhibition of serine transacetylase, and the other cysteine biosynthetic enzymes are then synthesized at a very slow rate. Strains carrying mutations in *cysE* are unable to synthesize *O*-acetyl-L-serine and require an exogenous supply of this compound for derepression.

Upon first consideration, the ability of L-cysteine to 'repress' enzymes of the biosynthetic pathway would appear to be through its action on serine transacetylase and *O*-acetyl-L-serine levels. This situation is more complicated than it seems, however, for exogenously supplied *O*-acetyl-L-serine does not derepress cells grown on L-cysteine. The explanation for this is not clear, but we believe it may be due to the desulphydration of cysteine (Kredich et al 1972) and the subsequent reaction of sulphide with *O*-acetyl-L-serine at a rate sufficient to keep intracellular levels of this inducer below those required for derepression.

The third element required for derepression, in addition to *O*-acetyl-L-serine and growth on a poor sulphur source, is the *cysB* gene (Jones-Mortimer 1968, Kredich 1971). The characterization of temperature-sensitive and amber mutations in *cysB* indicates that the gene product is a protein (Tully & Yudkin 1975), and our fine-structure genetic map of this region suggests that it codes for a single polypeptide chain (Cheney & Kredich 1975). Most known *cysB* mutations result in cysteine auxotrophy, and these *cysB*$^-$ strains cannot be appreciably derepressed for cysteine biosynthetic enzymes by any known nutritional means. A few prototrophic *cysB* mutants have been described which are designated *cysB*^{c} because they express high levels of cysteine biosynthetic enzymes even in the absence of *O*-acetyl-L-serine (in a *cysE*$^-$ background) and when grown on L-cysteine (Kredich 1971). A particularly interesting mutant is *cysB484*, which is a cysteine auxotroph and *cysB*$^-$ with respect to the pathway of sulphate reduction, but *cysB*^{c} for the expression of *O*-acetylserine sulphydrylase A. The product of this allele apparently is altered in such a way that it cannot effect the expression of most *cys* genes, but permits high-level synthesis of *O*-acetylserine sulphydrylase A even in the absence of *O*-acetyl-L-serine.

O-ACETYLSERINE SULPHYDRYLASE ENZYMES

I wish now to focus attention on the final step in this pathway, the reaction between *O*-acetyl-L-serine and sulphide to give L-cysteine. We originally observed this activity, termed *O*-acetylserine sulphydrylase, as one which co-purified with the enzyme serine transacetylase from *E. coli* (Kredich & Tomkins 1966). Subsequent studies in *S. typhimurium* showed conclusively that serine transacetylase and *O*-acetylserine sulphydrylase form a tight complex which we named 'cysteine synthetase' (Kredich et al 1969). It was also quite clear that large quantities of an unbound form of *O*-acetylserine sulphydrylase exist in crude extracts, and after purification it was found that the free and complexed enzymes are identical in all respects. This protein we call *O*-acetylserine sulphydrylase A (Becker et al 1969).

The genetic study of *O*-acetylserine sulphydrylase A has been aided by the ability of this enzyme to utilize certain analogues of its natural substrates *O*-acetyl-L-serine and sulphide (Fig. 3). The compound 1,2,4-triazole bears little obvious molecular

O-ACETYLSERINE + SULFIDE ⟶ CYSTEINE + ACETATE

O-ACETYLSERINE + 1,2,4-TRIAZOLE ⟶ 1,2,4-TRIAZOLE-1-ALANINE + ACETATE

O-DIAZOACETYLSERINE + SULFIDE ⟶ CYSTEINE + DIAZOACETATE

FIG. 3. Reactions catalysed by *O*-acetylserine sulphydrylase. Both sulphydrylases react with azaserine (*O*-diazoacetyl-L-serine), but we have not yet determined whether *O*-acetylserine sulphydrylase B has triazolylase activity.

resemblance to sulphide but is capable of substituting for the latter in what we call the 'triazolylase' activity of *O*-acetylserine sulphydrylase A (Kredich et al 1975). The product of this reaction, 1,2,4-triazole-1-alanine, is not toxic to *S. typhimurium*, but its formation from 1,2,4-triazole can sufficiently deprive cells of *O*-acetyl-L-serine to completely inhibit growth on minimal medium. The major effect of this form of *O*-acetyl-L-serine depletion appears to be on the derepression of the enzymes of sulphate reduction, for growth inhibition by 1,2,4-triazole can be overcome by adding a non-sulphate sulphur source such as sulphite or sulphide, as well as by cysteine, glutathione or *O*-acetyl-L-serine.

1,2,4-Triazole-resistant strains are easily obtained by positive selection on minimal plates containing this inhibitor, and almost all of these are found to carry mutations in *cysK*, the structural gene for *O*-acetylserine sulphydrylase A (Hulanicka et al 1974). Some mutant enzymes have preferentially lost triazolylase activity while retaining significant sulphydrylase activity. This situation allows sulphide to compete more effectively with 1,2,4-triazole for *O*-acetyl-L-serine. Others have lost so much of both activities that they create a metabolic block, causing *O*-acetyl-L-serine levels to rise and thereby ensuring adequate derepression of the enzymes of sulphate reduction. Of special interest are those strains which carry deletions in *cysK*, for although they synthesize no detectable *O*-acetylserine sulphydrylase A, they remain cysteine prototrophs. The explanation for this apparent paradox is the existence of another enzyme, designated *O*-acetylserine sulphydrylase B (Becker & Tomkins 1969), which is genetically and immunologically distinct from *O*-acetylserine sulphydrylase A.

Ordinarily one thinks of *S. typhimurium* as being too parsimonious to carry and express genes for two separate enzymes which carry out the same reaction. We originally felt that *O*-acetylserine sulphydrylase B might be a gratuitous activity of an enzyme such as cystathionase or tryptophan synthetase which normally would not function as a cysteine biosynthetic enzyme.

Additional insight into this problem came several years ago when Dr T. Mojica-A (personal communication 1976) at the Polish Academy of Sciences isolated several strains completely lacking *O*-acetylserine sulphydrylase B. He accomplished this by starting with a *cysK* strain completely lacking *O*-acetylserine sulphydrylase A activity and selecting for cysteine auxotrophs after mutagenesis and penicillin selection. Several such strains carried mutations in a new gene designated *cysM*, which is situated very close to *cysA*. The isolation of additional mutations in both *cysK* and *cysM* is now easily accomplished owing to our recognition that the compound azaserine, which is actually *O*-diazoacetyl-L-serine, is an excellent substrate for both sulphydrylases, giving L-cysteine and diazoacetate as products (Fig. 3). The formation by this reaction of the extremely reactive and toxic diazoacetate appears to account for much of the bacteriocidal action of azaserine, and we find that mutant strains either lacking *O*-acetylserine sulphydrylase A or unable to reduce sulphate to sulphide are resistant to azaserine. In the case of prototrophic *cysK* strains, selection for even greater resistance in the presence of sulphate and a sulphur source such as glutathione gives *cysM* mutants at high frequency as well as various kinds of sulphide auxotrophs (Kredich & Hulanicka, unpublished).

Cysteine auxotrophs of the type *cysK*$^-$ *cysM*$^-$ grow only on sulphur sources with a preformed cysteine moiety such as glutathione, L-djenkolate or L-cysteine itself, and assays of crude extracts of these mutants show no detectable *O*-acetylserine sulphydrylase activity. Furthermore, *cysK*$^+$ *cysM*$^-$ mutants, like their *cysK*$^-$ *cysM*$^+$ counterparts, are cysteine prototrophs, and no other nutritional requirements are noted in *cysM*$^-$ strains. Thus the genetic evidence indicates that either of the two *O*-acetylserine sulphydrylase enzymes is capable *in vivo* of carrying out the final step of cysteine biosynthesis and that *O*-acetylserine sulphydrylase B is an enzyme with no other obvious metabolic function.

The evidence implying a physiological role for *O*-acetylserine sulphydrylase A in cysteine biosynthesis is considerable. The expression of this enzyme is controlled by the sulphur source used for growth, *O*-acetyl-L-serine and *cysB*. Furthermore, although the significance of the binding of this enzyme to serine transacetylase is not understood, it seems unlikely that such an association would be purely accidental. We wish now to present evidence that the expression of *O*-acetylserine sulphydrylase B also is subject to control by the same elements which regulate *O*-acetylserine sulphydrylase A and other activities of the cysteine regulon.

Using an assay originally developed by Becker & Tomkins (1969) we have been able to separate *O*-acetylserine sulphydrylase A from *O*-acetylserine sulphydrylase B by DEAE-cellulose chromatography. Distinction between the two enzymes is also facilitated by the use of an antibody which specifically inactivates *O*-acetylserine sulphydrylase A. While adequate for quantifying *O*-acetylserine sulphydrylase B in extracts of repressed cells, where the ratio of *O*-acetylserine sulphydrylase A to its counterpart is only about 2, our methods are inadequate for analyses of derepressed cells in which this ratio is close to 100. This problem can

TABLE 1

Regulation of *O*-acetylserine sulphydrylase B activity

Strain	Genotype	*O-Acetylserine sulphydrylase B (units per mg protein)*	
		Grown on L-cysteine	*Grown on L-djenkolate*
Wild type	–	0.199	–
DW130	*cysK1751*	0.158	0.580
TK181	*trpC109 cysK1772*	0.045	0.713
cysB403	*cysB403*	0.072	0.079
DW42	*trpA160 cysB15*	0.062	0.122
DW393	*trpC109 cysB1773 cysK1772*	0.115	0.121
TK1190	*cysB1352 cysK1751*	0.542	0.492
TK1191	*cysB484 cysK1751*	0.097	0.092
DW18	*cysE2*	0.148	0.076[a]
DW383	*cysE2 cysK1751*	0.046	0.188[a] (0.414)[b]
DW384	*trpC109 cysM1770*	0.02	–
DW385	*trpC109 cysM1771*	0.02	–

Extracts of bacteria grown either on L-cysteine (to repress enzyme synthesis) or L-djenkolate (to derepress enzyme synthesis) were fractionated by DEAE-cellulose chromatography according to the method of Becker & Tomkins (1969). One unit of activity catalyses the synthesis of one μmol L-cysteine per min using the assay of Becker et al (1969). Units of *O*-acetylserine sulphydrylase B are expressed in terms of the total amount of protein applied to the column.
[a]Growth on 0.5 mM-glutathione.
[b]Growth on 0.5 mM-glutathione plus 2.0 mM-*O*-acetyl-L-serine.

be surmounted, however, by using *cysK* strains which synthesize wild-type quantities of *O*-acetylserine sulphydrylase A protein with little (*cysK1751*) or no (*cysK1772)* enzyme activity.

The experiments summarized in Table 1 show that in the *cysB*$^+$ strains DW130 and TK181, *O*-acetylserine sulphydrylase B activity is 4 to 16 times higher in sulphur-starved than in L-cysteine-grown cells. The *cysB*$^-$ strains *cysB403*, DW42 and DW393 have low levels of this enzyme whether grown on L-cysteine or on L-djenkolate. In contrast, the *cysB*^{c} strain TK1190 has high levels of *O*-acetylserine sulphydrylase B under conditions which in a *cysB*$^+$ strain would repress this enzyme. TK1191, which carries the *cysB*$^-$/*cysB*^{c} allele *cysB484*, appears to be *cysB*$^-$ with respect to *O*-acetylserine sulphydrylase B. Expression of *O*-acetylserine sulphydrylase B is variable in the two *cysE* strains, but is clearly stimulated in DW383 by the addition of *O*-acetyl-L-serine to a sulphur-deprived culture.

We conclude from these studies that both sulphydrylases are components of the cysteine regulon, and that either alone is sufficient for cysteine biosynthesis, at least under the laboratory conditions usually used for growth of *S. typhimurium*. Why then does this organism have two distinct enzymes catalysing the same reaction? Marcin Filutowicz, a graduate student at the Polish Academy of Sciences, has recently made an observation (personal communication, unpublished) which may eventually provide an answer to this question. He has found that although *cysK*$^{+}$ *cysM*$^{-}$ mutants are prototrophs during aerobic growth, they are cysteine bradytrophs when grown anaerobically. Wild-type and *cysK*$^{-}$ strains grow quite well anaerobically on minimal medium, but *cysM*$^{-}$ strains show a marked lag phase which can be overcome by L-cysteine but not by sulphite or sulphide. It seems therefore that *O*-acetylserine sulphydrylase B is somehow required or at least preferred for cysteine biosynthesis during anaerobic growth. We have no other data bearing on this phenomenon, but it is intriguing to speculate on whether the two sulphydrylases differ in their reactivities with different forms of sulphide. The studies of Tsang & Schiff (1976) on bound intermediates in sulphate reduction are worth considering in this regard, and perhaps we shall find that one of these two enzymes accepts a bound form of sulphide while the other is much more reactive with free sulphide. For the time being, however, the exact manner in which the two sulphydrylases function *in vivo* is unclear.

ACKNOWLEDGEMENTS

This work was supported by research grant AM12828 from the United States Public Health Service, and by the Polish Academy of Sciences with Project: 09.7.

References

Becker MA, Tomkins GM 1969 Pleiotropy in a cysteine-requiring mutant of *Salmonella typhimurium* resulting from altered protein-protein interaction. J Biol Chem 244:6023-6030

Becker MA, Kredich NM, Tomkins GM 1969 The purification and characterization of O-acetylserine sulfhydrylase A from *Salmonella typhimurium*. J Biol Chem 244:2418-2427

Cheney RW, Kredich NM 1975 Fine structure genetic map of the cysB locus in *Salmonella typhimurium*. J Bacteriol 124:1273-1281

Clowes RC 1958a Nutritional studies of cysteineless mutants of *Salmonella typhimurium*. J Gen Microbiol 18:140-153

Clowes RC 1958b Investigation of the genetics of cysteineless mutants of *Salmonella typhimurium* by transduction. J Gen Microbiol 18:154-172

Dreyfuss J 1964 Characterization of a sulfate and thiosulfate-transport system in *Salmonella typhimurium*. J Biol Chem: 239:2292-2297

Dreyfuss J, Monty KJ 1963 The biochemical characterization of cysteine-requiring mutants of *Salmonella typhimurium*. J Biol Chem 238:1019-1024

Hulanicka D 1972 Resistance to sulphate analogues in *Salmonella typhimurium*. Acta Biochim Pol 19:367-376

Hulanicka MD, Kredich NM, Treiman DM 1974 The structural gene for O-acetylserine sulfhydrylase A in *Salmonella typhimurium*. J Biol Chem 249:867-872

Jagura G, Hulanicka D, Kredich NM 1978 Analysis of merodiploids of the *cysB* region in *Salmonella typhimurium*. Mol Gen Genet 165:31-38

Jones-Mortimer MC 1968 Positive control of sulphate reduction in *Escherichia coli*. The nature of the pleiotropic cysteineless mutants of *E. coli K12*. Biochem J 110:597-602

Jones-Mortimer MC, Wheldrake JF, Pasternak CA 1968 The control of sulphate reduction in *Escherichia coli* by O-acetyl-L-serine. Biochem J 107:51-53

Kredich NM 1971 Regulation of L-cysteine biosynthesis in *Salmonella typhimurium*. J Biol Chem 246: 3474-3484

Kredich NM, Tomkins GM 1966 The enzymatic synthesis of L-cysteine in *Escherichia coli* and *Salmonella typhimurium*. J Biol Chem 241:4955-4965

Kredich NM, Becker MA, Tomkins GM 1969 Purification and characterization of cysteine synthetase, a bifunctional protein complex, from *Salmonella typhimurium*. J Biol Chem 244:2428-2439

Kredich NM, Keenan BS, Foote LJ 1972 The purification and subunit structure of cysteine desulfhydrase from *Salmonella typhimurium*. J Biol Chem 247:7157-7162

Kredich NM, Foote LJ, Hulanicka MD 1975 Studies on the mechanism of inhibition of *Salmonella typhimurium* by 1,2,4-triazole. J Biol Chem 250:7324-7331

Mizobuchi K, Demerec M, Gillespie DH 1962 Cysteine mutants of *Salmonella typhimurium*. Genetics 47:1617-1627

Murphy MJ, Siegel LM 1973 Siroheme and sirohydrochlorin. J Biol Chem 248:6911-6919

Siegel LM, Kamin H, Rueger DC, Presswood RP, Gibson QH 1971 An iron-free sulfite reductase flavoprotein from mutants of *Salmonella typhimurium* In: Kamin H (ed) Flavins and flavoproteins. University Park Press, Baltimore, p 523-554

Tsang ML-S, Schiff JA 1976 Sulfate-reducing pathway in *Escherichia coli* involving bound intermediates. J Bacteriol 125:923-933

Tsang ML-S, Schiff JA 1978 Assimilatory sulphate reduction in an *Escherichia coli* mutant lacking thioredoxin activity. J Bacteriol 134:131-138

Tully MD, Yudkin MD 1975 The nature of the product of the *cysB* gene of *Escherichia coli*. Mol Gen Genet 136:181-183

Discussion

Segel: Do either of the two enzymes work with *O*-succinylserine or *O*-acetylhomoserine?

Kredich: We haven't looked at sulphydrylase B for that yet; the A enzyme does not.

Jones-Mortimer: Under anaerobic conditions, is the A enzyme not synthesized, or is it inactivated?

Kredich: We plan to investigate this but haven't yet.

Whatley: While looking at the bacterium *Paracoccus denitrificans* growing on oxygen we found that one of the sulphur mutants that needed cysteine was lacking the enzyme system that synthesized *O*-acetylserine, though it appeared to be able to reduce sulphate (Paraskeva 1978). We have never seen anything to suggest that there is more than one enzyme for the synthesis of cysteine in this organism, presumably the A enzyme (Burnell & Whatley 1977). This would fit with your suggestion that the B enzyme is used under anaerobic conditions.

Meister: Perhaps the B enzyme is involved in homocysteine formation?

Kredich: There is a gene described for that already, however. That was one of our first thoughts, but it is now clear that the B enzyme is part of the cysteine regulon, for it is controlled by *cysB*, *O*-acetylserine, and the state of sulphur nutrition. The cell apparently regards both the A and B enzymes as being involved with cysteine biosynthesis.

Mudd: The double mutant grows solely on cysteine, so it would be hard to invoke a function in homocysteine formation.

Kredich: It certainly isn't an essential function.

Segel: Did you express some amazement that the cell was spending so much energy making two different enzymes with a similar function?

Kredich: I said that *Salmonella typhimurium* usually seems too parsimonious to carry the genes for two enzymes which catalyse the same reaction; there is a tremendous genetic conservatism on the part of such organisms. They tend to throw away a gene duplication if they don't need it. It's not a question of energetics, but simply the observation that such organisms don't keep genetic material around very long if they don't need it. The implication is that if a gene is there, it is doing something under some conditions.

Hamilton: When you grow the wild type anaerobically for many generations, do you get more B and less A?

Kredich: We haven't measured the two enzymes under such conditions.

Jones-Mortimer: Duplicated enzyme systems are commoner than you suggest, certainly in sugar transport in *Escherichia coli;* many sugars induce more than one transport system (Kornberg 1976).

Kredich: There is a lot of overlap for transport systems, but that seems to be an exception.

Hamilton: The advantage of duplication to the cell is that the two enzymes often have different affinities. Is there any difference in the affinity constants of these two enzymes?

Kredich: Our kinetic determinations show no appreciable differences in the apparent K_m for sulphide or for *O*-acetylserine.

Meister: Is the B enzyme inhibited by glutathione?

Kredich: We haven't looked at the effect of glutathione on either enzyme.

Postgate: Could you clarify for me the inducing effect of *O*-acetylserine, the regulatory effect of cysteine, and the nature of the regulation by the *cysB* product? Is regulation by cysteine not just a regulation of function, rather than a genetic regulation?

Kredich: Cysteine is a feedback inhibitor of serine transacetylase, while activation by *O*-acetylserine is genetic regulation.

Jones-Mortimer: In *E. coli* it controls the level of protein synthesis; we don't know whether it is at the level of transcription or translation.

Postgate: What does the *cysB* gene regulate?

Kredich: We don't know, but the presumption, by analogy with other systems of positive regulation, is that the *cysB* protein attaches to an initiator region and facilitates the attachment of the RNA polymerase to this region.

Postgate: In the cluster of structural genes *(CDHIJ)* shown in Fig. 2 (p 88) I gather that several operons are found, at least one of which is regulated by *O*-acetylserine, as well as by the cysteine product?

Kredich: Yes. It probably includes two operons, both regulated by *O*-acetylserine *and* by cysteine.

Segel: What is the black space between *D* and *H* on that diagram of the *S. typhimurium* gene?

Kredich: It is a silent region. There seems to be dispensible genetic information in that area. Dr Hulanicka has shown that the *HIJ* region is a single operon. *C* and *D* appear to be contiguous and probably constitute another operon.

Postgate: There is a paradox in that you say that the genes *C,D,H,I* and *J* are under the regulation of *O*-acetylserine itself and are also regulated by the *cysB* gene product, yet you can get point mutations in *cysB* which become constitutive *cysB*^{c}. How do those mutants respond to *O*-acetylserine?

Kredich: They don't require *O*-acetylserine. The implication is that the *cysB* gene product itself interacts with *O*-acetylserine and undergoes a conformational change which allows it to promote transcription. A *cysB*^{c} allele may specify a protein in which this conformational change is inherent even in the absence of *O*-acetylserine.

Dodgson: You described the association of serine transacetylase and the A enzyme. Is there any evidence of an association with the B enzyme? And what is the significance of the complex with the A form?

Kredich: We have no evidence that serine transacetylase and sulphydrylase B form a complex. Paul Cook and Randolph Wedding (1977,1978) have done very careful kinetic studies on the complex with the A enzyme that suggest that the *O*-acetylserine made by the complex diffuses off into the medium before returning to be converted into cysteine. If the complex doesn't function to funnel the product of the serine transacetylase reaction directly to the sulphydrylase, then we really don't know what its function is.

Roy: What happens when you dilute it down in the enzyme assay? Might the complex dissociate then?

Kredich: No. When sulphide is present at any concentration, the complex does not dissociate, however much *O*-acetylserine is present.

Whatley: Can you re-associate the complex by the reverse of that, i.e. by adding sulphide?

Kredich: I don't know, because in our assay sulphide and *O*-acetylserine are added simultaneously. If we add *O*-acetylserine a few seconds before sulphide the sulphydrylase A is dissociated and probably doesn't reassociate until *O*-acetylserine is completely consumed.

Bright: Do all your azaserine-resistant mutants have something to do with cysteine synthesis, or do you find another whole class of mutants to do with glutamine metabolism? Azaserine is an analogue of glutamine.

Kredich: It is in many systems. We find that cysteine auxotrophs that cannot reduce sulphate to sulphide are 5–10-fold more resistant to azaserine than wild type. Strains lacking the major sulphydrylase (the A enzyme) are also very resistant. Strains lacking both the A and B enzymes are even more resistant. We conclude that the major form of azaserine toxicity involves the formation of diazoacetic acid. Perhaps in the wild-type strain you never get enough azaserine into the cell to affect glutamine synthesis.

Meister: Azaserine doesn't affect glutamine synthesis; it affects its utilization.

Kredich: Yes. I suspect that by starting with mutants that are already blocked in cysteine biosynthesis one could obtain mutants in which glutamine functions are affected.

Dodgson: Has anything more been done on the sulphate permease system?

Jones-Mortimer: There are thought to be three cistrons in *cysA* (Mizobuchi et al 1962) and one can get mutants in any of them that lack the sulphate-binding protein. You can also get mutants in any of them that have the binding protein. None of the three is the structural gene for this protein (Ohta et al 1971). There is in blue lupins *(Lupinus augustifolia L.)* a similar enzyme system that makes cyanoalanine (Hendrickson & Conn 1969). Is cyanide a substrate for *O*-acetylserine sulphydrylase?

Kredich: I have looked, but found no such activity.

Le Gall: You said that the gene *cysG* is responsible for the biosynthesis of sirohaem. Does that gene affect the reduction of nitrite?

Kredich: I don't know whether sirohaem is a part of the nitrite reductase activity. It is interesting that the *cysG* gene appears not to be regulated by cysteine; it probably synthesizes the sirohaem-producing enzyme constitutively, which would go along with it having other functions.

Mudd: Where is the cystathionine synthase structural gene on the genetic map?

Kredich: It lies halfway between *cysE* and *leu*.

References

Burnell JN, Whatley FR 1977 Sulphur metabolism in *Paracoccus denitrificans*. Purification, properties, and regulation of serine transacetylase, *O*-acetylserine sulphydrylase and beta-cystathionase. Biochim Biophys Acta 481:246-265

Cook PF, Wedding RT 1977 Overall mechanism and rate equation for *O*-acetylserine sulfhydrylase. J Biol Chem 252:3459

Cook PF, Wedding RT 1978 Cysteine synthetase from *Salmonella typhimurium* LT-2. Aggregation, kinetic behaviour, and effect of modifiers. J Biol Chem 253:7874-7879

Hendrickson HR, Conn EE 1969 Cyanide metabolism in higher plants. IV. Purification and properties of the β-cyanoalanine synthase of blue lupins. J Biol Chem 244:2632-2640

Kornberg HL 1976 Genetics in the study of carbohydrate transport by bacteria. J Gen Microbiol 96:1-16

Mizobuchi K, Demerec M, Gillespie DH 1962 Cysteine mutants of *Salmonella typhimurium*. Genetics 47:1617-1627

Ohta N, Galsworthy PR, Pardee AB 1971 Genetics of sulfate transport by *Salmonella typhimurium*. J Bacteriol 105:1053-1062

Paraskeva C 1978 D.Phil. Thesis, Oxford University

The regulation of methionine biosynthesis and metabolism in plants and bacteria

S.W.J. BRIGHT, P.J. LEA and B.J. MIFLIN

Department of Biochemistry, Rothamsted Experimental Station, Harpenden, Herts., AL5 2JQ, UK

Abstract The amino acids biosynthetically derived from aspartate including methionine are all essential in the diet of monogastric animals. Most of this requirement is met by plant foods. The methionine biosynthetic pathways in plants and bacteria are outlined and compared. Regulation in bacterial systems is by a combination of repression and feedback inhibition whereas in plants repression is unimportant. Several enzymes in the branched pathway to methionine in plants are regulated by feedback inhibition; others are yet to be investigated. In plants many amino acid biosynthetic enzymes are localized in plastids and this is also likely for methionine biosynthesis. Methionine occupies an important position in cellular metabolism where the processes of one-carbon transfer via *S*-adenosylmethionine, protein synthesis, protein initiation and ethylene synthesis are interlocked. Attempts to increase the levels of free methionine have been made by selecting for plant mutants resistant to lysine plus threonine. One dominant mutation causes elevation of free amino acid levels in vegetative tissues but also has undesirable side-effects. The potential of such approaches is discussed.

Methionine, along with other members of the branched biosynthetic pathway derived from aspartate (lysine, threonine and isoleucine), is an essential amino acid in the diet of monogastric animals. The requirement for methionine may be met by homocysteine, provided that a supply of folic acid and vitamin B_{12} is available, but only in part by cysteine (Meister 1965). The major portion of this dietary requirement is obtained ultimately from plant sources. For this reason we propose to concentrate on methionine synthesis and metabolism in plants rather than on the extensively investigated areas of bacterial and fungal metabolism. Major differences between plants and microorganisms will be emphasized.

THE PATHWAY OF METHIONINE BIOSYNTHESIS

The pathway of methionine biosynthesis in plants is shown in Fig. 1. The methionine molecule can be considered to be made up of three separate parts:

Sulphur in biology
(Ciba Foundation Symposium 72) p 101-117

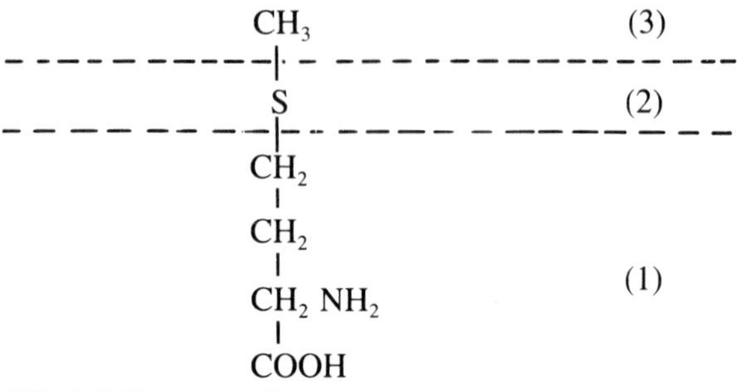

The four-carbon chain of part 1 is derived directly from aspartate. The sulphur atom (part 2) is derived from cysteine, and the methyl group (part 3) is derived either from formate or the C-3 atom of serine and is transferred to methionine via N^5-methyltetrahydrofolate. There is therefore a need to integrate and regulate the three elements in the biosynthesis of methionine.

The original experiments of Dougall & Fulton (1967) with plant cell cultures showed that homoserine, cystathionine and homocysteine are all intermediates in the

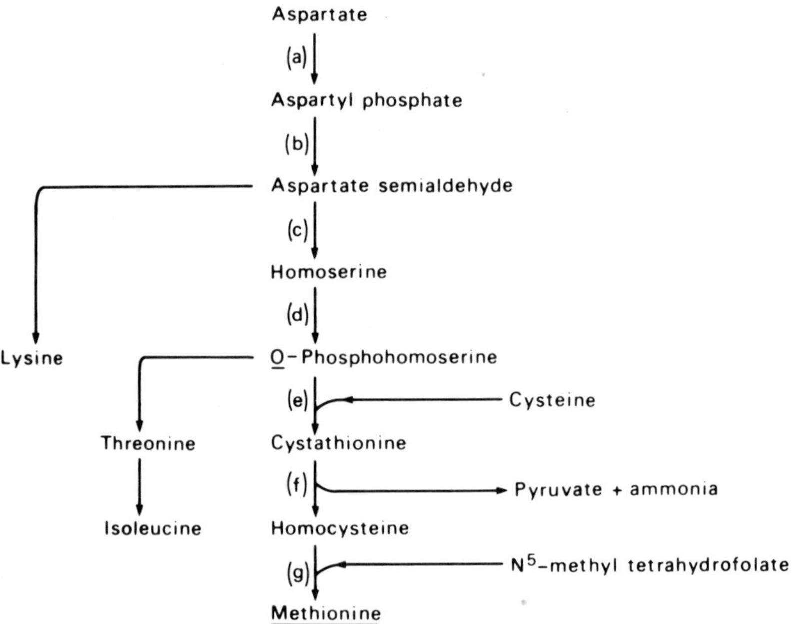

FIG. 1. The synthesis of methionine from aspartate in plants. (a) Aspartate kinase (EC 2.7.2.4); (b) Aspartate-semialdehyde dehydrogenase (EC 1.2.1.11); (c) Homoserine dehydrogenase (EC 1.1.1.3); (d) Homoserine kinase (EC 2.7.1.39); (e) Cystathionine γ-synthase (EC 4.2.99.9); (f) Cystathionine β-lyase (EC 4.4.1.8); (g) Tetrahydropteroyltriglutamate methyltransferase (EC 2.1.1.14).

synthesis of methionine, as they are in microorganisms (Flavin 1975). The enzymes of all the steps from aspartate have been demonstrated in plant tissues (Miflin et al 1979, Giovanelli et al 1980).

COOH		COOⓅ		CHO		CH_2OH		CH_2OⓅ
|	ATP	|	$NADH_2$	|	$NADH_2$	|	ATP	|
CH_2	$\longrightarrow$	CH_2	$\longrightarrow$	CH_2	$\longrightarrow$	CH_2	$\longrightarrow$	CH_2
$CHNH_2$		$CHNH_2$		$CHNH_2$		$CHNH_2$		$CHNH_2$
COOH		COOH		COOH		COOH		COOH
Aspartate		Aspartyl phosphate		Aspartate semialdehyde		Homoserine		*O*-Phospho-homoserine

Although the conversion of aspartate to homoserine in a light-dependent reaction has been demonstrated in pea chloroplasts (Lea et al 1979), other possible routes have been suggested (Bauer et al 1977).

There is a major difference between plants and bacteria in the branch point of the pathway after homoserine. In the bacteria homoserine is converted to the *O*-acetyl or *O*-succinyl derivative prior to condensation with cysteine (Flavin 1975). In plants homoserine is converted to *O*-phosphohomoserine prior to the condensation

CH_2 OⓅ		SH		CH_2 —	— S		
CH_2		CH_2		CH_2	CH_2		
$CHNH_2$	+	$CHNH_2$	$\longrightarrow$	$CHNH_2$	$CHNH_2$	+	P_i
COOH		COOH		COOH	COOH		
O-Phospho-homoserine		Cysteine		Cystathionine			

reaction with cysteine to yield cystathionine. *O*-Phosphohomoserine is also the precursor of threonine. Three lines of evidence support the hypothesis that *O*-phosphohomoserine is the activated form of homoserine involved in methionine synthesis in plants: (1) Most plants are only able to convert homoserine to the phosphate ester; synthesis of the *O*-acyl derivatives was detected only in *Pisum sativum* and *Lathyrus sativus*, which are known to accumulate *O*-acetylhomoserine and *O*-oxalylhomoserine respectively (Giovanelli et al 1974). (2) *O*-Phosphohomoserine was found to act as a substrate for cystathionine synthesis in plants, but not in bacteria (including the cyanobacteria) (Datko et al 1977). (3) In a search for endogenous α-aminobutyryl donors for cystathionine synthesis Datko et al (1974) showed that only one donor, *O*-phosphohomoserine, was present.

$$\begin{array}{cccccccc}
 & & & & SH & & & \\
 & & & & | & & & \\
CH_2 - & S & & & CH_2 & & & \\
| & | & & & | & & & \\
CH_2 & CH_2 & & & CH_2 & & CH_3 & \\
| & | & & & | & & | & \\
CHNH_2 & CHNH_2 & + H_2O & \longrightarrow & CHNH_2 & + & C{=}0 & + NH_3 \\
| & | & & & | & & | & \\
COOH & COOH & & & COOH & & COOH &
\end{array}$$

Cystathionine Homocysteine Pyruvate

The enzyme cystathionine β-lyase, which hydrolyses cystathionine to homocysteine, pyruvate and ammonia, has been demonstrated in a number of plants. There was no evidence that cystathionine could be cleaved at the γ-position to yield cysteine, a reaction that takes place in animals and fungi (Giovanelli & Mudd 1971). Evidence that cysteine is the SH donor for cystathionine synthesis has been provided in *Chlorella sorokiniana* by Giovanelli et al (1978). Radioactivity from $^{35}SO_4^{2-}$ fed to cells appeared initially in soluble cysteine, then in cystathionine and only later in homocysteine. After one second of incubation the ratio of ^{35}S cysteine to ^{35}S homocysteine was 127; this ratio decreased rapidly with time, so that after 10 seconds the ratio had reached the isotopic equilibrium value of 4. Cystathionine was soon labelled and reached isotopic equilibrium after 20 seconds, suggesting that this intermediate was rapidly turning over. The rate of labelling was at least sufficient to account for the rate of methionine synthesis. Analysis of the data suggested that a minimum of 97% of the sulphur in homocysteine was derived from cysteine rather than from direct sulphydration of *O*-phosphohomoserine.

$$\begin{array}{cccccc}
 & & & CH_3 & & \\
 & & & | & & \\
SH & & & S & & \\
| & & & | & & \\
CH_2 & & & CH_2 & & \\
| & & & | & & \\
CH_2 & + & N^5\text{-methyl-tetrahydro-folate} \longrightarrow & CH_2 & + & \text{tetrahydrofolate} \\
| & & & | & & \\
CHNH_2 & & & CHNH_2 & & \\
| & & & | & & \\
COOH & & & COOH & &
\end{array}$$

Homocysteine Methionine

The methyl group of methionine is derived from the triglutamyl derivative of N^5-methyltetrahydrofolic acid. The conversion of homocysteine to methionine is carried out in plants by a cobalamin-independent tetrahydropteroyltriglutamate methyltransferase (EC 2.1.1.14).

The supply of methyl groups via pteroylglutamate derivatives has been extensively studied in plants by Cossins and his colleagues (see Cossins 1980).

REGULATION OF METHIONINE BIOSYNTHESIS

In bacteria there are two levels of regulation of amino acid biosynthesis: (1) Repression of enzyme synthesis at the stage of transcription. (2) Modulation of enzyme activity by feedback regulation, often by end-products of the pathway. The two control systems have been extensively studied in *Escherichia coli* and *Salmonella typhimurium,* and a complex model of regulation has been discussed (Umbarger 1978). Methionine repression can be mediated by either methionine itself, *S*-adenosylmethionine, or the enzyme *S*-adenosylmethionine synthetase (methionine adenosyltransferase; EC 2.5.1.6). In the above bacteria it has been shown that there are three distinct isoenzymes of aspartate kinase and two of homoserine dehydrogenase. Only one of each set of isoenzymes is repressed by methionine, but none of them is inhibited by it. The other aspartate kinase isoenzymes are inhibited and repressed by threonine or lysine, and one homoserine dehydrogenase is inhibited and repressed by threonine (Umbarger 1978). The only methionine-inhibited enzyme is *O*-succinylhomoserine synthase, the first enzyme unique to methionine biosynthesis. The enzyme is inhibited by the concerted action of *S*-adenosylmethionine and methionine (Lee et al 1966). Despite the ability of methionine to regulate its own synthesis, it appears that carbon for methionine synthesis may still be supplied through lysine- and threonine-regulated enzymes (Umbarger 1978).

In plants there is no evidence for the repression of synthesis of any enzyme involved in amino acid biosynthesis (Miflin 1976). Aspartate kinase is subject to inhibition by lysine and/or threonine (see Miflin et al 1979, Bryan 1980 for reviews) and there is some evidence of a synergistic inhibitory effect of methionine at low concentrations of lysine, but never with methionine alone (Table 1). There is also evidence that the levels of the lysine-sensitive and threonine-sensitive enzymes vary with stage of tissue culture (Davies & Miflin 1978) and plant development (Table 1). Initial studies tend to suggest that the lysine-sensitive enzyme is predominant at times of rapid protein synthesis (Lea et al 1979). In carrot tissue cultures the alterations in isoenzyme levels do not correspond with any dramatic changes in free or protein amino acids (Bright et al 1979)*. Homoserine dehydrogenase is subject

* But see also Sakano K 1979 Plant Physiol (Bethesda) 63:583-585.

TABLE 1

The effect of end-product amino acids on aspartate kinase isolated from various sources

Amino acid addition	*Aspartate kinase activity (% of control with no additions)*			
	7-day-old barley leaves[a]	*9-day-old pea shoots*[b]	*Pods (no seeds) of pea plants 15 days after flowering*[c]	*Maturing seeds from pea plants 26 days after flowering*[c]
Lysine (Lys), 0.5 mM	67	58	98	62
Lys, 5mM	3	34	98	46
Threonine (Thr), 10mM	100	76	14	84
Methionine (Met), 10mM	100	90	101	100
Lys, 0.5mM + Met, 10mM	26	30	106	59
Lys, 5mM + Met, 10mM	3	28	103	45
Thr, 10mM + Met, 10mM	100	64	15	86
Lys, 10mM + Thr, 10mM	3	13	14	32
Lys, 10mM + Thr, 10mM + Met, 10mM	3	8	13	30

Aspartate kinase activity was assayed by the hydroxamate method as described in [a,b] in extracts purified by DEAE-cellulose chromatography.
[a]Shewry & Miflin (1977); [b]Lea et al (1979); [c]P.J. Lea & N. Gill, unpublished results.

to feedback inhibition by threonine, but there is also a significant proportion of insensitive enzyme which apparently increases during the growth of maize (Dicamelli & Bryan 1975), soybean (Matthews & Widholm 1979) and barley (P.J. Lea, unpublished results). The only inhibitory effect of methionine on its own biosynthesis can be seen in the inhibition of homoserine kinase by relatively high concentrations of *S*-adenosylmethionine in pea seedlings (Thoen et al 1978a). The inhibition of homoserine kinase in peas by *S*-adenosylmethionine, valine and isoleucine (not detected in barley) may account for the ability of young pea plants to accumulate large quantities of homoserine. Threonine synthase (EC 4.2.99.2), the last enzyme involved in threonine synthesis, is stimulated 30-fold by *S*-adenosylmethionine in sugar beet leaves (Madison & Thompson 1976). Similar stimulation of the enzyme has also been detected in barley (Aarnes 1976) and pea (Thoen et al 1978b). Thus in the presence of high levels of methionine, and presumably *S*-adenosylmethionine, carbon would be channelled away from methionine into threonine. As soon as sufficiently high levels of threonine had accumulated, homoserine dehydrogenase and aspartate kinase would be inhibited and the flow of carbon into methionine and threonine would cease. There is no evidence that methionine affects any other enzyme involved in the biosynthesis of amino acids from aspartate.

An understanding of the regulation of methionine biosynthesis *in vivo* can also be obtained by radioactive tracer studies on whole plants and by determining the inhibitory effects of amino acids on growth. A combination of lysine and threonine is inhibitory to growth in a number of plant systems (Dunham & Bryan 1969, Bright et al 1978a). This inhibition can be relieved by homoserine, homocysteine and methionine. The synthesis of methionine from $^{35}SO_4^{2-}$ is inhibited by lysine plus threonine in young wheat plants and the inhibition of methionine biosynthesis correlates quite well with the observed growth inhibition (Bright et al 1978b). In the young wheat plants the synthesis of methionine was inhibited 44% by 2mM-lysine and threonine whereas the synthesis of lysine and threonine from [^{14}C]acetate was inhibited 90% and 97% respectively by 1mM-lysine and threonine. The inhibition of methionine biosynthesis was interpreted as a sequential inhibition of aspartate kinase by lysine and homoserine dehydrogenase by threonine (Bright et al 1978b).

LOCALIZATION OF METHIONINE SYNTHESIS

Several enzymes of methionine biosynthesis have been shown to be located at least in part in the plastids of a cell. They include aspartate kinase, homoserine dehydrogenase and homocysteine-dependent tetrahydropteroyltriglutamate methyltransferase (see Miflin et al 1979 for a review). The enzymes required for cysteine synthesis are also located within the plastid (Giovanelli et al 1980). Further evidence that plastids are a site of methionine synthesis is incomplete but suggestive: (1) chloroplasts will rapidly incorporate aspartate in a light-dependent reaction into homoserine (Lea et al 1979); (2) light-driven protein synthesis in pea chloroplasts is inhibited by lysine and threonine; inhibition can be relieved by the addition of low levels of methionine (Mills & Wilson 1978); (3) sulphate reduction occurs in chloroplasts (Schwenn et al 1976); (4) cysteine synthesis occurs in chloroplasts (Ng & Anderson 1978); and (5) formyl and methyl tetrahydropteroylglutamates have been found in chloroplasts (Shah & Cossins 1970). These points indicate that at least some methionine biosynthesis takes place in the plastid but the quantitative contribution of the different cellular fractions remains to be determined. Mitochondria have been shown to be able to methylate homocysteine (Clandinin & Cossins 1974), but the ability of mitochondria to synthesize methionine from aspartate has not been determined.

METHIONINE METABOLISM

A summary of the metabolism of methionine is given in Fig. 2. The methyl group of methionine is activated by combination with ATP to form *S*-adenosylmethionine which is then able to donate the methyl group in a wide range of

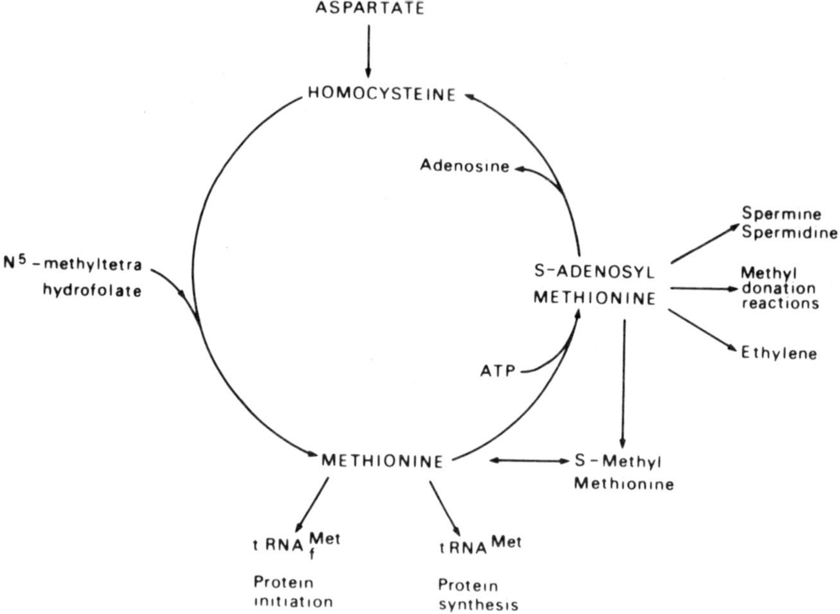

FIG. 2. Methionine metabolism.

reactions (for review see Cantoni 1977). Few of these methyl transfer reactions have been studied in plants. *S*-Adenosylhomocysteine is liberated in the methylation reaction and can be hydrolysed to homocysteine and adenine, thus allowing the recycling of parts 1 and 2 of the methionine molecule. Evidence for the functioning of this cycle has been provided by Dodd & Cossins (1968), who showed that ^{35}S-labelled *S*-adenosylmethionine turned over more slowly than [^{14}C]methyl-labelled *S*-adenosylmethionine in germinating pea seedlings.

In other reactions the carbon skeleton of *S*-adenosylmethionine is incorporated into the amines spermine and spermidine (Smith 1975) and the plant hormone ethylene (Adams & Yang 1977). The total methionine content of the tissue does not control the synthesis of ethylene in the ripening avocado (Baur et al 1971), although homocysteine synthesis was considered to be the rate-limiting step in ethylene synthesis during the senescence of morning glory petals (Hanson & Kende 1976). In this tissue it has been suggested that *S*-methylmethionine is the methyl donor for methionine synthesis during the senescence period (Hanson & Kende 1976).

Methionine is the first amino acid attached to the ribosome during the initiation stage of the synthesis of a peptide. In bacteria a specific tRNA molecule is involved which allows the formation of *N*-formylmethionine after the attachment of the amino acid, but before binding to the ribosome. The tRNA molecule $tRNA_f^{Met}$ is also found in plant chloroplasts and mitochondria (see Lea & Norris 1977 for a review). In the plant cytoplasm a specific $tRNA_i^{Met}$ is used as the initiator molecule, but the methionine is not formylated. In both cases different $tRNA^{Met}$ species are involved in the incorporation of methionine into other parts of the protein molecule. After the protein molecule has been formed the initiator methionine residue is removed and may be recycled.

SELECTION OF MUTANTS OF METHIONINE METABOLISM

Selection for resistance to methionine analogues in *S. typhimurium* and *E. coli* has led to the isolation of bacterial strains which accumulate or excrete methionine. Such mutants have aided considerably the elucidation of the regulatory patterns in these bacteria (Smith 1971). Ethionine-resistant tissue cultures of carrot have been selected with a 10- fold increase in soluble methionine (Widholm 1976) and two ethionine-resistant lines of *C. sorokiniana* accumulated or excreted increased amounts of methionine (Sloger & Owens 1974). Three-fold

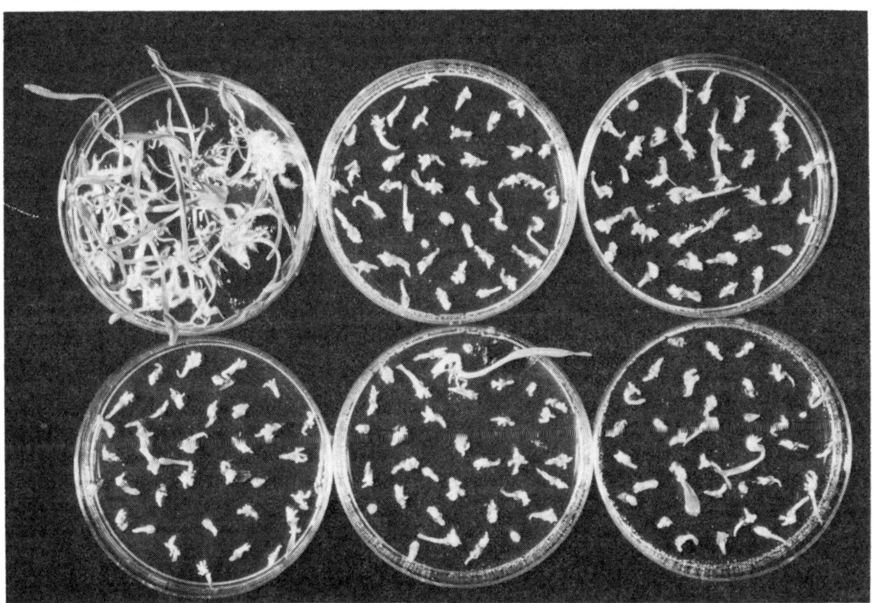

FIG. 3. Selection of mutant R2501. Barley embryos were incubated for 7 days under lights on nutrient medium without (top left petri dish) or with 2.5mM lysine and threonine (other dishes). One plant (R2501) has grown well in the presence of lysine and threonine and was selected for further study. (S.W.J. Bright & P. Norbury, unpublished results.)

TABLE 2

Inheritance of resistance to lysine plus threonine in progeny of four selected resistant barley plants after self-fertilization

Plant	*Progeny*		
	Resistant	*Sensitive*	*R/S*
R2501	68	46	1.5
R2506	27	15	1.8
R3004[a]	19	10	1.9
R3202[a]	121	43	2.8

[a]R3004 and R3202 also segregated for albino v. green.
R3004 18 green: 11 albino; R3202 125 green: 39 albino.

Dry mature barley embryos were surface-sterilized and incubated on sterile nutrient agar medium for two or more days at 25 °C under lights and then transferred to fresh medium containing lysine and threonine at 2mM. Resistant plants were green with long roots, long shoots and pale scutellum; sensitive plants were pale green or yellow with short shoots, short roots, dark scutellum and browning of the leaf base. (Unpublished results of S.W.J. Bright & J. Kueh.)

increases in the soluble methionine of tobacco plant leaves regenerated from methionine sulphoximine-resistant cells have been reported (Carlson 1973). In none of these cases has the regulation of methionine biosynthesis in the mutants been investigated. We have been interested in finding mutants of cereals and legumes with increased levels of aspartate-derived amino acids. A system selecting for resistance to the combined action of lysine and threonine has been used. The system has the advantage that the site of inhibition at aspartate kinase and homoserine dehydrogenase is clearly defined (Bright et al 1978b), and hence the selective pressure is directly applied to the feedback regulatory steps. The mode of action of inhibitory amino acid analogues is, on the other hand, often unknown. Other possible sites of action, such as in protein synthesis, must be considered (Lea & Norris 1976).

We have selected four barley mutants resistant to lysine plus threonine by screening 2×10^4 germinating barley embryos (Fig. 3). The embryos are dissected from the progeny of seed treated with the mutagen sodium azide. The four selected plants contain dominant mutations, as they produce sensitive and resistant progeny on self-fertilization (Table 2). The progeny of the self-fertilized plants do not all segregate in a 3:1 ratio as would be expected of a single dominant gene in the heterozygous form. It is possible that plants homozygous for a dominant gene are not viable or that other genes are involved.

One of the mutant plants, R2501, has been further characterized by analysis of free amino acids in progeny plants after growth for seven days (Table 3). The analysis

TABLE 3

Free amino acids in Bomi and progeny of R2501, a barley mutant resistant to lysine plus threonine

		Amino acid (μmol/g fresh weight)					
Plants	*Lys + Thr (mM)*	*Lysine*	*Threonine*	*Methionine*	*Isoleucine*	*Glycine*	*Alanine*
Bomi	0	0.40	0.62	0.07	0.26	0.93	0.83
		0.24	0.62	0.08	0.23	0.83	0.90
R2501	0	0.58	2.39	0.16	0.39	0.87	1.70
		0.95	3.64	0.25	0.55	1.16	2.55
Bomi	2 + 2	12.0	8.9	0.07	0.70	1.62	1.34
		10.3	12.5	0.07	0.86	1.68	1.76
R2501 resistant	2 + 2	13.5	16.8	0.12	0.84	0.92	1.30
		9.8	8.2	0.10	0.89	0.98	1.24
R2501 sensitive	2 + 2	8.7	7.5	0.08	0.74	1.27	2.59
		9.7	9.4	0.06	0.66	1.22	2.45

Three or four plants grown as in Table 2 were pooled for each replicate sample. Free amino acids were extracted in methanol: chloroform: water: mercaptoethanol (12:5:3:0.014), hydrolysed in 6*N*-HCl + 10mM-mercaptoethanol under nitrogen, and analysed on a technicon autoanalyser. Norleucine was added to the original extraction medium to act as internal standard.

(Unpublished results of S.W.J. Bright & S. Smith.)

is complicated by the heterozygosity in R2501, since plants grown on non-inhibitory medium are a mixture of sensitive and resistant types. There is a particularly large increase in the free threonine content of mutant plants and a smaller increase in methionine. Two unrelated amino acids, glycine and alanine, are included for comparison. Free alanine content is also increased in the mutant. The methionine content of resistant progeny grown in the presence of lysine and threonine is higher than in either the sensitive progeny or the parent variety. Some progeny of R2501 have undesirable agronomic characteristics, such as poor germination and altered morphology giving stunted plants with poor seed setting. These traits could be due to unrelated mutations in the original plant, which is derived from highly mutagenized material, or they could be pleiotropic effects of the selected gene. The former possibility is strengthened by the segregation of albino progeny from R3004 and R3202 and the production of only sensitive progeny plants after self-fertilization of one particular stunted plant derived from R2501. A comparison between the aspartate kinase and homoserine dehydrogenase of the parent and the resistant progeny of R2501 is now being made.

A maize tissue culture with an altered aspartate kinase less sensitive to feedback inhibition by lysine has been selected recently by its resistance to lysine plus threonine. The cultured cells have increases in free methionine (Hibberd et al 1978). These results imply that changes in the regulation of enzymes early in the biosynthesis of methionine can lead to changes in the free pool of this amino acid.

CONCLUSIONS

In plants, which are the main dietary source of methionine for man and animals, the pathway of methionine biosynthesis is now quite well established (see Giovanelli et al 1980 for a comprehensive review). The patterns of regulation of this pathway are understood only in a rudimentary fashion. Methionine is an important amino acid not only as a constituent of protein, but also as a methylating agent and a precursor of a plant hormone. The regulation of *S*-adenosylmethionine synthesis and its ability to methylate important molecules is little understood.

Methionine and lysine are the first nutritionally limiting amino acids in the seed proteins of legumes and cereals respectively (Woodham 1978). Mutants with alterations in the regulation of the branched pathway of methionine and lysine biosynthesis could be useful in improving nutritional quality and should also play an important part in improving the understanding of these pathways.

References

Aarnes H 1976 Homoserine kinase from barley seedlings. Plant Sci Lett 7:187-191

Adams DO, Yang SF 1977 Methionine metabolism in apple tissue. Implication of *S*-adenosylmethionine as an intermediate in the conversion of methionine to ethylene. Plant Physiol (Bethesda) 60:892-896

Bauer A, Joy KW, Urquart AA 1977 Amino acid metabolism of pea leaves. Plant Physiol (Bethesda) 59:920-924

Baur AH, Yang SF, Pratt HK, Biale JB 1971 Ethylene biosynthesis in fruit tissues. Plant Physiol (Bethesda) 47:696-699

Bright SWJ, Wood EA, Miflin BJ 1978a The effect of aspartate-derived amino acids (lysine, threonine, methionine) on the growth of excised embryos of wheat and barley. Planta (Berl) 139:113-117

Bright SWJ, Shewry PR, Miflin BJ 1978b Aspartate kinase and the synthesis of aspartate-derived amino acids in wheat. Planta (Berl) 139:119-125

Bright SWJ, Leggat MM, Miflin BJ 1979 Amino acids in carrot cell suspension culture: no correlation with aspartate kinase isoenzyme levels. Plant Physiol (Bethesda) 63 : 586-588

Bryan J 1980 The synthesis of the aspartate family and the branched chain amino acids. In: Miflin BJ (ed) The biochemistry of plants. Academic Press, New York, vol 5, in press

Cantoni GL 1977 In: Salvatore F et al (eds) The biochemistry of adenosylmethionine. Columbia University Press, New York, p 155-164

Carlson PS 1973 Methionine sulphoximine-resistant mutants of tobacco. Science (Wash DC) 180:1366-1368

Clandinin MT, Cossins EA 1974 Methionine biosynthesis in isolated *Pisum sativum* mitochondria. Phytochemistry (Oxf) 13:585-591

Cossins EA 1980 Metabolism and respiration. In: Davies DD (ed) The biochemistry of plants. Academic Press, New York, vol 2, in press

Datko AH, Giovanelli J, Mudd SH 1974 Homocysteine biosynthesis in green plants. J Biol Chem 249:1139-1155

Datko AH, Mudd SH, Giovanelli J 1977 Homocysteine synthesis in green plants. Studies of the homocysteine-forming sulfhydrylase. J Biol Chem 253:3436-3445

Davies HM, Miflin BJ 1978 Regulatory isoenzymes of aspartate kinase, and the control of lysine and threonine biosynthesis in carrot cell suspension culture. Plant Physiol (Bethesda) 62:536-541

DiCamelli CA, Bryan JK 1975 Changes in enzyme regulation during the growth of maize. Plant Physiol (Bethesda) 55:999-1005

Dodd WA, Cossins EA 1968 Biosynthesis of *S*-adenosylmethionine in germinating pea seeds. Phytochemistry (Oxf) 7:2143-2145

Dougall DK, Fulton MM 1967 Biosynthesis of protein amino acids in plant tissue culture. IV. Isotope competition experiments using glucose-U-^{14}C and potential intermediates. Plant Physiol (Bethesda) 42:1176-1178

Dunham VL, Bryan JK 1969 Synergistic effects of metabolically related amino acids on the growth of a multicellular plant. Plant Physiol (Bethesda) 44:1601-1608

Flavin M 1975 In: Greenberg DM (ed) Metabolic pathways, 3rd edn. Academic Press, New York, vol 7:457-503

Giovanelli J, Mudd SH 1971 Transsulfuration in higher plants. Partial purification and properties of β-cystathionase of spinach. Biochim Biophys Acta 227:654-670

Giovanelli J, Mudd SH, Datko A 1974 Homoserine esterification in plants. Plant Physiol (Bethesda) 54:725-736

Giovanelli J, Mudd SH, Datko AH 1978 Homocysteine biosynthesis in green plants. Physiological importance of the transsulfuration pathway in *Chlorella sorokiniana* growing under steady state conditions with limiting sulfate. J Biol Chem 253:5665-5677

Giovanelli J, Mudd SH, Datko AH 1980 Sulfur amino acids in plants. In: Miflin BJ (ed) The biochemistry of plants. Academic Press, New York, vol 5, in press

Hanson AD, Kende H 1976 Methionine metabolism and ethylene biosynthesis in senescent flower tissue of morning glory. Plant Physiol (Bethesda) 57:528-537

Hibberd KA, Green CE, Walter TJ, Gengenbach BG 1978 Characterisation of a methionine overproducer varient from maize callus. In: Thorpe TA (ed) Frontiers of plant tissue culture 1978. International Association of Plant Tissue Culture, Calgary, p 512

Lea PJ, Norris RD 1976 The use of amino acid analogues in studies on plant metabolism. Phytochemistry (Oxf) 15:585-595

Lea PJ, Norris RD 1977 tRNA synthetases in plants. In: Reinhold L, et al (eds) Progress in phytochemistry. Pergamon Press, Oxford, vol 4:121-168

Lea PJ, Mills WR, Miflin BJ 1979 The isolation of a lysine-sensitive aspartate kinase from pea leaves and its involvement in homoserine biosynthesis in isolated chloroplasts. FEBS (Fed Eur Biochem Soc) Lett 98:165-168

Lee L-W, Ravel JM, Shive W 1966 Multimetabolite control of a biosynthetic pathway by sequential metabolites. J Biol Chem 241:5479-5480

Madison JT, Thompson JF 1976 Threonine synthase from higher plants: stimulation by *S*-adenosylmethionine and inhibition by cysteine. Biochem Biophys Res Commun 71:684-691

Matthews BF, Widholm JM 1979 Regulation of homoserine dehydrogenase in developing organs of soybean seedlings. Phytochemistry (Oxf) 18:395-400

Meister A 1965 The biochemistry of the amino acids. Academic Press, New York, p 201-230

Miflin BJ 1976 The metabolic control of amino acid biosynthesis. In: Proceedings of international workshop on genetic improvement of seed proteins. National Research Council, Washington DC, p 135-158

Miflin BJ, Bright SWJ, Davies HM, Shewry PR, Lea PJ 1979 Amino acids derived from aspartate: their biosynthesis and its regulation. In: Hewitt E, Cutting CV (eds) Nitrogen assimilation in plants. Academic Press, London (Proc 6th Long Ashton Symp) p 335-357

Mills WR, Wilson KG 1978 Effects of lysine, threonine and methionine on light driven protein synthesis in isolated pea *(Pisum sativum* L.) chloroplasts. Planta (Berl) 142:153-160

Ng BH, Anderson JW 1978 Chloroplast cysteine synthases of *Trifolium repens* and *Pisum sativum*. Phytochemistry (Oxf) 17:879-885
Schwenn JD, Depka B, Hennies HH 1976 Assimilatory sulphate reduction in chloroplasts: evidence for the participation of both stromal and membrane bound enzymes. Plant Cell Physiol 17:165-176
Shah SPJ, Cossins EA 1970 Pteroylglutamates and methionine biosynthesis in isolated chloroplasts. FEBS (Fed Eur Biochem Soc) Lett 7:267-270
Shewry PR, Miflin BJ 1977 Properties and regulation of aspartate kinase from barley seedlings *(Hordeum vulgare* L.). Plant Physiol (Bethesda) 59:69-73
Sloger M, Owens LD 1974 Control of free methionine production in wild type and ethionine-resistant mutants of *Chlorella sorokiniana*. Plant Physiol (Bethesda) 53:469-473
Smith DA 1971 *S*-Amino acid metabolism and its regulation in *Escherichia coli* and *Salmonella typhimurium*. Adv Genet 16:141-165
Smith TA 1975 Recent advances in the biochemistry of plant amines. Phytochemistry (Oxf) 14:865-890
Thoen A, Rognes SE, Aarnes H 1978a Biosynthesis of threonine from homoserine in pea seedlings I. Homoserine kinase. Plant Sci Lett 13:103-112
Thoen A, Rognes SE, Aarnes H 1978b Biosynthesis of threonine from homoserine in pea seedlings II. Threonine synthase. Plant Sci Lett 13:113-119
Umbarger HE 1978 Amino acid biosynthesis. Annu Rev Biochem 47:533-606
Widholm JM 1976 Selection and characterisation of cultured carrot and tobacco cells resistant to lysine, methionine and proline analogues. Can J Bot 54:1523-1529
Woodham AA 1978 The nutritive value of mixed proteins. In: Friedman M (ed) Nutritional improvement of food and feed proteins. Plenum Press, New York, p 365-378

Discussion

Kredich: Is it possible to clone barley and grow a plant from a single cell in tissue culture?

Bright: No, not with barley or any other cereal. It has been achieved with a few forage legumes but not with pea or soybean. For isolating mutants of crop plants, in most cases whole plants are still the best system. Plants within the family Solanaceae are amenable to single-cell culture and cloning (Thomas et al 1979).

Postgate: In your general strategy of increasing the methionine content of the plant, where do you expect that extra methionine to go? It has either to be built into a protein or to become part of the amino acid pool, in which case it will be metabolized and eliminated.

Bright: It doesn't have to be eliminated; people have grown wheat plants and watered them with tryptophan and the seeds have higher levels of tryptophan, mostly in the free form (Singh & Widholm 1975).

Postgate: So you are increasing the pool of free amino acids?

Bright: Yes. The question is whether it makes any difference to the total seed amino acid content if you can get the plant itself to overproduce the amino acid. One can increase the free amino acid content of the seed by three- or four-fold naturally; a barley mutant defective in the synthesis of storage protein has increased levels of all the soluble amino acids, so there is a capacity for holding free amino acids in the seed (Brandt 1975). We want to discover whether this can be put to use, by finding mutants which overproduce particular amino acids.

Postgate: An idea put forward for upgrading the nutritional quality of crop plant protein was to introduce genes for polylysine formation, because there you don't ask the organism to make a new protein, or to make more of a natural lysine-containing protein. You would cause it to form a product that is not normally made but which is nutritionally useful when the crop is eaten.

Bright: Another approach is to insert lysine codons into the nuclear gene(s) for the major storage protein(s) and so modify the seed composition without requiring the expression of new genes. This approach requires the isolation and molecular cloning of the mRNA for these proteins. This is being attempted in a number of laboratories including ours at Rothamsted.

Segel: The comparative biochemistry of the cysteine–methionine pathway is a nightmare, because of the subtle differences between different species. You showed that cysteine inhibits homoserine dehydrogenase in some plants. Is this true in microorganisms also?

Kredich: We have evidence that that is so in *Salmonella typhimurium*.

Segel: Is there any cooperative, concerted, or cumulative effect between cysteine and threonine, or are there multifunctional enzymes (β-aspartyl kinase and homoserine dehydrogenase)?

Bright: Homoserine dehydrogenase is more complicated than aspartate kinase in maize in that there is one isoenzyme which is not feedback regulated at all and then a series of molecular forms, which may be interrelated, whose sensitivity to feedback regulation by threonine changes with the developmental age of the plant (see Bryan 1980). The subcellular localization of these forms is not known.

Whatley: You say that repression in plants in unimportant in the regulation of enzymes. Can you give examples of enzymes in plants which *are* repressed?

Mudd: Reuveny & Filner (1977) showed that in cultured tobacco plant cells, ATP sulphurylase is repressed, during growth with a normal sulphate source. One can derepress by starving the cells for sulphate or growing them on djenkolate.

Segel: I am confused by the many ways in which serine and homoserine are activated! Where does *O*-acetylhomoserine (or is it *O*-succinylhomoserine?) fit in? Which organisms use *O*-acetylhomoserine and which use *O*-succinylhomoserine?

Hamilton: Broadly speaking, the non-enteric bacteria and the fungi use *O*-acetylhomoserine, and the enteric bacteria use *O*-succinylhomoserine for cystathionine biosynthesis.

Segel: Do any bacteria form homocysteine directly from activated homoserine?

Mudd: Sulphurylation of an activated form of homoserine is probably a function of a minor alternative pathway in microorganisms which can come into play in special circumstances. In the higher plants, which use only *O*-phosphohomoserine, both direct sulphydration and cystathionine synthesis occur with *O*-phosphohomoserine as substrate, but in *Chlorella* at least 97% of total homocysteine comes from cystathionine, not by direct sulphydration; in *Lemna,* at least 95% does.

Kredich: Fungi of the *Aspergillus* type are much more liberal organisms with considerable freedom of choice as to whether they go to cystathionine!

Whatley: What appears to make methionine synthesis in plants irreversible? Is it just the thermodynamics? If so, what makes it reversible in animals? Do they have two different enzymes?

Mudd: Going via the transsulphuration pathway, cystathionine synthesis is functionally irreversible, and cystathionine cleavage in plants and most bacteria only goes in the direction of homocysteine. In animals, methionine synthesis is reversible, but only in the sense that one can regenerate methionine from homocysteine. The animal must have homocysteine.

Bright: One lacks the critical test in plants, namely a mutant which requires cysteine, to see whether it can use methionine to make cysteine.

Dodgson: Is any bacterium known that can make cysteine from methionine without going back, say, to sulphate?

Kredich: Not *Salmonella typhimurium* or *Escherichia coli,* certainly.

Dodgson: Dr Bright, what sort of plant hormone is ethylene?

Bright: Ethylene has a number of effects in plants, probably in interaction with other hormones. Particular actions are in fruit ripening, leaf abscission and auxin transport (Wareing & Phillips 1970).

Kredich: What is the current opinion on the lack of methionine in food plants? To what extent is it limiting, in comparison with amino acids like tryptophan and lysine? In other words, how much of a real problem in world nutrition is methionine deficiency?

Bright: Protein quality has been down-graded as a world problem over the past five years. Nutritionists are now stressing that there is not enough *food* available, in terms of calories mainly. Even the UN Protein Advisory Group has disbanded. At one time protein quality was considered a major problem but now only in more specific cases can you say that it's worthwhile to improve protein quality. Working with barley, for instance, we know that much barley in the UK is fed to pigs whose diet is accurately formulated (Fuller et al 1979). One can therefore estimate the benefit of increasing the lysine content of home-grown cereals rather than going on to the world market for soya beans, which is what happens now. In this particular case it might be worthwhile, but in terms of world nutrition, protein quality is low on the list of nutritional priorities at the moment.

Postgate: As you say, FAO in its statistics now completely disregards protein as a separate component of food, and protein quality is considered to be unimportant. At a recent meeting on food chains and human nutrition (Blaxter 1980) almost everybody felt that the pendulum had swung much too far in that direction and that in a few years protein intake and its quality, particularly of cereal proteins, is bound to become important again.

Bright: It illustrates the disadvantage of being too applied in research. On a time-scale of 10 or 20 years, one may be producing a new variety of crop plant and then find that no one wants it.

Postgate: If we are successful in increasing the world's food supplies the quality of life will go up and a demand for high-quality protein foods will also go up, so they will automatically become 'essential'. It is when one is considering the 10% of the world population who are undernourished that one can in principle disregard protein and concentrate on energy intake.

References

Blaxter K (ed) 1980 Food chains and human nutrition. Applied Science Publishers, London

Brandt A 1975 *In vivo* incorporation of [^{14}C]lysine into the endosperm proteins of wild type and high-lysine barley. FEBS (Fed Eur Biochem Soc) Lett 52:288-291

Bryan J 1980 The synthesis of the aspartate family and the branched chain amino acids. In: Miflin BJ (ed) The biochemistry of plants. Academic Press, New York, vol 5, in press

Fuller MF, Mennie I, Crofts RMJ 1979 The amino acid supplementation of barley for the growing pig. 22. Optimal additions of lysine and threonine for growth. Br J Nutr 41:333-340

Reuveny Z, Filner P 1977 Regulation of adenosine triphosphate sulfurylase in cultured tobacco cells. J Biol Chem 252:1858-1864

Singh M, Widholm JM 1975 Inhibition of corn, soybean and wheat seedling growth by amino acid analogues. Crop Sci 15:79-81

Thomas E, King PJ, Potrykus I 1979 Improvement of crop plants via single cells *in vitro* – an assessment. Z Pflanzenzücht 82:1-30

Wareing PF, Phillips IDJ 1970 The control of growth and differentiation in plants. Pergamon Press, Oxford

The oxidation of sulphite in animal systems

JEAN L. JOHNSON and K.V. RAJAGOPALAN

Department of Biochemistry, Duke University Medical Center, Durham, North Carolina 27710, USA

Abstract In animals the terminal step in the pathway for degradation of sulphur-containing amino acids is the oxidation of sulphite to sulphate. This reaction is catalysed by the enzyme sulphite oxidase. The enzyme contains molybdenum and a cytochrome b_5 type haem, is localized in the mitochondrial intermembrane space and transfers electrons from sulphite to cytochrome *c* on the inner membrane. The sulphite oxidase protein has a molecular weight of 110 000 (chicken) to 122 000 (human) and exists as a dimer of identical subunits. The haem and molybdenum cofactors are present on separate domains of the molecule. The structure of the molybdenum cofactor has not been worked out in detail, but this cofactor is known to be present in many other molybdoenzymes including xanthine oxidase and nitrate reductase. Three cases of genetic sulphite oxidase deficiency in humans have been reported. The three affected children displayed mental retardation, neurological abnormalities and dislocated ocular lenses. The biochemical basis for lack of enzyme activity in each case has been studied. All three have been shown to lack the sulphite oxidase protein, but in one case this appears to be secondary to a defect in synthesis of the molybdenum cofactor. Sulphite oxidase deficiency has been produced in the rat by administration of high levels of tungsten. Sulphite oxidase-deficient animals are particularly susceptible to the toxic effects of sulphite and atmospheric sulphur dioxide.

In animal tissues sulphite arises from the degradation of sulphur-containing amino acids as outlined in Fig. 1. Ingestion of food products containing bisulphite as a preservative and exposure to atmospheric sulphur dioxide also contribute to the sulphite load but under ordinary circumstances probably have a minor role. The oxidation of sulphite to sulphate is catalysed by the enzyme sulphite oxidase (sulphite: ferricytochrome *c* oxidoreductase, EC 1.8.2.1). The sulphate which is produced by this reaction is used for numerous biosynthetic processes and when produced in excess of this requirement is excreted in the urine. Sulphate is also supplied normally in the diet. Sulphite oxidase has been purified to homogeneity from several animal sources and its properties have been characterized in some detail. The essential nature of the enzyme-catalysed oxidation of sulphite has been

Sulphur in biology
(Ciba Foundation Symposium 72) p 119-133

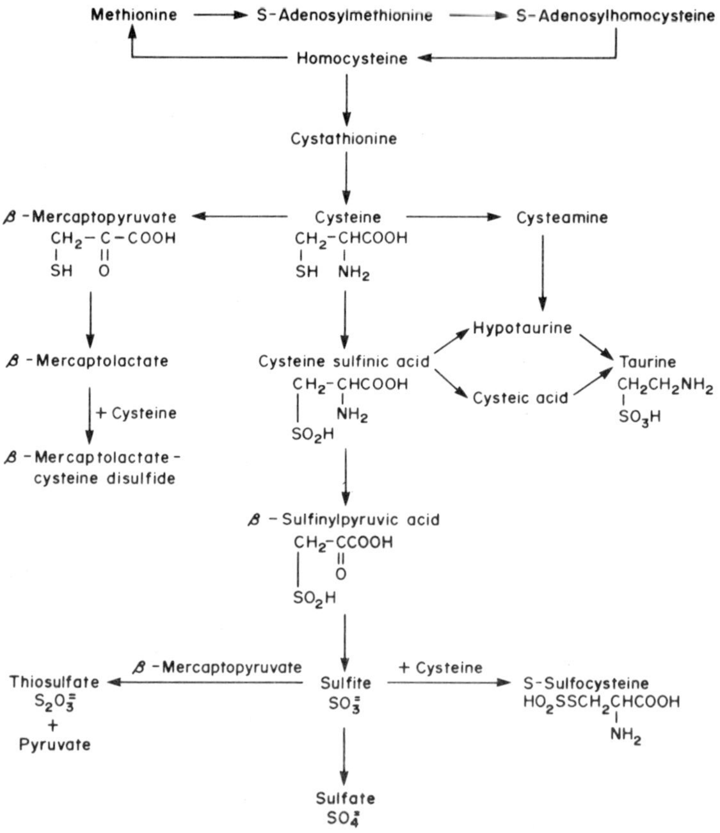

FIG. 1. Metabolism of sulphur-containing compounds in animals.

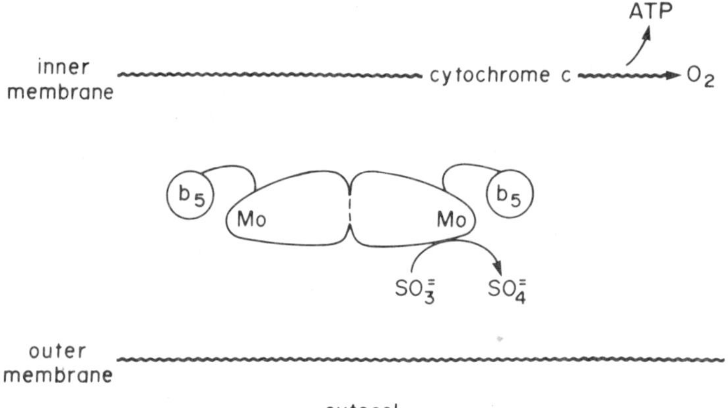

FIG. 2. Schematic representation of sulphite oxidase in the mitochondrial intermembrane space.

demonstrated by the reports of three instances of human sulphite oxidase deficiency (Irreverre et al 1967, Shih et al 1977, Duran et al 1979a,b). The properties of sulphite oxidase and its role in sulphur metabolism as determined from studies of sulphite oxidase deficiency in humans and in an animal model are presented below.

PROPERTIES OF SULPHITE OXIDASE

Nearly all animal tissues which have been assayed contain measurable amounts of sulphite oxidase, but the highest levels appear in the liver. The enzyme has been purified from beef, chicken, rat and human liver and shown to contain a cytochrome b_5 type haem and molybdenum (Cohen et al 1971, Kessler & Rajagopalan 1972, Johnson & Rajagopalan 1976a). The protein molecular weight ranges from 110 000 (chicken) to 122 000 (human), and the molecule is always isolated as a dimer of identical subunits. A schematic model of the enzyme structure is shown in Fig. 2. The molybdenum and haem binding regions of the protein have been shown to be present on separate domains of the molecule (Johnson & Rajagopalan 1977). Mild treatment of the rat enzyme with trypsin or other proteolytic enzymes effects a cleavage in the hinge region and allows isolation of a dimeric molybdenum fragment (molecular weight 94 000) and monomeric haem fragments (molecular weight 10 000). Both the native molecule and the molybdenum fragment are quite asymmetrical (Southerland et al 1978).

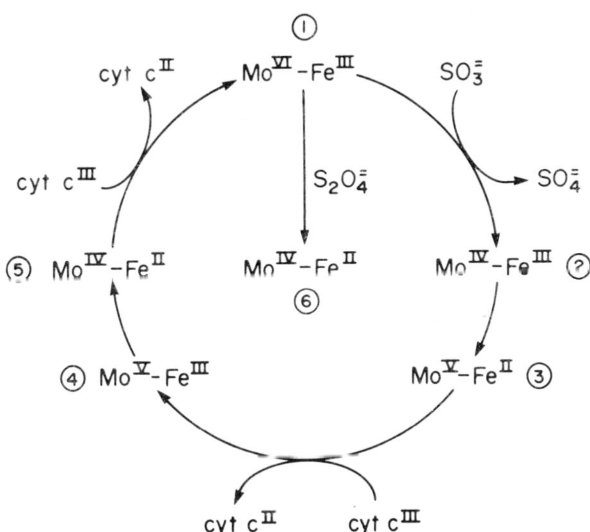

FIG. 3. Proposed catalytic cycle for sulphite oxidase.

The oxidation of sulphite to sulphate takes place at the molybdenum centre with an oxygen atom derived from water. The two electrons taken up by the enzyme can be transferred to ferricyanide or molecular oxygen, but the physiological electron acceptor is cytochrome *c*. The enzyme is located in the mitochondrial intermembrane space where it has free access to the electron transport chain on the inner membrane. It has been shown that the oxidation of sulphite does produce one molecule of ATP (Cohen et al 1972). Sulphite oxidation is efficiently catalysed by the molybdenum fragment of the enzyme using ferricyanide as electron acceptor, but the interaction with cytochrome *c* requires the presence of the b_5 haem fragment which must be in covalent linkage to the molybdenum domain (Johnson & Rajagopalan 1977).

The proposed catalytic mechanism for sulphite oxidase is shown in Fig. 3. The oxidized resting enzyme ① contains molybdenum VI and ferric iron. Two-electron reduction by a molecule of sulphite yields species ②with molybdenum in the +4 state. Rapid transfer of one electron from molybdenum to haem produces an enzyme ③poised for interaction with cytochrome *c*. When the ferrous iron in the enzyme molecule has been oxidized by cytochrome *c* the second electron can move from the molybdenum to the enzyme haem (④→⑤), and another molecule of cytochrome *c* can be reduced. Complete reduction of the electron carriers in sulphite oxidase can be achieved by dithionite, yielding species ⑥.

Molybdenum is a trace metal required in extremely low amounts in the diet and present in a very limited number of enzymes. All of these are in some way involved in nitrogen metabolism. Nitrogenase and nitrate reductase act on inorganic forms of nitrogen to provide the proper reduction state for biosynthetic reactions, whereas xanthine dehydrogenase and sulphite oxidase function in degradative pathways. The metal in the various molybdoenzymes has been shown to be present as part of a molybdenum cofactor which can be separated from the enzyme protein and made to function in other molybdoenzymes. The concept of a molybdenum cofactor common to several enzymes was first proposed by Nason et al (1971). This proposal has been confirmed and extended by many laboratories, but recent work from Pienkos et al (1977) has indicated that the nitrogenase cofactor contains iron and is unrelated to the cofactor present in other molybdoenzymes. The cofactor in sulphite oxidase, xanthine dehydrogenase and nitrate reductase is thought to be composed of a small peptide, a fluorescent molecule of undetermined structure and the metal (Hainline et al 1979). The cofactor has been isolated from the various enzymes by procedures that involve denaturation of the enzyme protein. The cofactor prepared in this manner is extremely oxygen sensitive, losing activity and molybdenum very readily. A more stable form of the molybdenum cofactor has been identified in various organisms and is presumed to represent newly synthesized molecules in storage for use in the biosynthesis of molybdoenzymes. In animals this cofactor is present on the mitochondrial outer membrane (Johnson et al 1977). Assay systems for quantifying molybdenum cofactor are described in detail elsewhere. Two which have been extensively used in the studies described below involve the reconstitution of sulphite oxidase activity in preparations of the demolybdoenzyme purified from tungsten-treated rats (Johnson et al 1977) and the reconstitution of nitrate reductase activity in extracts of the cofactor mutant of *Neurospora crassa, nit*-1 (Lee et al 1974).

HUMAN SULPHITE OXIDASE DEFICIENCY

Three cases of sulphite oxidase deficiency in children have been identified. The first instance was discovered in 1967 by Dr S. Harvey Mudd and coworkers at the National Institute of Mental Health (Irreverre et al 1967). The patient was studied at $2^1/_2$ years of age and displayed neurological abnormalities, severe mental retardation and dislocated ocular lenses. During the course of the study he was excreting extremely low amounts of sulphate but high levels of sulphite, thiosulphate and an unusual amino acid identified as *S*-sulphocysteine (see Table 1). Post-mortem analysis of brain, kidney and liver revealed the absence of sulphite oxidase activity (Mudd et al 1967). Biochemical studies on the parents revealed no abnormalities, but a genetic basis for the disease was suggested from the report that three of the patient's seven siblings had died in infancy of an undiagnosed disorder with similar symptoms.

TABLE 1

Sulphite oxidase deficiency: clinical symptoms and excretion of sulphur compounds

Patient	*NIMH*	*Boston*	*Utrecht*
Sex	Male	Male	Female
Clinical symptoms			
time of onset	Birth	17 months	Birth
neurological problems	+	+	+
poor mental development	+	+	+
dislocated lenses	+	+	+
Excretion of sulphur compounds			
sulphite[a]	+	+	+
S-sulphocysteine[a]	+	+	+
$S_2O_3^{2-}$[a]	Elevated	Elevated	Elevated
SO_4^{2-}	Very low	Slightly reduced	Very low
taurine	Slightly elevated	Slightly elevated	Elevated

[a]Excreted in extremely low amounts by control subjects.

TABLE 2

Sulphite oxidase deficiency: biochemical findings

Patient	*NIMH*	*Boston*	*Utrecht*
Sulphite oxidase activity	0	0 (fibroblasts)	0
Sulphite oxidase CRM[b]	0	Decreased (fibroblasts)	0
Xanthine oxidase activity	Normal	–[a]	0
Xanthine oxidase CRM	–	–[a]	Normal
Molybdenum	Normal	–	Normal (serum) 0 (liver)
Molybdenum cofactor	Normal	–	0

Zero indicates none detected; dash indicates not measured. Assays were done on samples of liver tissue except as indicated.

[a]Presumed normal from measurements of serum urate concentrations.
[b]Cross-reacting material.

With the emergence of new information defining the role of molybdenum in sulphite oxidase, liver tissue from this patient was re-examined to see whether it was possible to determine the biochemical basis for the lack of enzyme activity (Johnson & Rajagopalan 1976b). The results of this study are included in Table 2. As shown, the hepatic molybdenum level and activities of xanthine oxidase and molybdenum cofactor were comparable to those in control samples. It should be noted that several of the controls were also provided by Dr Mudd and had been retained from the original study. Immunological analysis of sulphite oxidase cross-reacting material failed to detect any inactive enzyme protein in the liver sample. Recent studies in this laboratory (W. Waud, personal communication 1979) using a sensitive radioimmunoassay for sulphite oxidase have confirmed the lack of enzyme protein in this liver sample.

A second case of sulphite oxidase deficiency was identified by Dr Vivian Shih and coworkers at Massachusetts General Hospital in Boston (Shih et al 1977). This child appeared normal at birth but at 17 months of age had a sudden onset of neurological problems. Sulphite oxidase deficiency was not suspected until at age four years bilateral subluxation of the lenses was discovered. This patient, like the one at NIMH, excreted high levels of sulphite, thiosulphate and *S*-sulphocysteine. Sulphate excretion was decreased but not nearly to the extent seen in the first case (see Table 1). Liver tissue from the Boston patient was not available for study but sulphite oxidase activity was assayed directly in cultured skin fibroblasts. As shown in Table 2, no activity could be detected. Radioimmunoassay of fibroblast extracts indicated that sulphite oxidase cross-reacting material was considerably reduced as well (W. Waud, personal communication 1979). Xanthine oxidase activity and overall molybdate metabolism in this individual appeared to be unimpaired, since serum uric acid levels were close to normal on two occasions. Intermediate levels of sulphite oxidase activity were found in fibroblasts cultured from the parents, further suggesting a genetic basis for this deficiency disease. It should be noted that dietary therapy consisting of restricted intake of sulphur-containing amino acids has been effective in correcting the biochemical aberrations in this patient but has not led to significant clinical improvement.

A third case of sulphite oxidase deficiency has recently been identified by Drs S.K. Wadman and M. Duran at Rijksuniversiteit Utrecht, The Netherlands (Duran et al 1979a). This case differs dramatically from the first two in that the child is also xanthinuric. Analysis of liver tissue obtained by needle biopsy showed that the patient was lacking both sulphite oxidase and xanthine oxidase activities (Duran et al 1979b). Neurological problems, poor mental development and dislocated lenses were evident very early in life, and the pattern of excretion of sulphur compounds was as described for the first two patients (Table 1). Analysis in this laboratory of liver tissue obtained by surgical biopsy revealed that nearly every factor related to molybdenum metabolism was decreased or undetectable in this individual. As shown in Table 2, sulphite oxidase and xanthine oxidase activities, sulphite oxidase protein and molybdenum were all below detectable levels. Xanthine oxidase protein was present and immunologically identical to native enzyme. From these data it seems likely that the underlying defect in this patient is related to molybdenum metabolism, with a defect in molybdenum cofactor synthesis a distinct possibility. In the absence of functional cofactor to bind and store molybdate for use in molybdoenzymes, a lack of accumulation of the metal in the liver is not unexpected. The lack of sulphite oxidase protein is surprising, however, and might suggest that levels of this enzyme are regulated by the molybdenum cofactor. This would be in contrast to the molybdoenzymes nitrate reductase and nitrogenase which are fully expressed in fungal and bacterial cofactor mutants (Pateman et al 1964, Nason et al 1971, Nagatani et al 1974).

SULPHITE OXIDASE DEFICIENCY IN THE RAT

An animal model of sulphite oxidase deficiency has been produced in the rat by administering high levels of tungsten (Johnson et al 1974a). Animals maintained on a low molybdenum diet supplemented with 100 p.p.m. sodium tungstate undergo a time-dependent loss in sulphite oxidase and xanthine oxidase activities. It has been shown (Johnson et al 1974b) that under these conditions the demolybdo sulphite oxidase protein continues to be synthesized and accumulates partially as a tungsten derivative and partially in a form free of cofactor and metal. (The haem portion of the molecule is unaltered.) The cofactor-free population can be reconstituted by molybdenum cofactor and has been used extensively as a convenient and quantitative assay system (Johnson et al 1977).

The importance of sulphite oxidase in the metabolism of injected bisulphite and of sulphite derived from atmospheric sulphur dioxide has been documented (Cohen et al 1973). Sulphite oxidase-deficient animals were found to be much more susceptible to the toxic effects of intraperitoneal injections of bisulphite, with an LD_{50} approximately one-half that of controls. As shown in Table 3, absence of enzyme activity was also detrimental to animals exposed to certain levels of sulphur dioxide. At 2350 and 925 p.p.m. SO_2, survival times were considerably shorter for tungsten-treated rats. At extremely high levels of SO_2 (50 000 and 500 000 p.p.m.) death was extremely rapid and a protective effect of the enzyme was not observed. At 590 p.p.m., protection by sulphite oxidase was also not apparent, and it was concluded that under these conditions toxicity resulted from a direct effect of the gas on the respiratory tract before it was absorbed and made available to the enzyme.

TABLE 3

Effect of various concentrations of inhaled SO_2 on survival time of rats

Concentration of SO_2 (p.p.m.)	*Survival time (min ± S.E.)*	
	Control	*Tungsten-treated*
590	1866 ± 210	1542 ± 210
925	750 ± 54	366 ± 36
2350	176 ± 9	63 ± 3
50 000 (5%)	<10	<10
500 000 (50%)	<2	<2

DISCUSSION

The complex nature of the sulphite oxidase molecule has led to the suggestion that a deficiency in the enzyme activity could result from a defect at any one of several genetic loci (Johnson & Rajagopalan 1976b). Proper expression of this activity in animal systems requires that molybdenum be absorbed, transported to those tissues which require it and activated to the level of a functional molybdenum cofactor. The sulphite oxidase protein must be synthesized correctly and be supplied with its cytochrome and molybdenum cofactors, and must be incorporated into its mitochondrial intermembrane locale. So far, a very limited number of deficiency cases have been identified, but already they represent more than one class of genetic defect. An additional lesion leading to absence of sulphite oxidase activity, lack of adequate supplies of molybdenum, has been generated experimentally in the rat by tungstate competition.

The lack of sulphite oxidase activity appears to have been extremely detrimental to the development and well-being of the three affected children. The essential nature of the enzyme could derive from its function in preventing tissue accumulation of sulphite, its role in supplementing sulphate which may not be adequately provided in the diet, or both. Percy et al (1968) examined brain and kidney tissue and urine from the NIMH patient to determine whether lack of sulphate might have led to defective synthesis of organic sulphate esters. The tissue sulphatides were shown to be of expected composition and present in normal concentrations, and the urinary excretion of tyrosine-*O*-sulphate and indoxyl-3-sulphate was found to be comparable to controls. The total brain weight, however, was reduced by about 50%. It was not possible to determine from these studies whether the low brain weight was a direct result of lack of sulphate or secondary to the sulphite oxidase deficiency.

In the Boston patient, accumulation of sulphite was considered to be a more severe problem than lack of sulphate (Shih et al 1977). This child appeared normal at birth and for several months thereafter and at all times tested was excreting reduced but significant amounts of sulphate. It appeared that sulphate was present in excess of biosynthetic requirements, either from dietary sources or provided by very low levels of sulphite oxidase activity which could not be detected by the assay system employed. On the other hand, a reduced or totally deficient ability to remove accumulated sulphite could be responsible for the clinical pattern seen in this patient.

Sulphate excretion in the Utrecht patient was seen to be very low and close to values measured in the NIMH case. An additional similarity between these two cases is the observation that symptoms which appear to be associated with sulphite oxidase deficiency were detected at birth or shortly thereafter. The clinical manifestations in the Utrecht patient reflect the combined effects of the deficiency of two molybdoenzymes. However, it may be concluded that the lack of sulphite oxidase presents a more severe problem than does the xanthine oxidase deficiency. Symptoms associated with xanthinuria are generally not life-threatening, and in many cases detection of the metabolic defect is incidental (Wyngaarden 1978). In contrast, the lack of mental development and the neurological problems seen in this individual are most probably ramifications of the sulphite oxidase deficiency. This would suggest that therapy directed at alleviating sulphite toxicity and/or sulphate deficiency would be most beneficial.

The animal model of sulphite oxidase deficiency has provided very little information on the function of the enzyme in sulphur amino acid degradation but rather has stressed the role of sulphite oxidase in detoxifying sulphite administered exogenously. This information is of importance when one evaluates the effects of the atmospheric pollutant sulphur dioxide or the preservative sodium bisulphite. While these compounds in high amounts are unquestionably detrimental to living systems, the presence of sulphite oxidase in all but a few deficient individuals limits to a large extent the tissues which are vulnerable and the specific sites of chemical attack which need be considered.

ACKNOWLEDGEMENTS

The authors wish to acknowledge the generous contributions of Dr S. Harvey Mudd of the National Institute of Mental Health, Dr Vivian Shih of Massachusetts General Hospital, Boston, and Drs S.K. Wadman and M. Duran of Rijksuniversiteit Utrecht, The Netherlands. Tissue samples from the three sulphite oxidase-deficient patients made available to us by these individuals have been invaluable to the studies described.

This work was supported by grant GM 00091 from the National Institutes of Health, US Public Health Service.

References

Cohen HJ, Fridovich I, Rajagopalan KV 1971 Hepatic sulfite oxidase. A functional role for molybdenum. J Biol Chem 246:374-382

Cohen HJ, Betcher-Lange S, Kessler DL, Rajagopalan KV 1972 Hepatic sulfite oxidase. Congruency in mitochondria of prosthetic groups and activity. J Biol Chem 247:7759-7766

Cohen HJ, Drew RT, Johnson JL, Rajagopalan KV 1973 Molecular basis of the biological function of molybdenum. The relationship between sulfite oxidase and the acute toxicity of bisulfite and SO_2. Proc Natl Acad Sci USA 70:3655-3659

Duran M, Beemer FA, vd Heiden C, Korteland J, de Bree PK, Brink M et al 1979a Combined deficiency of xanthine oxidase and sulphite oxidase: a defect of molybdenum metabolism or transport? J Inher Metab Dis, in press

Duran M, Korteland J, Beemer FA, vd Heiden C, deBree PK, Brink M et al 1979b Variability of sulfituria: combined deficiency of sulfite oxidase and xanthine oxidase. In: Hommes FA (ed) Models of the studies of inborn errors of metabolism. Elsevier/North-Holland Biomedical Press, Amsterdam, p 103-107

Hainline BE, Johnson JL, Rajagopalan KV 1979 Isolation and properties of the molybdenum cofactor of xanthine oxidase and sulfite oxidase. Fed Proc 38:314

Irreverre F, Mudd SH, Heizer WD, Laster L 1967 Sulfite oxidase deficiency: studies of a patient with mental retardation, dislocated ocular lenses, and abnormal urinary excretion of *S*-sulfo-L-cysteine, sulfite, and thiosulfate. Biochem Med 1:187-217

Johnson JL, Rajagopalan KV 1976a Purification and properties of sulfite oxidase from human liver. J Clin Invest 58:543-550

Johnson JL, Rajagopalan KV 1976b Human sulfite oxidase deficiency. Characterization of the molecular defect in a multicomponent system. J Clin Invest 58:551-556

Johnson JL, Rajagopalan KV 1977 Tryptic cleavage of rat liver sulfite oxidase. Isolation and characterization of molybdenum and heme domains. J Biol Chem 252:2017-2025

Johnson JL, Rajagopalan KV, Cohen HJ 1974a Molecular basis of the biological function of molybdenum. Effect of tungsten on xanthine oxidase and sulfite oxidase in the rat. J Biol Chem 249:859-866

Johnson JL, Cohen HJ, Rajagopalan KV 1974b Molecular basis of the biological function of molybdenum. Molybdenum-free sulfite oxidase from livers of tungsten-treated rats. J Biol Chem 249:5046-5055

Johnson JL, Jones HP, Rajagopalan KV 1977 *In vitro* reconstitution of demolybdosulfite oxidase by a molybdenum cofactor from rat liver and other sources. J Biol Chem 252:4994-5003

Kessler DL, Rajagopalan KV 1972 Purification and properties of sulfite oxidase from chicken liver. Presence of molybdenum in sulfite oxidase from diverse sources. J Biol Chem 247:6566-6573

Lee K-Y, Pan S-S, Erickson R, Nason A 1974 Involvement of molybdenum and iron in the *in vitro* assembly of assimilatory nitrate reductase utilizing *Neurospora* mutant *nit*-1. J Biol Chem 249:3941-3952

Mudd SH, Irreverre F, Laster L 1967 Sulfite oxidase deficiency in man: demonstration of the enzymatic defect. Science (Wash DC) 156:1599-1602

Nagatani HH, Shah VK, Brill WJ 1974 Activation of inactive nitrogenase by acid-treated component I. J Bacteriol 120:697-701

Nason A, Lee K-Y, Pan S-S, Ketchum PA, Lamberti A, DeVries T 1971 *In vitro* formation of assimilatory reduced nicotinamide adenine dinucleotide phosphate: nitrate reductase from a *Neurospora* mutant and a component of molybdenum-enzymes. Proc Natl Acad Sci USA 68:3242-3246

Pateman JA, Cove DJ, Rever BM, Roberts DB 1964 A common co-factor for nitrate reductase and xanthine dehydrogenase which also regulates the synthesis of nitrate reductase. Nature (Lond) 201:58-60

Percy AK, Mudd SH, Irreverre F, Laster L 1968 Sulfite oxidase deficiency: sulfate esters in tissues and urine. Biochem Med 2:198-208

Pienkos PT, Shah VK, Brill WJ 1977 Molybdenum cofactors from molybdoenzymes and *in vitro* reconstitution of nitrogenase and nitrate reductase. Proc Natl Acad Sci USA 74:5468-5471

Shih VE, Abroms IF, Johnson JL, Carney M, Mandell R, Robb RM et al 1977 Sulfite oxidase deficiency. Biochemical and clinical investigations of a hereditary metabolic disorder in sulfur metabolism. N Engl J Med 297:1022-1028

Southerland WM, Winge DR, Rajagopalan KV 1978 The domains of rat liver sulfite oxidase. Proteolytic separation and characterization. J Biol Chem 253:8747-8752

Wyngaarden JB 1978 Hereditary xanthinuria. In: Stanbury JB et al (eds) The metabolic basis of inherited disease, 4th edn. McGraw-Hill, New York, p 1037-1044

Discussion

Idle: You mentioned the deaths in infancy of three siblings of the first case and the intermediate levels of sulphite oxidase in fibroblasts from the parents of the Boston case. In your interviews with the families did you obtain any evidence of similar conditions further back in the pedigree, which would strengthen the idea that this is a recessive genetic disease and not an idiosyncrasy?

Johnson: There were isolated instances of mental deficiency in the family history of the NIMH case (Irreverre et al 1967) but nothing that could be directly linked to sulphite oxidase deficiency. The family histories of the other two patients were unremarkable. We are now culturing fibroblasts from parents and siblings of the NIMH case and parents of the Utrecht case. Perhaps we shall be able to demonstrate intermediate levels of sulphite oxidase in these individuals also.

Mudd: The parents of the first case (and the other surviving siblings) did not excrete sulphite, and appeared to be normal, so there is no unequivocal proof that they are heterozygotes.

Dodgson: You said that the sulphatides of the brain of the first patient were normal in both amount and composition. Was the connective tissue studied also?

Mudd: Unfortunately not. The child died before we had any idea of the diagnosis.

Dodgson: There is some confusion in the literature about the contribution that dietary inorganic sulphate makes to the formation of sulphate esters. It is sometimes argued that sulphate is really derived, within the cell, from the oxidation of cysteine and that this provides a more effective supply of sulphate than that obtained by transporting sulphate into the cell. The sort of metabolic disease studied by Dr Johnson might perhaps throw some light on the unresolved question of whether the sulphate required for the synthesis of 3′-phosphoadenosine 5′-phosphosulphate (PAPS) is coming from sulphate entering cells, or made inside the cell.

Kredich: Several years ago we were concerned by the fact that researchers studying mucopolysaccharide synthesis in cultured fibroblasts use a relatively sulphate-free medium containing ^{35}S-labelled sulphate. They rarely measure the absolute amount of polysaccharide synthesized but simply measure the radioactivity appearing in the medium as acid-insoluble material. In the same kinds of experiments we isolated the chondroitin sulphates, hydrolysed them and determined the specific activity of the sulphur to evaluate the importance of isotope dilution effects. It appeared that 50–90% of the sulphate incorporated into newly made mucopolysaccharides originated from exogenous inorganic sulphate, indicating that sulphate does get into cultured fibroblasts in sufficient quantity to satisfy the sulphate requirement.

Roy: There are similar problems with the uptake of sulphate from the mammalian gut. There are conflicting reports on whether it is absorbed poorly or well.

Rose: The sulphate which is utilized by enzyme systems involved in the synthesis of macromolecules such as chondroitin sulphate in guinea-pig liver microsomes was once claimed to come from homocysteic acid (McCully 1971). It would be interesting to know if this has been confirmed and extended.

Mudd: Perhaps the Utrecht patient will provide a better answer to this. Is the child excreting virtually no inorganic sulphate?

Johnson: Urinary excretion of sulphate was found to be 0.74–1.85 mmol/l, or approximately 10% of normal (Duran et al 1979).

Mudd: Is there any indication of connective tissue defects?

Johnson: Not at present, but that is something to be looked at.

Roy: When you look at the amounts of sulphate esters I wonder whether you could expect much change, because there are two opposing effects of the enzyme deficiency: firstly, you have a deficiency of sulphate, which will lower the amounts of sulphate esters; secondly, you have a vast excess of sulphite, which is a powerful inhibitor of sulphatases, and might therefore increase the amount of sulphate ester. These effects might balance out.

Johnson: I agree that this could be a problem in such studies. It is clearly going to be very difficult if not impossible to directly correlate poor mental development with sulphate deficiency in these patients. Perhaps a more detailed assessment of connective tissue in the two living patients will provide a clearer definition of the role of sulphite oxidase in providing sulphate for biosynthetic reactions.

Kredich: Although you have no positive evidence for a connective tissue defect, the lens dislocation noted in all three patients is one of the earliest and most subtle defects found in other inborn errors of connective tissue metabolism. For example, patients with Marfan's syndrome are usually born with dislocated lenses, in the absence of any of the other abnormalities which develop later. Of these three cases of sulphite oxidase deficiency, one is dead and the other two are still quite young. I would suspect a connective tissue defect, unless there is some other explanation for the dislocated lens.

Johnson: I agree that the dislocated lenses seen in all three patients *may* be an early symptom of a more generalized connective tissue disorder resulting from sulphate deficiency. However, dislocated lenses are also frequently seen in homocystinuric patients who lack cystathionine synthase (Mudd & Levy 1978). In these patients connective tissue disorders are not apparent, nor is sulphate deficiency likely to be a problem. In addition, the dislocated lenses and connective tissue defects seen in Marfan's syndrome are most probably unrelated to sulphate deficiency.

Rose: We might expect that in these sulphite oxidase deficient patients, abnormally sulphated mucopolysaccharides would be produced, whereas in Marfan's syndrome where sulphate metabolism is not affected this would not occur. The result might be an abnormal connective tissue matrix which in turn affects the collagen.

Le Gall: Dr Johnson, you have shown a nice homology between sulphite oxidase and yeast lactate dehydrogenase. Is there any homology between the mammalian enzyme and the bacterial sulphite oxidase from *Thiobacillus*? And does the homology between sulphite oxidase and lactate dehydrogenase haem domains imply that there could be something in common between the molybdenum binding site of sulphite oxidase and the flavin binding site of lactate dehydrogenase?

Johnson: I don't believe there is any homology between the flavin domain of lactate dehydrogenase and the molybdenum domain of sulphite oxidase. What is more likely is that a gene for the b_5 domain has fused with a gene coding for a flavin-binding component in the case of lactate dehydrogenase or with a gene coding for a molybdenum-binding component in the case of sulphite oxidase. One might further speculate that this primordial gene for a b_5 domain has been joined to genes coding for a flavin and for a molybdenum domain to yield a protein structure characteristic of assimilatory nitrate reductase. Sulphite oxidase has been partially purified from *Thiobacillus thioparus* and shown to exhibit a molybdenum e.p.r. spectrum (Kessler & Rajagopalan 1972). Lyric & Suzuki (1970) have also studied a partially purified enzyme preparation and reported the presence of one non-haem iron, no acid-labile sulphide and no cytochrome component. I am not familiar with any more recent studies on the enzyme from this source.

Segel: Is sulphite oxidase present in plants and other bacteria or is it mainly an animal enzyme?

Johnson: There have been reports of sulphite oxidase activity in pea internodes (Arrigoni 1959), in wheat roots (Fromageot et al 1960) and in oats (Tager & Rautanen 1955). We tried to assay it in preparations of spinach, mung bean, yeast and *Aspergillus nidulans* without success.

Ziegler: I was interested in your comment that sulphite oxidase is especially abundant in rat liver. What are the relative amounts in other species?

Johnson: Gunnison et al (1977) have measured sulphite oxidase in various tissues from rat, rhesus monkey and rabbit and reported values for liver of 5.45, 1.42 and 0.36 units per mg protein, respectively. Chicken liver contains levels comparable to those reported for the rat.

Ziegler: The relative activities of various pathways in sulphur metabolism may show considerable species variation. The cysteine sulphinic acid pathway appears more significant in the rat than in some other species. Is the level of sulphite oxidase related to amounts of sulphite produced in different species?

Johnson: Sulphite production by various species has never been measured. Perhaps it could be quantified if sulphite oxidase activity was first eliminated by tungstate administration. Even then, however, I'm afraid properly controlled studies would be difficult.

Kägi: You indicated that cytochrome *c* is the natural electron acceptor for this enzyme. Are other electron acceptors known?

Johnson: Yes; ferricyanide and molecular oxygen can both function as electron acceptors for the enzyme.

Kägi: Do you know how cytochrome *c* interacts with sulphite oxidase? Is it through the molybdenum centre or the cytochrome b_5 domains?

Johnson: Electrons from sulphite enter the enzyme at the molybdenum centre. Reoxidation by ferricyanide is at the molybdenum, whereas cytochrome *c* reduction is via the b_5-type cytochrome on the enzyme. The oxidase activity of the enzyme is very low. When we cleave the enzyme between the haem and molybdenum domains we see a partial loss of oxidase activity. This may suggest that this activity reflects, in part, autooxidation of both the molybdenum and haem centres of the enzyme (Johnson & Rajagopalan 1977).

Le Gall: What is the redox potential of the b_5 in sulphite oxidase?

Johnson: The potential of the b_5 has been measured in the bovine liver sulphite oxidase and found to be +60 mV (Barber & Siegel 1979).

References

Arrigoni O 1959 The enzymatic oxidation of sulphite in mitochondrial preparations of pea internodes. Ital J Biochem 8:181-186

Barber MJ, Siegel LM 1979 Redox potentials of molybdenum and heme in bovine liver sulfite oxidase. Abstracts XIth International Congress of Biochemistry p 265 (abstr 04-2-582)

Duran M, Beemer FA, van der Heiden C, Korteland J, deBree PK, Brink M et al 1979 Combined deficiency of xanthine oxidase and sulphite oxidase: a defect of molybdenum metabolism or transport? J Inher Metab Dis, in press

Fromageot P, Vaillant R, Perez-Milan H 1960 Oxidation du sulfite en sulfate par la racine d'avoine. Biochim Biophys Acta 44:77-85

Gunnison AF, Bresnahan CA, Palmes ED 1977 Comparative sulfite metabolism in the rat, rabbit and rhesus monkey. Toxicol Appl Pharmacol 42:1-11

Irreverre F, Mudd SH, Heizer WD, Laster L 1967 Sulfite oxidase deficiency: studies of a patient with mental retardation, dislocated ocular lenses, and abnormal urinary excretion of *S*-sulpho-L-cysteine, sulfite, and thiosulfate. Biochem Med 1:187-217

Johnson JL, Rajagopalan KV 1977 Tryptic cleavage of rat liver sulfite oxidase. Isolation and characterization of molybdenum and heme domains. J Biol Chem 252:2017-2025

Kessler DL, Rajagopalan KV 1972 Purification and properties of sulfite oxidase from chicken liver. Presence of molybdenum in sulfite oxidase from diverse sources. J Biol Chem 247:6566-6573

Lyric RM, Suzuki I 1970 Enzymes involved in the metabolism of thiosulfate by *Thiobacillus thioparus*. I. Survey of enzymes and properties of sulfite:cytochrome *c* oxidoreductase. Can J Biochem 48:334-343

McCully KS 1971 Homocysteine metabolism in scurvy, growth and arteriosclerosis. Nature (Lond) 231:391

Mudd SH, Levy HL 1978 Disorders of transsulfuration. In: Stanbury FA et al (eds) The metabolic basis of inherited disease, 4th edn. McGraw-Hill, New York, p458-503

Tager JM, Rautanen N 1955 Sulphite oxidation by a plant mitochondrial system. I. Preliminary observations. Biochim Biophys Acta 18:111-121

New aspects of glutathione metabolism and translocation in mammals

ALTON MEISTER, OWEN W. GRIFFITH, ABRAHAM NOVOGRODSKY and SURESH S. TATE

Department of Biochemistry, Cornell University Medical College, 1300 York Avenue, New York, NY 10021, USA

Abstract An appreciable fraction of the sulphur present in the mammal occurs in the form of glutathione, whose concentration in various tissues ranges from about 0.8 to about 8 mM; the extracellular concentration of glutathione (largely present as the disulphide) is in the micromolar range. The synthesis of glutathione and its utilization take place by the reactions of the γ-glutamyl cycle, which include those catalysed by γ-glutamylcysteine and glutathione synthetases, γ-glutamyl transpeptidase, cysteinylglycinase, γ-glutamyl cyclotransferase, and 5-oxoprolinase. γ-Glutamyl transpeptidase catalyses transpeptidation (with amino acids and dipeptides) and hydrolysis reactions with both glutathione and its disulphide. The transpeptidase is membrane-bound, apparently to the outer surface of the cell, and is found in certain epithelial cells in anatomical sites that are involved in transport and secretory activities (e.g., renal tubule, jejunal villi, choroid plexus, ciliary body). Evidence that the reactions of the γ-glutamyl cycle take place *in vivo* has come from studies with labelled metabolites and selective enzyme inhibitors, and on inborn errors of metabolism associated with specific enzyme deficiencies. Inhibition *in vivo* of γ-glutamyl cyclotransferase and 5-oxoprolinase leads, respectively, to decreased and increased renal levels of 5-oxoproline. Administration of a specific inhibitor of γ-glutamylcysteine synthetase, such as buthionine sulphoximine, leads to a rapid decline in the glutathione level of the kidney and other tissues, reflecting the appreciable rate of glutathione utilization. When γ-glutamyl transpeptidase is inhibited *in vivo* by injection of L- or D-γ-glutamyl-(o-carboxy)phenylhydrazide, there is extensive glutathionuria and the blood plasma level of glutathione increases. Studies in which inhibitors of glutathione synthesis and transpeptidation were given to mice showed that transport of intracellular glutathione to membrane-bound transpeptidase is a discrete step in the γ-glutamyl cycle, and that the level of plasma glutathione reflects (a) synthesis of glutathione and its export by liver, muscle, and other tissues and (b) utilization of glutathione by kidney and other tissues. Studies on several lymphoid cell lines show that these cells also actively translocate glutathione out of the cell. A summary scheme is given for the metabolism of glutathione in which glutathione is translocated to the cell membrane where it may be utilized as such or oxidized to glutathione disulphide. Oxidation is inhibited, and transpeptidation is promoted by the presence of amino acids that are substrates of the transpeptidase. Glutathione exported from cells that have membrane-bound transpep-

Sulphur in biology
(Ciba Foundation Symposium 72) p 135-161

tidase may be recovered by the cell by transport of γ-glutamyl amino acids and free amino acids. Translocation of glutathione may also reflect the function of a general cellular mechanism that protects and maintains the integrity of cell membranes.

A significant amount of the sulphur present in mammalian tissues occurs in the tripeptide glutathione. Thus, the concentration of glutathione found in several mouse organs varies from about 0.8 to 8 millimoles per kilogram of fresh tissue (Table 1). The tissue concentrations of glutathione are considerably higher than

TABLE 1

Glutathione levels of mouse tissues before and after administration of buthionine sulphoximine

Tissue	*Control level of glutathione (μmol/g)*	*Glutathione level after giving buthionine sulphoximine*	
		μmol/g	*% of control*
Brain	2.08 ± 0.15	1.93 ± 0.11	93
Heart	1.35 ± 0.10	1.19 ± 0.14	88
Lung	1.52 ± 0.13	1.42 ± 0.09	94
Spleen	3.43 ± 0.35	3.46 ± 0.07	101
Liver	7.68 ± 1.22	2.67 ± 1.15	35
Pancreas	1.78 ± 0.31	0.81 ± 0.30	46
Kidney	4.13 ± 0.15	0.75 ± 0.16	18
Small intestine	2.94 ± 0.16	2.40 ± 0.36	82
Colon	2.11 ± 0.19	1.83 ± 0.16	87
Skeletal muscle	0.78 ± 0.05	0.52 ± 0.11	67
Plasma	0.0284 ± 0.0062	0.009 ± 0.0038	33

Glutathione (GSH + GSSG) was determined by the glutathione reductase procedure. Mice injected subcutaneously with buthionine sulphoximine (Griffith & Meister 1979a) at a dose of 4 mmol/kg were killed after two hours and the glutathione levels of their tissues were determined.

those of cysteine, cystine, and methionine, and of the γ-glutamyl derivatives of these amino acids. This suggests that cysteine sulphur is stored in the form of glutathione. Virtually all cells contain glutathione, and the intracellular concentration of glutathione is far greater than the extracellular level, which is in the micromolar range. It is significant that only a few percent at most of the total intracellular glutathione is present in the form of glutathione disulphide; on the other hand, a substantial fraction, perhaps as much as 75%, of the extracellular glutathione (e.g., that present in blood plasma) can be accounted for as glutathione disulphide. Glutathione is thus the major intracellular thiol compound in most mammalian tissues; glutathione and glutamine are the most prevalent intracellular γ-glutamyl compounds.

Studies in this laboratory elucidated the metabolic pathway, the γ-glutamyl cycle, that accounts for the synthesis and the utilization of glutathione. These investigations began with *in vitro* studies on the individual enzymes of the cycle. Later, evidence was obtained that these reactions take place *in vivo*. Much of this information has come from studies involving the use of selective inhibitors of the γ-glutamyl cycle enzymes. This work has also revealed several new facets of glutathione metabolism; these findings will be presented here together with a brief review of earlier results. (Reviews of earlier work have appeared [Meister & Tate 1976, Meister 1975]; these sources cite the relevant literature more fully than is possible here.)

ENZYMES OF THE γ-GLUTAMYL CYCLE

Glutathione is synthesized by the sequential action of two enzymes, γ-glutamylcysteine synthetase and glutathione synthetase, which catalyse reactions (1) and (2), respectively.

$$\text{L-Glutamate} + \text{L-cysteine} + \text{ATP} \rightleftharpoons \text{L-}\gamma\text{-glutamyl-L-cysteine} + \text{ADP} + P_i \quad (1)$$

$$\text{L-}\gamma\text{-Glutamyl-L-cysteine} + \text{glycine} + \text{ATP} \leftrightharpoons \text{glutathione} + \text{ADP} + P_i \quad (2)$$

These reactions were first recognized by Bloch and his collaborators (Snoke & Bloch 1954). The enzymes have been obtained in highly purified form from several sources and the reaction mechanisms, which involve acyl phosphate intermediates (γ-glutamyl phosphate and γ-glutamylcysteinyl phosphate, respectively), have been elucidated (Meister 1974a). The most highly purified preparations of γ-glutamylcysteine and glutathione synthetases currently available are those isolated from rat kidney (Sekura & Meister 1977, Oppenheimer et al 1979). The broad distribution of glutathione synthesis activity in mammalian tissues suggests that each tissue may synthesize the glutathione it requires. However, as discussed below, one cannot exclude the possibility that glutathione may be translocated into cells.

The major pathway of glutathione breakdown involves the action of γ-glutamyl transpeptidase, which catalyses reactions in which the γ-glutamyl moiety of glutathione (and of other γ-glutamyl compounds) is transferred to an acceptor substrate (e.g., certain L-amino acids and dipeptides) or to water (reactions (3) and (4)):

$$\text{Glutathione} + \text{L-amino acid} \rightleftharpoons \text{L-}\gamma\text{-glutamyl-L-amino acid} + \text{L-cysteinylglycine} \quad (3)$$

$$\text{Glutathione} + H_2O \rightarrow \text{L-glutamate} + \text{L-cysteinylglycine} \quad (4)$$

The activity of the transpeptidase was recognized many years ago; Hanes et al (1952) were the first to demonstrate reaction (3). γ-Glutamyl transpeptidase is a membrane-bound enzyme found in a variety of epithelial cells in anatomical locations thought to be involved in transport and secretory phenomena, such as jejunal villus tip cells, proximal renal tubules, choroid plexus, ciliary body, visual receptor cells, retinal epithelium, cerebral astrocytes or their capillaries, bile duct, seminal vesicles, epididymis. (See Tate 1975 and Meister et al 1976 for recent reviews.) The enzyme is also found on the membrane of certain lymphoid cells (Novogrodsky et al 1976). There is substantial evidence that a major fraction of the membrane-bound enzyme is located on the outer surface of the cell membrane (Meister & Tate 1976). Thus, the enzyme is accessible to externally supplied substrates.

Purified preparations of the transpeptidase have been obtained from several sources; perhaps the most highly purified preparation is that from rat kidney (Tate & Meister 1975). The enzyme is composed of two dissimilar glycoprotein subunits, the smaller of which has the binding site for the γ-glutamyl moiety (Tate & Meister 1976, 1977). The mechanism of the reaction involves interaction between the enzyme and γ-glutamyl donor to form a γ-glutamyl-enzyme,which may react with an acceptor substrate or with water (Tate & Meister 1974, Thompson & Meister 1977).

γ-Glutamyl transpeptidase may act on glutathione as well as glutathione disulphide. *In vitro* studies have shown that the transpeptidation reaction with glutathione (reaction (3)) is about 20–30 times more rapid than the hydrolytic reaction (reaction (4)) (Tate & Meister 1974, 1975, Tate et al 1979). The hydrolyses of glutathione and of glutathione disulphide, as judged by the rates of glutamate release, occur at similar rates. Transpeptidation reactions occur with both glutathione and glutathione disulphide; the reaction with glutathione disulphide takes place at about 20% of that found with glutathione. Under these conditions there is appreciable autotranspeptidation, i.e., formation of γ-glutamyl-glutathione and γ-glutamyl-γ-glutamyl-glutathione. Such reactions occur at about two- and five-fold greater rates than those of the corresponding hydrolytic reactions with glutathione disulphide and glutathione.

The cysteinylglycine (or the corresponding disulphide) formed by the action of the transpeptidase may be hydrolysed by the action of membrane-bound dipeptidase (Hughey et al 1978) (reaction (5)):

$$\text{L-Cysteinylglycine} + H_2O \rightarrow \text{L-cysteine} + \text{glycine} \qquad (5)$$

Similar enzyme activity is also found intracellularly.

The γ-glutamyl amino acids formed by the action of the transpeptidase (reaction (3)) may serve as substrates for additional hydrolytic or transfer reactions catalysed by the transpeptidase, or they may be converted intracellularly by the action of γ-glutamyl cyclotransferase to 5-oxo-L-proline and the corresponding amino acids (reaction (6)):

$$\text{L-}\gamma\text{-Glutamyl-L-amino acid} \rightarrow \text{5-oxo-L-proline} + \text{L-amino acid} \qquad (6)$$

The formation of 5-oxoproline as a product of glutathione metabolism was first noted by Woodward & Reinhart (1942); Connell & Hanes (1956) subsequently demonstrated reaction (6). The enzyme has been found in many tissues. The most highly purified preparation of γ-glutamyl cyclotransferase has been obtained from rat kidney (Taniguchi & Meister 1978).

Studies in which labelled 5-oxo-L-proline was administered to mice showed that this compound is rapidly metabolized (Richman 1969, Van Der Werf 1970). Subsequently, the enzyme 5-oxo-L-prolinase was discovered (Van Der Werf et al 1971) and purified (Van Der Werf et al 1975, Williamson 1979). The reaction catalysed by this enzyme (reaction (7)) is of interest in that it involves coupling between the exergonic cleavage of ATP and the endergonic decyclization of 5-oxoproline (Van Der Werf & Meister 1975):

$$\text{5-oxo-L-proline} + \text{ATP} + 2\ H_2O \rightarrow \text{L-glutamate} + \text{ADP} + P_i \qquad (7)$$

FUNCTION OF THE γ-GLUTAMYL CYCLE *IN VIVO*

The metabolic pathway described by reactions (1) to (7), which has been termed the γ-glutamyl cycle, accounts for the synthesis and the breakdown of glutathione. Evidence that this series of reactions occurs *in vivo* has come from a variety of experimental approaches including experiments with labelled metabolites and studies with selective inhibitors of the enzymes of the cycle. Observations have also been made on patients with inborn errors of metabolism associated with specific enzyme deficiencies. It is notable that the enzymes of the cycle are widely distributed and that the rate of turnover of glutathione in a given tissue is, in general, related to the magnitude of the enzyme activities. Thus, the kidney, which has a rapid turnover of glutathione, has very high activities of the cycle enzymes (Sekura & Meister 1974). The rate of incorporation of administered labelled 5-oxoproline into glutathione in the mouse kidney is close to that found after administration of glutamate, and labelled 5-oxoproline is found in kidney and liver after administration of labelled glutamate to rats (Sekura et al 1976).

In vivo inhibition of 5-oxoprolinase has been observed after injection into mice of L-2-imidazolidone-4-carboxylate, a competitive inhibitor of this enzyme (Van Der Werf et al 1973, 1974). Mice treated with the inhibitor exhibited decreased ability to metabolize 5-oxoproline, and they accumulated 5-oxoproline in their tissues. Such accumulation was greater after administration of both inhibitor and amino acids.

When β-glutamyl-α-aminobutyrate, a competitive inhibitor of γ-glutamyl cyclotransferase, was given to mice there was a marked decrease in the steady-state concentration of 5-oxoproline in the kidney (Griffith et al 1978). This inhibitor also decreased the accumulation of 5-oxoproline found after *in vivo* inhibition of 5-

oxoprolinase (Table 2). The studies with L-2-imidazolidone-4-carboxylate and β-glutamyl-α-aminobutyrate thus support the conclusion that γ-glutamyl cyclotransferase and 5-oxoprolinase are, respectively, major *in vivo* catalysts for the formation and utilization of 5-oxoproline.

TABLE 2

Effect of β-aminoglutaryl-α-aminobutyrate (β-Glu-αAba) on kidney levels of 5-oxoproline

Compound given	*5-Oxoproline (nmol/g)*	*β-Glu-αAba (μmol/g)*
None (control)	36 ± 5	- - -
β-Glu-αAba	18 ± 4	36 ± 10
ICA	110 ± 9	- - -
ICA + β-Glu-αAba	60 ± 20	46 ± 7

β-Glu-α-[^{14}C]Aba (10 mmol/kg) and L-2-imidazolidone-4-carboxylate (ICA) (2.4 mmol/kg) were given to mice intraperitoneally. After 30 min the doses were repeated subcutaneously; 30 min later the animals were killed. Data are shown as mean ± S.D. (From Griffith et al 1978.)

The steady-state level of glutathione in a tissue, according to the γ-glutamyl cycle, would be expected to reflect a balance between the synthesis of glutathione and its utilization. Thus, significant perturbations of the rates of synthesis or use of glutathione would be expected to affect the tissue level of glutathione.

Administration to rats of methionine sulphoximine (a convulsant which inhibits both γ-glutamylcysteine synthetase and glutamine synthetase [Meister & Tate 1976]), led to markedly decreased levels of glutathione in the kidney and liver (Palekar et al 1975). When mice were given prothionine sulphoximine and buthionine sulphoximine (which are specific inhibitors of γ-glutamylcysteine synthetase and are not convulsants), the levels of glutathione in the kidney also decreased rapidly (Griffith et al 1979b, Griffith & Meister 1979b). Indeed, the levels of glutathione in most tissues of the mouse declined two hours after administration of buthionine sulphoximine (Table 1, p 136) (Griffith & Meister 1979). The marked decrease in glutathione found in the liver, kidney, pancreas and skeletal muscle indicates a relatively rapid turnover of glutathione in these tissues. Although the concentration of glutathione is lowest in skeletal muscle, the total amount of muscle glutathione must evidently be of the same order as that present in liver. This finding and the relatively rapid turnover of muscle glutathione suggest that this organ may play a significant role in the overall metabolism of body glutathione (see below). The marked decrease in plasma glutathione after buthionine sulphoximine administration is also considered below. Brain and spleen evidently turn over glutathione rather slowly. Administration of buthionine sulphoximine to tumour-bearing mice led to markedly decreased levels of glutathione in the tumours (Anderson 1979). This observation is of potential interest in relation to the possibility of modifying the glutathione content of tumours during therapy. It seems clear that the rapid fall of tissue glutathione levels after inhibition of synthesis must reflect the appreciable rate at which glutathione is normally utilized.

Glutathione levels also decrease when there is an increase in transpeptidation. This was first observed in studies in which rats were injected with glycylglycine, an excellent γ-glutamyl transpeptidase acceptor substrate. In these studies rats given 3.34 mmoles of glycylglycine per kilogram showed about a 50% decline in renal glutathione one hour after injection; it is notable that there was no decline in the glutathione level of the liver, which has relatively little transpeptidase (Palekar et al 1975). Such a decline in renal glutathione did not occur after giving equivalent doses of glycine or other amino acids, presumably because after small doses of amino acids, the amount of glutathione used for transpeptidation can be effectively regenerated by synthesis. The extensive transpeptidation that occurs in the kidney after giving glycylglycine evidently uses glutathione at a considerably faster rate than that of its resynthesis; the level of glutathione therefore falls.

When very large amounts of L-methionine or other amino acids were given to mice there was a considerable decline in the level of renal glutathione (Griffith et al 1978). Thus, after giving 32 mmoles per kilogram of L-methionine the renal glutathione level decreased to 50% of the control value in 30 minutes, and this effect was accompanied by a marked increase in the tissue level of 5-oxoproline. Although increased levels of 5-oxoproline were found after giving large amounts of other amino acids, the level of 5-oxoproline did not increase after administration of glycylglycine, the γ-glutamyl derivative of which is not a substrate of γ-glutamyl cyclotransferase.

The substantial decline in renal glutathione found after giving a large dose of methionine was not seen when the animals were also treated with a mixture of L-serine and sodium borate (Fig. 1); serine plus borate effectively inhibits γ-glutamyl transpeptidase, apparently by forming a complex with the enzyme that mimics the transition state (Tate & Meister 1978). It was also found that *in vivo* inhibition of the transpeptidase by serine plus borate (as well as by other inhibitors of the transpeptidase) decreased the rate of glutathione disappearance in the kidneys of mice treated with inhibitors of γ-glutamylcysteine synthetase, such as prothionine sulphoximine.

The findings reviewed above on the effects of *in vivo* inhibition of glutathione synthesis, γ-glutamyl cyclotransferase, 5-oxoprolinase and γ-glutamyl transpeptidase are consistent with and support the view that the reactions of the γ-glutamyl cycle occur *in vivo*. Other evidence that supports this conclusion has come from

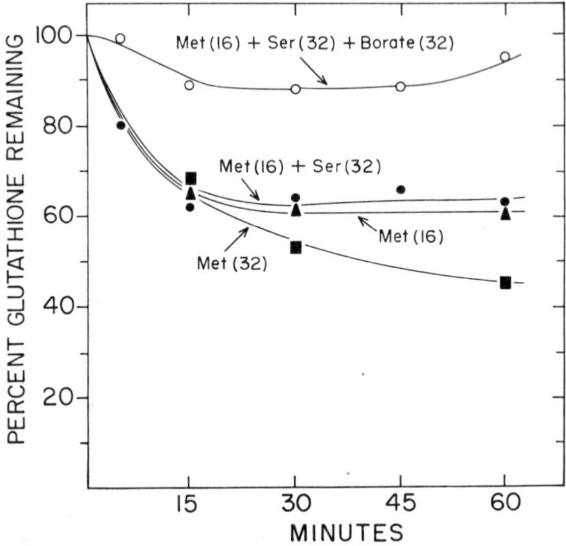

FIG. 1. Inhibition of methionine-induced depletion of renal glutathione in mice by serine and borate. The values given in parentheses indicate doses (mmol/kg). (From Griffith et al 1978.)

studies on patients who have 5-oxoprolinuria, an inborn metabolic error associated with a marked deficiency of glutathione synthetase (Wellner et al 1974, Meister 1974b, 1978a), and from other investigations (see below). Patients with 5-oxoprolinuria overproduce γ-glutamylcysteine which is effectively converted to 5-oxoproline. The amount of 5-oxoproline formed exceeds the capacity of 5-oxoprolinase and the excess of 5-oxoproline (about one-third of that formed) accumulates in the blood plasma, cerebrospinal fluid, and tissues, and there is massive excretion of 5-oxoproline in the urine. Glutathione normally regulates its own biosynthesis by feedback inhibition of γ-glutamylcysteine synthetase (Richman & Meister 1975), so that in the presence of very low levels of glutathione, γ-glutamylcysteine is produced in large amounts. Thus, in 5-oxoprolinuria there is a futile cycle of γ-glutamylcysteine synthesis followed by its conversion to 5-oxoproline and cysteine. Administration of amino acids to a patient with this disease led to increased 5-oxoproline excretion (Jellum et al 1970), a finding consistent with increased transpeptidation followed by a corresponding increase in conversion of γ-glutamyl amino acids to 5-oxoproline and overflow of the latter into the urine (Meister 1978a).

TRANSLOCATION OF GLUTATHIONE AS A DISCRETE STEP IN THE γ-GLUTAMYL CYCLE

There is now considerable evidence indicating that a significant fraction of γ-glutamyl transpeptidase is located on the outer surface of the cell membrane. Thus, it was found that suspensions of lymphoid cells act on γ-glutamyl substrates that cannot enter the cell, and disruption of such cells does not lead to an increase in transpeptidase activity (Novogrodsky et al 1976). In experiments in which a closed intact perfused rat kidney system was used, glutathione added to the perfusate was rapidly degraded (Maack et al 1974, Meister & Tate 1976). Glutathione injected into animals is known to disappear rapidly; apparently the kidney plays an important role in this process (Hahn et al 1978). Perfusion of kidneys with papain leads to liberation of a significant fraction of the transpeptidase present in renal tubular cells (Kuhlenschmidt & Curthoys 1975). These studies show that the transpeptidase is accessible to substrates that are supplied extracellularly, but they do not exclude the possibility that the enzyme is also accessible to intracellular glutathione. It is possible that the enzyme is also located within the cell membrane and that it occurs, at least in part, intracellularly as well. The findings discussed above indicate that intracellular glutathione is rapidly utilized. If the transpeptidase is the major enzyme involved in the utilization of glutathione, and if this enzyme is mainly (if not entirely) located on the surface of the cell membrane, it would seem that there must be a mechanism for making intracellular glutathione available to the enzyme. This problem was recognized and considered previously (Meister et al 1977, Meister 1978a, Griffith et al 1978), and it was proposed that a mechanism exists for the translocation of intracellular glutathione to the membrane-bound enzyme.

Definitive evidence for the translocation of intracellular glutathione has come from studies in which highly effective competitive inhibitors of γ-glutamyl transpeptidase were administered to mice (Griffith & Meister 1979b). In this work it became desirable to establish the extent of *in vivo* inhibition of the transpeptidase after administration of an inhibitor. This was accomplished by giving the mice a standard dose of D-γ-glutamyl-L-α-amino[^{14}C]butyrate; this compound is a substrate of γ-glutamyl transpeptidase, but not of γ-glutamyl cyclotransferase (Griffith & Meister 1977). Thus, the conversion of the L-α-amino[^{14}C]butyryl moiety to respiratory $^{14}CO_2$ requires the action of the transpeptidase, and a decrease in $^{14}CO_2$ formation after giving an inhibitor indicates *in vivo* inhibition of the transpeptidase. By this assay, several inhibitors were found to be effective; these included 6-diazo-5-oxo-L-norleucine, L-serine plus borate, and L- and D-γ-glutamyl-(*o*-carboxy)phenylhydrazide. The hydrazides were by far the most potent inhibitors *in vitro* and *in vivo*. Although the L-isomer of the hydrazide is somewhat more active than the D-isomer, we chose to use the D-isomer in many studies because of its lower toxicity. After administration of the hydrazide there was substantially

TABLE 3

Glutathionuria induced by inhibition of γ-glutamyl transpeptidase

Expt	*Compound given (mmol kg^{-1})*	*Glutathionuria*	
		Concn (μM)	*Amount (nmol)*
1	None (Control)	2-5	0.4-1.3
2	L-γ-Glutamyl-(*o*-carboxy)phenylhydrazide (1.0)	3600-5640	360-846
3	D-γ-Glutamyl-(*o*-carboxy)phenylhydrazide (2.5)	1770-3940	208-788
4	DL-γ-(γ-Methyl)glutamyl-(*o*-carboxy)phenylhydrazide (2.5)	74-144	22-26
5	L-γ-Glutamyl-(*p*-carboxy)phenylhydrazide (1.0)	88-148	11-22
6	*0*-Hydrazinobenzoate (0.2)	2-5	0.4-1.0
7	L-Serine (32) + borate (32)	0.5-4	0.4-0.75
8	6-Diazo-5-oxo-L-norleucine (2.0)	9-35	1.8-7.0

Mice (not starved) were injected intraperitoneally with the compounds indicated. Small volumes (5–50 μl) of urine were then collected directly from each mouse every 10–15 min. After 60 min the animals were killed and the bladder urine was collected. The pooled urine (75–300 μl) of individual mice was assayed by the glutathione reductase method. In Expt 7 the period of urine collection was extended to 4–6 h because very little urine is produced for several hours after injection of serine plus borate.

In Expt 8 the animals were injected with hippurate (10 mmol kg^{-1}) and 6-diazo-5-oxo-L-norleucine (2 mmol kg^{-1}) 12 h before urine collection. The ranges given were determined in studies on four to six animals for each experiment. (From Griffith & Meister 1979b.)

complete inhibition *in vivo* of the transpeptidase for about one hour — that is, no $^{14}CO_2$ was formed from the administered dose of D-γ-glutamyl-L-α-amino[^{14}C]butyrate; subsequently $^{14}CO_2$ was formed as the inhibitor was excreted in the urine. The most notable finding in these studies was the appearance in the urine of animals treated with the hydrazide of substantial amounts of glutathione (Table 3). Much less glutathionuria was observed after giving another hydrazide, L-γ-glutamyl-(*p*-carboxy)phenylhydrazide, 6-diazo-5-oxo-L-norleucine, or L-serine plus borate; all of these compounds are also less effective as inhibitors of transpeptidase *in vitro*. About 90% of the total urinary glutathione found after giving L- or D-γ-glutamyl(*o*-carboxy)phenylhydrazide was present as glutathione disulphide. It is notable that, in addition to glutathione disulphide, the urine of such mice also contained about 30-60% as much γ-glutamylcysteine (determined as the vinylpyridine derivative after treatment of the urine with dithiothreitol) as glutathione. This observation suggests that there was some conversion of glutathione to γ-glutamylcysteine (or an analogous conversion of the corresponding disulphides).

Studies were also made on the levels of glutathione present in the blood plasma of mice after treatment with D-γ-glutamyl-(*o*-carboxy)phenylhydrazide and with the γ-glutamylcysteine synthetase inhibitor, prothionine sulphoximine. It was found that administration of the hydrazide led to about a three-fold increase in the level of glutathione in the blood plasma (control, 24 ± 1.3 nmol/ml; after the hydrazide was given, 73 ± 15 nmol/ml). On the other hand, administration of prothionine sulphoximine led to a decrease in the blood plasma glutathione level to 8.3 ± 1.4 nmol/ml (see also Table 1, p 136). Studies in which both prothionine sulphoximine and the hydrazide were given in sequence are consistent with the conclusion that a substantial fraction of the urinary glutathione arises from the kidney; however, it is likely that some of the urinary glutathione comes via the plasma from the liver and other tissues as well. It was also found that administration of the hydrazide decreases the rate at which the glutathione level declines after the administration of prothionine sulphoximine. It is notable, however, that injection of prothionine sulphoximine produced a decrease in kidney glutathione even when the transpeptidase was markedly inhibited. The substantial decrease in intracellular glutathione and the extensive glutathionuria that occurs after inhibition both of glutathione synthesis and of the transpeptidase indicate that translocation of glutathione takes place as an event separate from its utilization by the transpeptidase. Thus, it appears that the translocated glutathione normally interacts with the transpeptidase, but when the transpeptidase is inhibited, the glutathione appears in the urine.

These observations show that there is a mechanism for translocation of intracellular glutathione which can function independently of the transpeptidase. The experimental findings are in general accord with observations reported on a patient with an apparent inborn deficiency of γ-glutamyl transpeptidase who was found to exhibit glutathionuria and glutathionaemia (Schulman et al 1975). The translocation of glutathione in this patient and in the hydrazide-treated mice thus seems to reflect a step that occurs normally in glutathione metabolism. This then explains how an enzyme, which appears to be chiefly extracellular in location, can interact with a substrate that is formed intracellularly.

The translocation of glutathione is, however, not restricted to cells such as those of the renal tubule, which are equipped with γ-glutamyl transpeptidase. The increase in the blood plasma level of glutathione after inhibition of transpeptidase may probably be ascribed to decreased utilization of glutathione by extrarenal tissues that have the transpeptidase. On the other hand, the substantial fall in the glutathione level of the plasma after administration of an inhibitor of glutathione synthesis seems to closely parallel the decreases in tissue glutathione that occur under these conditions. Thus, two hours after administration of prothionine sulphoximine to mice, the levels of glutathione decreased by 75%, 30%, and 65%, respectively, in the kidney, liver, and plasma. Similar results were obtained in studies with buthionine sulphoximine (Table 1, p 136). It seems likely that the liver, which contains a large fraction of the total body glutathione, is a major, but probably not the only, source of plasma glutathione. Thus, it appears that muscle, which has very little γ-glutamyl transpeptidase, may also serve as a significant source of plasma glutathione. The rapid fall in plasma glutathione that occurs after inhibition of glutathione synthesis indicates that there is a substantial flow of glutathione into the plasma, presumably largely from tissues such as liver and muscle, that have very little transpeptidase. The very low concentration of glutathione in normal blood plasma indicates that glutathione is effectively removed from the circulation. Our findings on the origin and utilization of plasma glutathione are consistent with the observations of Bartoli & Sies (1978) who found a substantial efflux of glutathione from isolated perfused liver preparations into a non-haemoglobin-containing perfusate, and also with those of Hahn et al (1978), who found that glutathione administered to rats is rapidly cleared from the blood.

It has been possible to estimate the relative amounts of glutathione cleared by the kidneys and by extra-renal tissues by studying anephric rats treated with a transpeptidase inhibitor (D-γ-glutamyl-(*o*-carboxy)phenylhydrazide) (Griffith & Meister 1979a). Nephrectomized rats on the day after surgery were found to have plasma glutathione levels that were about twice as high as those of sham-operated controls (33 ± 7 μM compared to 17 ± 4). This increase is much less than found when anephric animals are given a transpeptidase inhibitor. This indicates that there is a substantial amount of extra-renal transpeptidase, which limits the increase in the concentration of plasma glutathione after nephrectomy. Sham-operated animals treated with inhibitor had plasma glutathione levels of 63 ± 7 μM, whereas anephric animals similarly treated had levels of 169 ± 31 μM. The difference between these values, about 100 μM, would seem to be a measure of the glutathione cleared by the kidneys. This suggests that about two-thirds of the plasma glutathione is normally utilized by the kidneys, and that the remainder is utilized by extra-renal transpeptidase. It is interesting to note that rats that were surgically deprived of one kidney and then treated with inhibitor had plasma glutathione levels of 102 ± 1 μM, or a value about midway between similarly treated controls and anephric animals.

In a further study of the phenomenon of glutathione translocation we have examined the consequences of inhibiting the γ-glutamyl transpeptidase activity of several lymphoid cell lines grown in tissue culture (Griffith et al 1979a) (Table 4). Three types of cells were used: (a) a cell line with a high γ-glutamyl transpeptidase activity (8000-10 000 units/mg protein), (b) cell lines with very low transpeptidase activity (60–230 units/mg protein), and (c) a cell line with an intermediate level of transpeptidase activity (about 1500 units/mg protein). The cells were suspended in a solution containing various concentrations of a transpeptidase inhibitor. In studies on the lymphoid cell line with an intermediate level of activity, we found no glutathione in the suspending medium. However, as

TABLE 4

Translocation of glutathione from lymphoid cells that have markedly different transpeptidase activities

Cell line	*Inhibitor* (*μM*)	*γ-Glutamyl transpeptidase* (*nmol mg*$^{-1}$ *h*$^{-1}$)	*Glutathione* (*nmol mg*$^{-1}$)
RPMI 6237	None	1750	0
	10	1490	0.48
	50	1130	1.2
	500	276	2.9
	1000	184	3.4
RPMI 8226	None	8700	0
	50	4840	0
	100	3110	0
	500	899	0.60
	1000	555	0.92
CEM	None	60	1.2
	1000	16	1.3
MOLT	None	232	0.61
	1000	20	1.4

The cells were suspended in solutions containing various concentrations of the inhibitor, L-γ-glutamyl-(*o*-carboxy)phenylhydrazide; after incubation with shaking for 1 h the cells were removed by centrifugation and the glutathione present was determined by the glutathione reductase procedure (Tietze 1969). (From Griffith et al 1979a.)

the concentration of L-γ-glutamyl-(*o*-carboxy)phenylhydrazide in the suspending solution was increased, and the transpeptidase was inhibited 15% or more, the amount of glutathione found increased significantly. In experiments on the cell line with high transpeptidase activity, no glutathione was found in the medium until the concentration of inhibitor was relatively high and thus sufficient to inhibit the transpeptidase by more than 90%. In the studies on cell lines that exhibit very little transpeptidase, a substantial amount of glutathione was detectable in the medium in the absence of inhibitor, and there was only a moderate increase in the glutathione found when inhibitor was added. It thus appears that glutathione is translocated to the medium by all of the cell lines studied. However, glutathione does not accumulate in the medium unless the cellular transpeptidase activity is very low or substantially inhibited. Translocation of glutathione by such cells is therefore not directly related to the activity of γ-glutamyl transpeptidase. These lymphoid cell lines are, at least in one respect, similar to other mammalian cells such as those of the liver and kidney which contain, respectively, very low and very high transpeptidase activity. In one type of cell, glutathione is translocated to the medium, where it is not appreciably metabolized; in the other, it is utilized and the products of such reactions may be returned to the cell. It is notable that the glutathione translocated to the medium by lymphoid cells was found to be more than 90% in the reduced form; the small amount of glutathione disulphide present may well have been due to non-enzymic oxidation. This indicates that glutathione, rather than glutathione disulphide, is the form that is translocated.

EXTRACELLULAR OXIDATION OF GLUTATHIONE TO GLUTATHIONE DISULPHIDE

More than half of the total glutathione found in the blood plasma is in the form of glutathione disulphide, and a large fraction of the glutathione found in the urine of animals treated with the potent transpeptidase inhibitor, γ-glutamyl-(*o*-carboxy)phenylhydrazide, and of the patient with transpeptidase deficiency was also found to be in the form of glutathione disulphide. These observations are consistent with the existence of one or more systems that mediate the oxidation of extracellular glutathione. A number of years ago Elvehjem and his colleagues (Ames & Elvehjem 1945, Ziegenhagen et al 1947) found that homogenates of mouse kidney catalyse the oxygen-dependent conversion of glutathione to glutathione disulphide. The oxidation was inhibited by cyanide. Although considerable oxidase activity was found in kidney homogenates, much less was found in homogenates of liver. Subsequent studies in Orrenius' laboratory (Moldeus et al 1978, Jones et al 1978) on the metabolism of glutathione and of a glutathione conjugate by isolated rat kidney cells have confirmed these findings, and have provided evidence for the stepwise breakdown of glutathione disulphide to its constituent amino acids; these reactions are presumably catalysed by the combined actions of γ-glutamyl transpeptidase and dipeptidase. In these experiments, as in the earlier ones, little or no activity was found with preparations obtained from liver, and the oxidation reaction catalysed by kidney cells was inhibited by cyanide and by replacement of oxygen with nitrogen.

The conversion of glutathione to glutathione disulphide was examined in our laboratory in studies in which the specificity of the reaction and its distribution in various mammalian tissues were examined (Tate et al 1979). Oxidase activity was found in the kidney, epididymis, jejunum, choroid plexus, and retina; no activity was found in liver. It is notable that the oxidase activity followed the same pattern of distribution as that of γ-glutamyl transpeptidase. Thus, renal cortex exhibited very high oxidase activity. In epididymis, there was substantially more oxidase in the caput than in the cauda, as is the case with γ-glutamyl transpeptidase (DeLap et al 1977). A gradient of oxidase activity was found in fractions obtained by the micrometer planing method (Imondi et al 1969) from jejunal mucosa, with the highest oxidase present in the villus tip cells, as was previously found for γ-glutamyl transpeptidase (Cornell & Meister 1976). Appreciable activity was found in the choroid plexus, whereas no activity was detected in cerebral cortex.

The oxidase activity was found to be associated with the particulate fractions of the tissues. Subcellular fractionation of rat kidney carried out as described by Grau et al (1979) led to a brush-border membrane fraction in which both the transpeptidase and the oxidase were enriched about 18-fold. Both activities were solubilized (and remained associated) after treatment with papain; neither activity was affected by treatment with detergents. These observations suggest that both activities are located on the luminal surface of the cell membrane.

The oxidase appears to be specific for glutathione; little or no activity was found with *N*-acetylcysteine, cysteinylglycine, cysteine, and dithiothreitol. No activity was found when the reaction was studied under nitrogen, and marked inhibition was found by cyanide and by EDTA. Hydrogen peroxide could not be detected as a product of the reaction. The oxidase seems to be a glycoprotein.

It is of interest that the ratios of transpeptidase to oxidase activities found in various mammalian tissues are very similar. However, examination of two lymphoid cell lines that have substantial levels of transpeptidase failed to reveal the presence of oxidase activity; it may be noted (see above) that most of the glutathione translocated from such cells was in the form of glutathione rather than glutathione disulphide.

Studies in which the brush-border membrane fraction (containing both oxidase and transpeptidase activities) was examined showed that the oxidation of glutathione and the degradation of total glutathione (presumably catalysed by transpeptidase and

TABLE 5

Rates of oxidation and utilization of glutathione by rat renal brush-border membranes

Additions (mM)	*GSSG formed (μmol/15 min)*	*Disappearance of total glutathione (GSH + GSSG) (μmol/15 min)*
None (control)	0.45	0.55
EDTA	0.03	0.62
L-Methionine (5)	0.08	0.83
Glycylglycine (5)	0.02	1.59
L-γ-Glutamyl-(*o*-carboxy)-phenylhydrazide (0.5)	0.39	0.14

The reaction mixtures (final volume, 1 ml) contained 2 mM-GSH, renal brush-border membranes (20 μg of P_5B protein), Hepes-phosphate buffer (pH 7.4), and other additions as shown. After incubation at 37 °C with aeration, samples (0.2 ml) were withdrawn at 5 min intervals and mixed with 0.02 ml of 2 M-acetic acid. Total glutathione (GSH + GSSG) and glutathione disulphide (GSSG) (after removal of glutathione (GSH) by reaction with 2-vinylpyridine) were determined enzymically with glutathione reductase. (From Tate et al 1979.)

dipeptidase) proceed at similar rates, suggesting that the oxidation of glutathione by the renal brush border may be a significant metabolic process (Table 5). Under these conditions EDTA inhibited oxidation without affecting disappearance of total glutathione. γ-Glutamyl-(*o*-carboxy)phenylhydrazide did not affect oxidase activity, but inhibited glutathione utilization. Relatively low concentrations of acceptor substrates of γ-glutamyl transpeptidase (e.g., L-methionine, glycylglycine) increased glutathione utilization and decreased oxidation. It seems likely that glutathione translocated from cells would be accessible to both transpeptidation and oxidation. The pathway of glutathione metabolism would probably be influenced by the presence of amino acids, which would tend to inhibit oxidation and promote transpeptidation. As discussed further below, it seems probable that the translocation of intracellular glutathione, the extracellular oxidation of glutathione to glutathione disulphide, and the reactions of the γ-glutamyl cycle are closely related and perhaps coordinated phenomena.

DISCUSSION

The chart given in Fig. 2 presents an attempt to integrate the findings now available about glutathione metabolism and translocation. Glutathione is synthesized intracellularly from its constituent amino acids, and is translocated to the surface of the cell membrane. Glutathione may be utilized as such or oxidized to glutathione disulphide. Both glutathione and glutathione disulphide are substrates of γ-glutamyl transpeptidase and may therefore participate in hydrolysis and transpeptidation reactions leading, in various ways, to the formation of γ-glutamyl amino acids, glutamate, glycine, cysteine, cystine, cysteinylglycine, cystinyl-bis-glycine, and mixed disulphides between glutathione and cysteine, glutathione and cysteinylglycine, and cysteine and cysteinylglycine. There are evidently transport pathways for the uptake by the renal cell of glycine, glutamate, cysteine and cystine, and there is also evidence (Griffith et al 1978) that γ-glutamyl amino acids can be transported into the kidney. As indicated in Fig. 2, γ-glutamyl amino acids may be converted to the corresponding amino acids and 5-oxoproline, followed by decyclization of the latter to glutamate by 5-oxoprolinase.

Translocation of glutathione appears to be a general property of many, perhaps most, mammalian cells. Such translocation was revealed clearly in the studies in which intact animals and also cells grown in tissue culture were treated with inhibitors of γ-glutamyl transpeptidase, and is also apparent from the observations made on a patient with γ-glutamyl transpeptidase deficiency. It is also evident from the studies on lymphoid cells that have low levels of transpeptidase activity.

It would appear that cells such as those of the renal tubule, epithelial cells of the

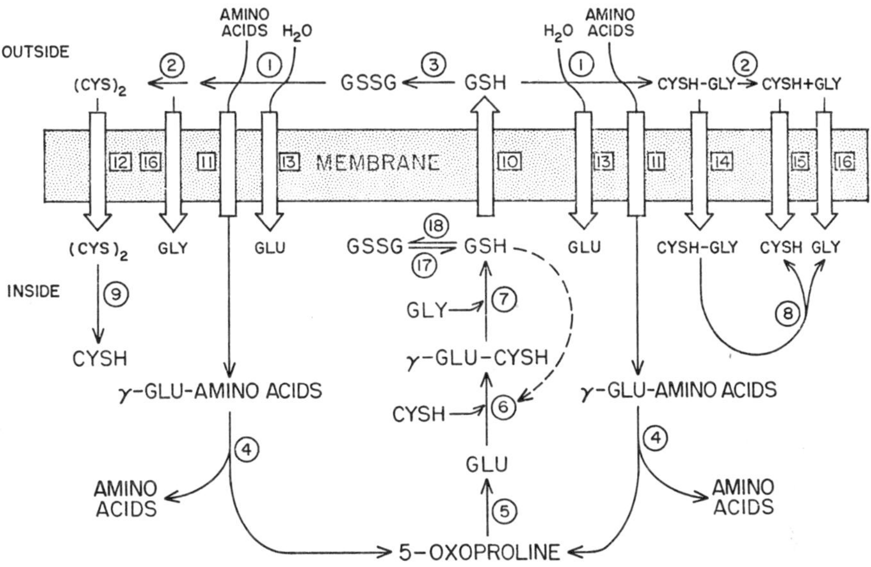

FIG. 2. Overall scheme of the metabolism and translocation of glutathione (see the text). 1. γ-Glutamyl transpeptidase. 2. Dipeptidase (membrane). 3. Oxidation of glutathione. 4. γ-Glutamyl cyclotransferase. 5. 5-Oxo-L-prolinase. 6. γ-Glutamylcysteine synthetase. 7. Glutathione synthetase. 8. Dipeptidase (cytoplasmic). 9. Coupled to glutathione reductase. 10. Transport of glutathione. 11. Transport of γ-glutamyl amino acids. 12-16. Transport of cystine, glutamate, cysteinylglycine, cysteine, and glycine. 17. Glutathione reductase. 18. Intracellular oxidation of glutathione. The dashed line indicates feedback inhibition of γ-glutamylcysteine synthetase by glutathione. L-Amino acids promote transpeptidation (reaction 1) and inhibit oxidation of glutathione (reaction 3).

jejunum, ciliary body, choroid plexus, and so forth, that are equipped with transpeptidase, may recover the translocated glutathione as γ-glutamyl amino acids or as free amino acids. Cells such as those of the liver, which lack appreciable transpeptidase activity, evidently translocate glutathione to the blood plasma, from which it ultimately may reach and be utilized by cells equipped with transpeptidase.

Although translocation of glutathione and the activity of γ-glutamyl transpeptidase can be distinguished as separate phenomena, several observations suggest that they are interrelated. Thus, it was found that administration of L-serine and borate prevents the marked fall in the level of renal glutathione induced by giving amino acids or glycylglycine (Griffith et al 1978). In addition, inhibitors of transpeptidase decrease the rate at which the renal glutathione level falls after administration of an inhibitor of glutathione synthesis. These effects may indicate that there is some inhibition of intracellular glutathione breakdown. Another possibility is that the rate of translocation of glutathione may be affected by the concentration of glutathione within the renal tubule. If there is a substantial concentration of glutathione in the renal tubule, a decrease of this level would be expected to facilitate glutathione efflux and thus tend to reduce the intracellular level of glutathione. An increase in the concentration of tubular glutathione, produced by inhibition of transpeptidase, would tend to decrease the translocation of and thus the utilization of intracellular glutathione.

The mechanism by which glutathione is translocated requires study. Since the intracellular concentration of glutathione is so much higher than the extracellular concentration, one might conclude that translocation could occur by diffusion and that energy would not be required. In addition to the utilization of extracellular glutathione by transpeptidation, its oxidation to glutathione disulphide may also serve to increase the rate of translocation by reducing the extracellular glutathione level. Other factors need to be examined in relation to the translocation phenomenon. Thus, in the studies on lymphoid cells differences were found in the rates of glutathione translocation in different experiments on the same cell line. The possibilities that such differences are related to the metabolic status of the cells or to an influence associated with the stage of the cell cycle need to be explored. It will also be of interest to examine directly the possibility that intact glutathione may be taken up by certain cells.

Although it seems likely that translocation, transpeptidation, and oxidation of glutathione are related, the details of such a relationship are not yet clear. The availability of transpeptidase acceptor substrates probably influences the relative extents of the two pathways of extracellular glutathione metabolism shown in Fig. 2. It seems of significance that, although intracellular oxidation of glutathione may be readily reversed by the action of glutathione reductase, no such mechanism exists extracellularly, so that the action of transpeptidase is required for recovery of the amino acid constituents of glutathione disulphide. Oxidation may promote translocation of glutathione, as discussed above, or it may be a part of a membrane electron transport system that plays a role in transport. It may possibly serve in a protective mechanism that prevents a reduction of essential disulphide bonds in membrane proteins by removing excess glutathione.

The translocation of glutathione by cells that lack transpeptidase may provide a source of cysteine for cells that are capable of degrading glutathione. The translocation of glutathione may also function to protect the cell membrane itself against oxidation. It has been thought that glutathione functions to protect the erythrocyte membrane, and it has been reported that glutathione is required to permit or preserve assembly of microtubules (Oliver et al 1976). Thus, the finding that many mammalian cells translocate glutathione may reflect the operation of a rather general mechanism that protects and maintains the integrity of cell membranes.

In summary, there is considerable evidence that the γ-glutamyl cycle functions *in vivo*, and that translocation of glutathione is a step in the cycle. When the synthesis of glutathione is inhibited, the level of glutathione in the kidney and many other tissues declines. The rate of this decline is slowed by administration of inhibitors of γ-glutamyl transpeptidase. Similarly, the renal glutathione level declines after administration of amino acids and this effect is decreased by inhibition of transpeptidation. Administration of amino acids was found to increase intracellular 5-oxoproline levels, a finding that, along with others, is in accord with the hypothesis that γ-glutamyl amino acids are transported into the cell. The observation that administered γ-glutamyl amino acids may be transported into the kidney also supports the transport hypothesis. Although further studies on the translocation of γ-glutamyl amino acids are needed, the available data are in accord with the proposal that the γ-glutamyl cycle functions as *one* of the systems that mediate amino acid transport. There are undoubtedly other systems, and these may overlap in specificity with the γ-glutamyl cycle. The quantitative aspects of amino acid transport clearly need further study.

The present findings do not exclude the very likely possibility that the reactions of the γ-glutamyl cycle also function to hydrolyse glutathione, conjugates of glutathione (as in mercapturic acid formation), and other γ-glutamyl compounds. Glutathione appears to serve as a storage form of cysteine, and it is evident that the γ-glutamyl cycle functions in the uptake of cysteine into glutathione, and in its release from glutathione. It is thus probable that glutathione is of major significance in the overall metabolism of the sulphur-containing amino acids.

The new findings presented here show that whereas the synthesis of glutathione occurs intracellularly, its degradation is initiated extracellularly, and indeed a significant fraction of the translocated glutathione may be completely degraded to its constituent amino acids extracellularly. Such reactions may well be of importance in facilitating the recovery by cells of the amino acid components of glutathione, which might otherwise be lost by excretion of the tripeptide. These studies also indicate that the translocation of glutathione provides a means for delivering a thiol compound to the surface and perhaps to other regions of the cell membrane, where it may participate in phenomena associated with transport, catalysis, and protection of the cell membrane. Although these studies have elucidated several new aspects of glutathione metabolism and translocation, it is clear that they have also raised new questions which need to be answered. The additional new finding of γ-glutamylcysteine in the urine of animals treated with a potent transpeptidase inhibitor raises yet another new question about glutathione metabolism, since this observation suggests the possibility that there may be an alternative mechanism of glutathione utilization initiated by cleavage of the glycine moiety.

References

Ames SR, Elvehjem CA 1945 Enzymatic oxidation of glutathione. J Biol Chem 159:549-562

Anderson ME 1979 Glutathione metabolism in tumors. Fed Proc 38:1803

Bartoli GM, Sies H 1978 Reduced and oxidized glutathione efflux from liver. FEBS (Fed Eur Biochem Soc) Lett 86:89-91

Connel GE, Hanes CS 1956 Enzymic formation of pyrrolidone carboxylic acid from γ-glutamyl peptides. Nature (Lond) 177:377-378

Cornell JS, Meister A 1976 Glutathione and γ-glutamyl cycle enzymes in crypt and villus tip cells of rat jejunal mucosa. Proc Natl Acad Sci USA 73:420-422

DeLap LW, Tate SS, Meister A 1977 γ-Glutamyl transpeptidase and related enzyme activities in the reproductive system of the male rat. Life Sci 20:673-680

Grau EM, Marathe GV, Tate SS 1979 Rapid purification of rat kidney brush borders enriched in γ-glutamyl transpeptidase. FEBS (Fed Eur Biochem Soc) Lett 98:91-95

Griffith OW, Meister A 1977 Selective inhibition of γ-glutamyl-cycle enzymes by substrate analogs. Proc Natl Acad Sci USA 74:3330-3334

Griffith OW, Meister A 1979a Metabolism and transport of glutathione and γ-glutamyl amino acids; *in vivo* studies with selective inhibitors of the γ-glutamyl cycle enzymes. Fed Proc 38:99

Griffith OW, Meister A 1979b Translocation of intracellular glutathione to membrane-bound γ-glutamyl transpeptidase as a discrete step in the γ-glutamyl cycle: glutathionuria after inhibition of transpeptidase. Proc Natl Acad Sci USA 76:268-272

Griffith OW, Bridges RJ, Meister A 1978 Evidence that the γ-glutamyl cycle functions *in vivo* using intracellular glutathione: effects of amino acids and selective inhibition of enzymes. Proc Natl Acad Sci USA 75:5405-5408

Griffith OW, Novogrodsky A, Meister A 1979a Translocation of glutathione from lymphoid cells that have markedly different γ-glutamyl transpeptidase activities. Proc Natl Acad Sci USA 76: 2249-2252

Griffith OW, Anderson ME, Meister A 1979b Inhibition of glutathione biosynthesis by prothionine sulphoximine (*S-n*-propyl homocysteine sulphoximine), a selective inhibitor of γ-glutamylcysteine synthetase. J Biol Chem 254:1205-1210

Hahn R, Wendel A, Flohe L 1978 The fate of extracellular glutathione in the rat. Biochim Biophys Acta 539:324-337

Hanes CS, Hird FJR, Isherwood FA 1952 Enzymic transpeptidation reactions involving γ-glutamyl peptides and α-aminoacyl peptides. Biochem J 51:25-35

Hughey RP, Rankin BB, Elce JS, Curthoys NP 1978 Specificity of a particulate rat renal peptidase and its localization along with other enzymes of mercapturic acid synthesis. Arch Biochem Biophys 186:211-217

Imondi AR, Balis ME, Lipkin M 1969 Changes in enzyme levels accompanying differentiation of intestinal epithelial cells. Exp Cell Res 58:323-330

Jellum E, Kluge T, Borresen HC, Stokke O, Eldjarn L 1970 Pyroglutamic aciduria– a new inborn error of metabolism. Scand J Clin Lab Invest 26:327-335

Jones DP, Stead AH, Moldeus P, Orrenius S 1978 Glutathione and glutathione conjugate metabolism in isolated liver and kidney cells. In: Sies H, Wendel A (eds) Functions of glutathione in liver and kidney. Springer, Berlin (Proc Life Sci) p 194-200

Kuhlenschmidt T, Curthoys NP 1975 Subcellular localization of rat kidney phosphate-independent glutaminase. Arch Biochem Biophys 167:519-524

Maack T, Johnson V, Tate SS, Meister A 1974 Effects of amino acids on the function of the isolated perfused rat kidney. Fed Proc 33:305

Meister A 1974a Glutathione synthesis. In: Boyer PD (ed) The enzymes, 3rd edn. Academic Press, New York, vol 10:671-697

Meister A 1974b The γ-glutamyl cycle; diseases associated with specific enzymatic deficiencies. Ann Intern Med 81:247-253

Meister A 1975 Biochemistry of glutathione. In: Greenberg DM (ed) Metabolic pathways, 3rd edn. Academic Press, New York, vol 7: Metabolism of sulfur compounds, p 101-188

Meister A 1978a Current status of the γ-glutamyl cycle. In: Sies H, Wendel A (eds) Functions of glutathione in liver and kidney. Springer, Berlin (Proc Life Sci), p 43-59

Meister A 1978b 5-Oxoprolinuria (pyroglutamic aciduria) and other disorders of glutathione biosynthesis. (Chapter 16) In: Stanbury JB et al (eds) The metabolic basis of inherited diseases, 4th edn. McGraw-Hill, New York, ch 16, p 328-336

Meister A, Tate SS 1976 Glutathione and related γ-glutamyl compounds: biosynthesis and utilization. Annu Rev Biochem 45:559-604

Meister A, Tate SS, Ross LL 1976 Membrane bound γ-glutamyl transpeptidase. In: Martinosi A (ed) The enzymes of biological membranes. Plenum, New York, vol 3:315-347

Meister A, Tate SS, Thompson G 1977 On the function of the γ-glutamyl cycle in the transport of amino acids and peptides. In: Peptide transport and hydrolysis. Excerpta Medica, Amsterdam (Ciba Found Symp 50) p 141

Moldeus P, Jones DP, Ormstad K, Orrenius S 1978 Formation and metabolism of a glutathione-S-conjugate in isolated rat liver and kidney cells. Biochem Biophys Res Commun 83:195-200

Novogrodsky A, Tate SS, Meister A 1976 γ-Glutamyl transpeptidase, a lymphoid cell-surface marker; relationship to blastogenesis, differentiation, and neoplasia. Proc Natl Acad Sci USA 73:2414-2418

Oliver JM, Albertini DF, Berlin RD 1976 Effects of glutathione-oxidizing agents on microtubule assembly and microtubule-dependent surface properties of human neutrophils. J Cell Biol 71:921-932

Oppenheimer L, Wellner VP, Griffith OW, Meister A 1979 Glutathione synthetase; purification from rat kidney and mapping of the substrate binding sites. J Biol Chem 254: 5184-5190

Palekar AG, Tate SS, Meister A 1975 Decrease in glutathione levels of kidney and liver after injection of methionine sulphoximine into rats. Biochem Biophys Res Commun 62:651-657

Richman PG 1969 γ-Glutamylcysteine synthetase and related studies on the metabolism of the γ-glutamyl group. Doctoral Dissertation, Cornell University Graduate School of Medical Sciences, New York

Richman P, Meister A 1975 Regulation of γ-glutamyl-cysteine synthetase by non-allosteric feedback inhibition by glutathione. J Biol Chem 250:1422-1426

Schulman JD, Goodman SI, Mace JW, Patrick AD, Tietze F, Butler EJ 1975 Glutathionuria: inborn error of metabolism due to tissue deficiency of γ-glutamyl transpeptidase. Biochem Biophys Res Commun 63:68-74

Sekura R, Meister A 1974 Glutathione turnover in the kidney; considerations relating to the γ-glutamyl cycle and the transport of amino acids. Proc Natl Acad Sci USA 71:2969-2972

Sekura R, Meister A 1977 γ-Glutamylcysteine synthetase: further purification, "half of the sites" reactivity, subunits, and specificity. J Biol Chem 252:2599-2605

Sekura R, Van Der Werf P, Meister A 1976 Mechanism and significance of the mammalian pathway for elimination of D-glutamate; inhibition of glutathione synthesis by D-glutamate. Biochem Biophys Res Commun 71:11-18

Snoke JE, Bloch K 1954 In: Colowick S et al (eds) Glutathione. Academic Press, New York, p 129-141

Taniguchi N, Meister A 1978 γ-Glutamyl cyclotransferase from rat kidney; sulfhydryl groups and isolation of a stable form of the enzyme. J Biol Chem 253:1799-1806

Tate SS 1975 γ-Glutamyl transpeptidase: properties in relation to its proposed physiological role. In: Markert CL (ed) Isozymes, physiology and function. Academic Press, New York, vol 2:743-765

Tate SS, Meister A 1974 Interaction of γ-glutamyl transpeptidase with amino acids, dipeptides, and derivatives and analogs of glutathione. J Biol Chem 249:7593-7602

Tate SS, Meister A 1975 Identity of maleate-stimulated 'glutaminase' with γ-glutamyl transpeptidase in rat kidney. J Biol Chem 250:4619-4624

Tate SS, Meister A 1976 Subunit structure and isozymic forms of γ-glutamyl transpeptidase. Proc Natl Acad Sci USA 73:2599-2603

Tate SS, Meister A 1977 Affinity labeling of γ-glutamyl transpeptidase and location of the γ-glutamyl binding site on the light subunit. Proc Natl Acad Sci USA 74:931-935

Tate SS, Meister A 1978 Serine-borate complex as a transition-state inhibitor of γ-glutamyl transpeptidase. Proc Natl Acad Sci USA 75:4806-4809

Tate SS, Grau EM, Meister A 1979 Conversion of glutathione to glutathione disulfide by cell membrane-bound oxidase activity. Proc Natl Acad Sci USA 76: 2715-2719

Thompson GA, Meister A 1977 Interrelationships between the binding sites for amino acids, dipeptides, and γ-glutamyl donors in γ-glutamyl transpeptidase. J Biol Chem 252:6792-6797

Tietze F 1969 Enzymatic method for quantitative determination of nanogram amounts of total and oxidized glutathione: applications to mammalian blood and other tissues. Anal Biochem 27:502-522

Van Der Werf P 1970 Metabolism of 5-oxo-L-proline; conversion to L-glutamate in an ATP-dependent reaction catalyzed by 5-oxoprolinase. Doctoral Dissertation, Cornell University Graduate School of Medical Sciences, New York

Van Der Werf P, Meister A 1975 The metabolic formation and utilization of 5-oxo-L-proline (L-pyroglutamate, L-pyrrolidone carboxylate). Adv Enzymol Relat Areas Mol Biol 43:519-556

Van Der Werf P, Orlowski M, Meister A 1971 Enzymatic conversion of 5-oxo-L-proline (L-pyrrolidone carboxylate) to L-glutamate coupled with ATP cleavage to ADP: a reaction in the γ-glutamyl cycle. Proc Natl Acad Sci USA 68:2982-2985

Van Der Werf P, Stephani RA, Orlowski M, Meister A 1973 Inhibition of 5-oxoprolinase by 2-imidazolidone-4-carboxylic acid. Proc Natl Acad Sci USA 70:759-761

Van Der Werf P, Stephani RA, Meister A 1974 Accumulation of 5-oxoproline in mouse tissues after inhibition of 5-oxoprolinase and administration of amino acids; evidence for function of the γ-glutamyl cycle. Proc Natl Acad Sci USA 71:1026-1029

Van Der Werf P, Griffith OW, Meister A 1975 5-oxo-L-prolinase (L-pyroglutamate hydrolase); purification and catalytic properties. J Biol Chem 250:6686-6692

Wellner VP, Sekura R, Meister A, Larsson A 1974 Glutathione synthetase deficiency, an inborn error of metabolism involving the γ-glutamyl cycle in patients with 5-oxoprolinuria (pyroglutamic aciduria). Proc Natl Acad Sci USA 71:2505-2509

Williamson J 1979 Structure-function relationships of 5-oxo-L-prolinase. Fed Proc 38:2629

Woodward GE, Reinhart FE 1942 The effect of pH on the formation of pyrrolidone-carboxylic acid and glutamic acid during enzymatic hydrolysis of glutathione by rat kidney extract. J Biol Chem 145:471-480

Ziegenhagen AJ, Ames SR, Elvehjem CA 1947 Enzymatic oxidation and hydrolysis of glutathione by different tissues. J Biol Chem 161:129-133

Discussion

Idle: Do fetal mammals have 5-oxoprolinase activity? The exact biochemical lesion produced by thalidomide has never been worked out, and some of its spontaneous hydrolysis products are rather analogous to the inhibitors of 5-oxoprolinase and γ-glutamyltranspeptidase that you use.

Meister: We have not looked at fetal tissues for 5-oxoprolinase.

Hamilton: You mentioned that glutathione translocation may help to maintain the integrity of the cell membrane. How do you visualize the glutathione export effect operating? And what do you mean exactly by protection of the membrane?

Meister: The translocation of glutathione may not require energy since the concentration inside the cell is so much higher than the concentration outside. But the translocation may be regulated in ways we don't yet understand. The suggestion that it may protect the cell membrane is neither original nor supported by much evidence. It is thought that intracellular glutathione protects the erythrocyte membrane.

Perhaps glutathione protects the membranes of other cells or plays a role in cell membrane structure and function. Translocation of glutathione seems to be involved in this. I envisage protection as primarily against oxidation: for example, protection of membrane proteins that have SH groups or other groups susceptible to oxidation.

Ziegler: Your results with cultured lymphocytes indicate that 90% of the translocated glutathione appeared as the free thiol in the medium. This is somewhat surprising, since it is generally assumed that only glutathione thioethers or glutathione disulphides readily perfuse from cells. For example, Sies et al (1972) have shown that glutathione disulphide is rapidly expelled by liver tissue.

Meister: As far as we have looked, we have found nothing else.

Ziegler: The liver cell expels GSSG very rapidly, but it retains GSH.

Meister: Sies et al (1973) perfused liver with peroxides and found that glutathione leaves the liver as the disulphide. However, when no peroxide was added, they did not find much translocation of the disulphide.

Segel: How does the combination of serine and borate act as an inhibitor of γ-glutamyl transpeptidase?

Meister: We think that serine and borate form a complex with the transpeptidase that resembles the γ-glutamyl enzyme and the transition state intermediate formed in the reaction (Tate & Meister 1978). The serine–borate compound which is formed may have a structure such as:

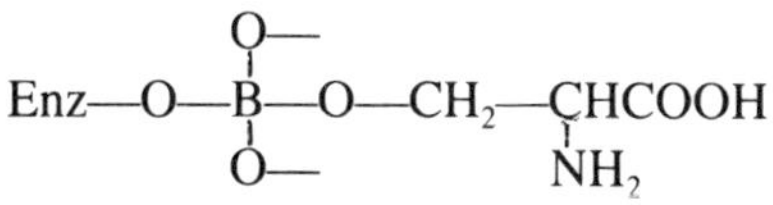

$$\begin{array}{c} \qquad\qquad \text{O—} \\ \qquad\qquad | \\ \text{Enz—O—B—O—CH}_2\text{—CHCOOH} \\ \qquad\qquad | \qquad\qquad\quad | \\ \qquad\qquad \text{O—} \qquad\quad \text{NH}_2 \end{array}$$

Segel: So borate is not simply chelating with serine, but forms a covalent bond? A chelation of a serine hydroxyl and something else is what first comes to mind.

Meister: Yes, the serine hydroxyl and probably a hydroxyl group on the enzyme.

Dodgson: Some years ago, Dr Thomas in our laboratory examined the serum SH content in cases of coal miner's pneumoconiosis and certain other diseases (Thomas & Evans 1975). The patients were shown to have significantly decreased amounts of protein sulphydryl in serum. You say that glutathione levels are low in plasma as compared with intracellular levels. What is the present status of the view that some glutathione is attached to the free SH group on albumin?

Meister: It is possible that some glutathione binds to albumin and other proteins.

Dodgson: Do your measurements of glutathione in plasma include glutathione bound to SH groups of plasma proteins such as albumin?

Meister: The determinations of plasma glutathione I presented here do not include bound glutathione. I do not know whether there is an increase in plasma glutathione levels in disease. We have done only a few determinations on human blood plasma, and have studied only healthy people. Their levels were 1–3 μM, considerably lower than those found in the plasma of the mouse or rat. The estimation is tricky, however, because haemolysis has to be avoided; there is considerable glutathione in red cells.

Kägi: What is the half-life of glutathione in the body?

Meister: In mouse kidney the average half-life is about 20 minutes; in the liver it is a bit longer. However, there may be two pools in the liver, one turning over almost as fast as in the kidney and the other more slowly. I mentioned (p 140) the relatively rapid turnover in skeletal muscle. Skeletal muscle must contain an amount of glutathione that is as great as that of the liver. Pancreatic glutathione turns over quite rapidly too.

Kägi: Is there any extracellular glutathione peroxidase?

Meister: No; I believe that it is all intracellular. The oxidation of glutathione in renal brush border membranes does not seem to lead to the formation of hydrogen peroxide.

Ziegler: Jones (1979) finds that renal brush border membranes contain a very active glutathione oxidase. This suggests that it would be extremely difficult to detect an increase in serum glutathione, since it would be oxidized to the disulphide during each pass through the kidney.

Meister: There is glutathione oxidase activity in the renal brush border, in the jejunal villi, and in other locations that also have γ-glutamyl transpeptidase (Tate et al 1979).

Kredich: Most mammalian cell lines have a cysteine requirement when grown in synthetic media but many lose that requirement when the cells reach high density. The implication is that cysteine leaks out of the cell. Do you think that leakage of glutathione, or perhaps γ-glutamylcysteine, might contribute to the apparent loss of cysteine?

Meister: We are looking at this now. If glutathione can leak out of a cell one might wonder if it can also leak into a cell. Obviously, with a lot of glutathione in the cell it's easier to conceive of it leaking out than going back in, but there may be conditions in which the translocation process is reversed, especially in cells without very much transpeptidase. It's possible that glutathione, cysteine, cystine or γ-glutamylcysteine could be transferred intact from one cell in tissue culture to another.

Kredich: With respect to the loss of either cysteinylglycine or glutathione from the cell, since mammalian cells are usually grown under aerobic conditions, these compounds would readily form mixed disulphides with proteins which then wouldn't be able to re-enter the cell. Could one overcome the apparent cysteine requirement by using one of your transpeptidase inhibitors?

Meister: I don't think so. The transpeptidase of the lymphoid cell is on the surface of the cell. The enzyme is accessible to externally supplied substrate. The problem you raise is presumably whether one cell could provide another cell with cysteine?

Kredich: Not really, because they are all the same kind of cell and if you starve them all for cysteine you can't expect one to be making cysteine for another.

Meister: Some of the cell lines we have studied behave quite differently at different times, and we don't yet know why. The rate of translocation of glutathione from a particular cell line may vary from one day to the next or one experiment to the next. There may be metabolic control over the rate of translocation, or it may have something to do with the cell cycle. We wonder also whether there is translocation of cysteine or glutathione, or some related product, between cells.

Bright: From your model, you seem to be saying that each cell must have a cysteine transport system and a glycine transport system, plus the transpeptidase. Can you demonstrate transport of any other amino acid into the cultured cell in the presence of a transpeptidase inhibitor, or only cysteine and glycine?

Meister: We haven't looked in our laboratory at these transport systems in the way you suggest, but there is much evidence from other studies to indicate that many amino acids are transported into these and similar cells.

Whatley: You mentioned experiments in which you gave the sulphoximine, which reduced glutathione levels, while other subjects were given the phenylhydrazide inhibitor of transpeptidase, which diminished the rate at which the glutathione urinary level fell. What happens to an animal when you treat it successively with both these reagents? Does it revert towards some sort of normality — can it stand this dual treatment? Or does it perhaps fail because it is no longer able to move things around in the γ-glutamyl cycle?

Meister: These are relatively short-term experiments. In the mouse we gave γ-glutamyl-(*o*-carboxy)phenylhydrazide in large doses, almost LD50 doses. Our experiments were complete within a couple of hours, so we didn't have much chance to observe effects on the animals. This inhibitor stops the breakdown of glutathione. When glutathione synthesis is inhibited for short periods we don't see any unusual behaviour. We are now doing some longer-term experiments to determine a number of things.

Mice that have had high doses of buthionine sulphoximine for two weeks seem normal. We have not used the transpeptidase inhibitors over such long periods, but the one we use breaks down to give a toxic hydrazine. We want to make less toxic inhibitors. Then we may be able to answer the question you ask, and to determine whether glutathione or transpeptidase (or both) are essential for happiness or life!

Ziegler: How much of the phenylhydrazine inhibitor is present in the liver as a conjugate? If you block the transpeptidase, it probably wouldn't be broken down to the cysteine conjugate and converted to a mercapturic acid.

Meister: It is excreted in the urine as the γ-glutamyl derivative; you are quite right that it is not converted to a typical mercapturic acid. In these experiments we found not only glutathione in the urine but also the disulphide of γ-glutamylcysteine, which suggests that there may be a pathway of glutathione breakdown in which glycine is split off first. We normally don't see this, but only when we inhibit the transpeptidase.

Kredich: Carboxypeptidase does that also.

Meister: Yes. Pancreatic carboxypeptidase acts on glutathione disulphide to give γ-glutamylcystine.

Kredich: It only works on the disulphide; isn't that correct?

Meister: Yes, presumably because the SH compound interacts with the zinc on the enzyme.

Mudd: Can you say more about the clinical state of individuals deficient in one or other of these γ-glutamyl cycle enzymes?

Meister: Patients with 5-oxoprolinuria have a deficiency in glutathione synthetase. The first reported patient (Jellum et al 1970) was a 19-year-old boy of normal height and weight who had been mentally retarded since childhood. At one stage he had a severe acidosis which was treated with bicarbonate. This patient had signs of organic cerebral damage, spastic quadriparesis and cerebellar disturbances. Two other patients who developed severe metabolic acidosis during the first few days of life have been successfully treated with bicarbonate. Both these patients seem to be relatively normal (Meister 1978). Several other patients with 5-oxoprolinuria have been described more recently including one who apparently had no neurological disease at 18 months of age. However, others have had some mental retardation and other central nervous system disease.

Glutathione synthetase deficiency without 5-oxoprolinuria has also been reported in a somewhat larger number of patients. In contrast to patients with 5-oxoprolinuria, there are no indications in the published reports that any of these patients was ill during the neonatal period. There is also no evidence for acidosis. In this form of glutathione synthetase deficiency the genetic lesion may lead to synthesis of an unstable glutathione synthetase molecule. The turnover of such a defective, but active, enzyme may be sufficiently rapid in most tissues to compensate for the defect; however, such compensation is not possible in the erythrocyte, which does not synthesize protein. Consequently, patients with this form of glutathione synthetase deficiency show only a tendency toward haemolysis.

Two patients have been found to have γ-glutamylcysteine synthetase deficiency. These patients exhibit haemolytic anaemia, spinocerebellar degeneration, peripheral neuropathy, myopathy and aminoaciduria. One patient is known to have a deficiency, apparently generalized, of γ-glutamyl transpeptidase. This patient appears to be mentally retarded (Schulman et al 1975).

References

Jellum E, Kluge T, Borrensen HC, Stokke O, Eldjarn L 1970 Pyroglutamicaciduria — a new inborn error of metabolism. Scand J Clin Lab Invest 26:327-335

Jones DP 1979 Metabolism of glutathione and a glutathione conjugate by isolated kidney cells. J Biol Chem 254:2787-2792

Meister A 1978 5-Oxoprolinuria and other disorders of glutathione biosynthesis. In: Stanbury JB et al (eds) The metabolic basis of inherited diseases (4th edn). McGraw-Hill, New York, p 328-336

Schulman JD, Goodman SI, Mace JW, Patrick AD, Tietze F, Butler E 1975 Glutathionuria: inborn error of metabolism due to tissue deficiency of γ-glutamyl transpeptidase. Biochem Biophys Res Commun 65:68-74

Sies H, Gerstenecker C, Menzel H, Flohe L 1972 Oxidation in the NADP system and release of GSSG from hemoglobin-free perfused rat liver during peroxidatic oxidation of glutathione by hydroperoxides. FEBS (Fed Eur Biochem Soc) Lett 27:171-175

Sies H, Gerstenecker C, Summer KH, Menzel H, Flohé L 1973 Glutathione. Georg Thieme, Stuttgart, p 261-275

Tate SS, Meister A 1978 Serine–borate complex as a transition-state inhibitor of γ-glutamyl transpeptidase. Proc Natl Acad Sci USA 75:4806-4809

Tate SS, Grau EM, Meister A 1979 Conversion of glutathione to glutathione disulfide by cell membrane-bound oxidase activity. Proc Natl Acad Sci USA 76: 2715-2719

Thomas J, Evans PH 1975 Serum protein changes in coal workers' pneumoconiosis. Clin Chim Acta 60:237-247

Observations on the biological roles of sulphatases

KENNETH S. DODGSON and FREDERICK A. ROSE

Department of Biochemistry, University College, P.O. Box 78, Cardiff CF1 1XL, UK

Abstract Until recently little was known about the biological roles played by sulphatase enzymes, owing in part to the selection of assay substrates that were convenient but only remotely related to the natural substrates. Once this was recognized the elucidation of function proceeded more rapidly. Microbial sulphatases appear to have roles to play in the nutrition of individual microorganisms whilst collectively they enable sulphur, returned to soils and waters in the form of sulphate esters, to be made available for recycling. In contrast, with one or two important exceptions, mammalian sulphatases are concerned, in association with other enzymes, with the turnover of macromolecules.

Still defying understanding are the roles of sulphatases acting on adenosine 5′-phosphosulphate (APS) and 3′-phosphoadenosine 5′-phosphosulphate (PAPS). APS sulphatases have now been purified from ox-liver lysosomes and cytosol and from a strain of *Comamonas terrigena*. The lysosomal enzyme shows wide specificity and can hydrolyse ATP, ADP, FAD and pyrophosphate. The cytosol enzyme is apparently specific and may be active only when cellular concentrations of ATP are low. The bacterial enzyme is also specific and has properties and a cellular localization that suggest the possibility of its involvement in sulphate transport.

For many years little interest was shown in sulphuric acid esters. Much of this lack of interest arose from the failure to appreciate adequately that sulphate esters were not necessarily inert end-products of metabolism or evolutionary relics of no physiological significance. Time has adequately dispelled that view and we have seen the discovery of an astonishingly diverse variety of sulphate esters, distributed widely through the living world. Closely related to these esters are other compounds in which sulphate is linked through oxygen to a nitrogen atom (e.g. mustard oil glycosides) or to a phosphorus atom (adenosine 5′-phosphosulphate, APS; 3′-phosphoadenosine 5′-phosphosulphate, PAPS) or by an N–S bond (e.g. sulphamate residues in heparin). All will be regarded as relevant to the present theme.

The biosynthesis of sulphated compounds involves the production of PAPS and

Sulphur in biology
(Ciba Foundation Symposium 72) p 163-176

transfer of its sulphate group to acceptors via the agency of sulphotransferases, a large group of distinct enzymes, few of which have been studied in detail. Also widely distributed are hydrolytic enzymes (the sulphatases that are able to remove sulphate groups from sulphated compounds; see Dodgson & Rose 1975, Roy 1976 for general reviews). In the cases of those enzymes acting on true sulphate esters (C–O–S linkage) it has always been assumed, from studies on arylsulphatases (Spencer 1958), that hydrolysis involves fission of the O–S bond. This would accord with the accepted view that esterases rupture the bond on the acid side of the ester linkage. This view must now be modified following our findings that primary and secondary alkylsulphatases involved in the microbial degradation of sulphated detergents operate by fission of the C–O bond of the C–O–S linkage (Bartholomew et al 1977, Cloves et al 1977). Furthermore, the secondary alkylsulphatases that we have examined exhibit stereospecificity and fission of the C–O bond is accompanied by bond inversion; so that, for example, L-octan-2-ol is produced from D-octan-2-yl sulphate. These alkylsulphatases thus appear to be unique in their properties, although the question must now be asked whether any other sulphatases operate by C–O bond fission. Early studies by Suzuki et al (1957) raised the possibility that sulphatase enzymes were capable of transferring sulphate to appropriate acceptors. We made many abortive attempts to explore this possibility further but concluded that the enzymes played purely hydrolytic roles. However, recent observations (Burns et al 1977) that one of three arylsulphatases produced by *Aspergillus oryzae* is able to transfer sulphate from certain aryl sulphates to tyramine perhaps indicate that the matter is not closed.

For many years sulphatases were studied because some of them possessed intriguing kinetic properties, but little attention was paid to their biological significance. The breakthrough in this search for function came from two rather different directions: firstly, the discovery that mammalian arylsulphatase A (assay substrate, nitrocatechol sulphate) was a cerebroside sulphatase (cerebroside 3-sulphate 3-sulphohydrolase, EC 3.1.6.8), active towards the D-galactose 3-*O*-sulphate component of the cerebroside substrate (Mehl & Jatzkewitz 1968); and secondly, the demonstration of the relationship between the role of choline sulphate as a fungal sulphur store and the production of choline sulphatase, the action of which enables that store to be broached in times of sulphur insufficiency (Scott & Spencer 1968). Two things were clearly revealed by these studies, namely the danger of assuming that ability to hydrolyse a particular type of sulphate ester necessarily means that one has identified the natural substrate for a given enzyme, and, secondly, that sulphatases can have a vital physiological role to play. From such beginnings, interest in establishing biological function has grown, although there is still a long way to go before all the answers are known. Most of the available information relates to microbial and mammalian enzymes and attention will therefore be restricted to these.

MICROBIAL SULPHATASES

Although absolute proof is not available it is accepted that microbial sulphatases play a major role in the recycling of sulphur. Sulphate esters are returned to the soils and waters of our planet in enormous quantities from microbial, plant and animal sources and there is much evidence to indicate that organically bound sulphate groups form a major component of the sulphur content of soils (see Fitzgerald 1976 for review). We have shown that soils are capable of releasing sulphate from a variety of sulphate esters, including choline sulphate and aryl, alkyl and carbohydrate sulphates (Houghton & Rose 1976). In some cases sulphate release begins immediately, suggesting the presence of preformed enzyme, whilst in others there is a lag period suggestive of enzyme induction and/or enhanced microbial growth. Interestingly, the ester sulphate groups in soil humic acids could not be removed by treatment with alkyl-, aryl- or glycosulphatase preparations from bacterial or molluscan sources and the problem of whether this material is ever turned over, or simply accumulates and removes sulphate from circulation, remains to be resolved.

At present the sulphated compounds shown in Table 1 are known to serve as substrates for sulphatases present in soil microorganisms. For a complex polysulphated compound like heparin, several different sulphatases are involved in the overall biodegradation of the polymer. Most of these compounds occur in substantial amounts in nature or are closely related in structure to naturally occurring compounds. In any event, one becomes increasingly convinced that there must be very few naturally occurring sulphate esters that cannot serve as substrates for microbial sulphatases. Unfortunately, little is known about the ability of fresh- or salt-water microorganisms to degrade sulphate esters. However, years ago we made a fairly detailed study of the arylsulphatase of a strain of *Alcaligenes metalcaligenes* isolated from intertidal mud, whilst Chandramohan et al (1974) have

TABLE 1

Some sulphated compounds known to serve either directly or indirectly as substrates for different microbial sulphatases

Aryl sulphates	Chondroitin 4-sulphate
Choline sulphate	Chondroitin 6-sulphate
Primary alkyl sulphates	Heparan sulphate
Secondary alkyl sulphates	Keratan sulphate
3:5-Dichlorophenoxyethyl sulphate	Heparin
D-Glucose 6-sulphate	Sulphamates
D-Gluconate 6-sulphate	Adenosine 5′-phosphosulphate
D-Glycerate 3-sulphate	3′-Phosphoadenosine 5′-phosphosulphate

shown that the soils of mangrove swamps possess arylsulphatasc activity. Moreover, biodegradation of alkyl sulphate detergents by microorganisms present in river waters is well established and there is every reason to suppose that sulphate esters of many types, entering aqueous environments via excretion, secretion or decay processes, will undergo desulphation via microbial sulphatases.

The question might now be posed of the function of microbial sulphatases in the physiology of individual microorganisms. Here, apart from the obvious example of fungal choline sulphatase (choline-sulphate sulphohydrolase, EC 3.1.6.6), we must speculate to some degree. Even with choline sulphatase we must reckon with the anomalous situation of the bacterial enzyme. There is no evidence that choline sulphate plays a storage role in bacteria, as in fungi and plants. Indeed, Fitzgerald (1973 and personal communication) has established that the ester is readily formed from choline by *Pseudomonas* C12B, *Comamonas terrigena* and *Escherichia coli* and then released almost completely into the medium. Curiously, *Pseudomonas* C12B grows well when choline sulphate is used as sulphur source but attempts to detect choline sulphatase activity at any phase of growth have completely failed (see Fitzgerald et al 1977). Fitzgerald believes that some unique mechanism may exist for the transfer of sulphate from the ester without the intervention of a sulphatase. In those cases where a bacterial choline sulphatase has been clearly demonstrated (see Fitzgerald & Scott 1974 for references) the ester has had to be present in the growth media as a source of carbon or nitrogen or sulphur. Collectively, the results from several independent studies suggest that the enzyme is produced in response to a need for sulphur and other nutrients.

Of the other microbial sulphatases only arylsulphatases have been studied in detail, although some useful information has been gleaned on the alkylsulphatases. Space does not permit full discussion of factors influencing enzyme production but the overall impression, based mainly on work on fungal and bacterial arylsulphatases, is that of microorganisms scavenging for sulphur. This is generally reflected in the repression of sulphatase formation by SO_4^{2-} and, in most cases, by other components of the cysteine biosynthetic pathway, including cysteine, although the nature of the actual repressor is usually uncertain. In bacteria, repression is relieved by growth on methionine (see Adachi et al 1975, Fitzgerald & Payne 1972 for examples), presumably reflecting the inability or poor ability to convert cysteine to methionine. This is expressing an over-simple view of the systems involved and in some instances the situation is complicated by other important factors (see e.g. Murooka et al 1978, Burns & Wyn 1977).

The few studies on alkylsulphatases have been concerned with primary and secondary enzymes in *Pseudomonas* C12B and the inducible primary alkylsulphatase of *Pseudomonas aeruginosa*. In the former organism, three of the enzymes (one primary and two secondaries) appear to be constitutive while the others (one primary, one secondary) are inducible. In all these cases enzyme production or induction is unaffected by the sulphur supply but appears to be regulated by the supply of carbon. With both the alkyl- and arylsulphatases of bacteria the available evidence shows them to be localized at the cell periphery, readily released by osmotic shock or during spheroplast formation. Only one bacterial sulphatase (APS-sulphatase) has yet been shown to have an intracellular localization (see later). The peripheral localization of sulphatases may be related to their scavenging role and to the problems of transporting highly charged esters through the plasma membrane. Transport of sulphate esters — with the exception of the charge-compensated choline sulphate — has not yet been studied in detail but is an important aspect of bacterial metabolism that is ripe for investigation.

In accord with the views expressed here, fungal choline sulphatases appear to have an intracellular localization. This would be appropriate in the light of their role in releasing sulphate from the intracellular store of choline sulphate.

MAMMALIAN SULPHATASES

Not too many years ago it was fairly well accepted that mammals possessed four sulphatases only, three being arylsulphatases (aryl-sulphate sulphohydrolase, EC 3.1.6.1) and the other a 3-β-steroid sulphatase (sterol-sulphate sulphohydrolase, EC 3.1.6.2). Of the three arylsulphatases, two (arylsulphatases A and B) were lysosomal in origin, while the other (arylsulphatase C) was localized, together with steroid sulphatase, in the endoplasmic reticulum of the cell. It is now clear that enzymes A and B are not true arylsulphatases, but, fortuitously, are active towards nitrocatechol sulphate, an ester that has long served as a chromogenic assay substrate. In fact, both enzymes attack sulphate groups attached to carbohydrate residues and, *in vivo,* A is certainly concerned with the turnover of cerebroside sulphate (see Roy 1976) and B with the desulphation of *N*-acetylgalactosamine 4-sulphate residues during glycosaminoglycan turnover (Gorham & Cantz 1978). In our view it is also virtually certain that arylsulphatase C (usual assay substrate, 4-acetylphenyl sulphate) is an oestrogen sulphatase concerned with regulating oestrogen levels, and there is mounting evidence to support this (see Dodgson & Rose 1975). However, in spite of efforts to resolve the problem, we are still uncertain whether the 3-β-steroid sulphatase is a distinct enzyme.

All four enzymes provide good examples of the way in which research can be thrown off course by the fortuitous selection of substrates that bear only partial structural resemblance to the true physiological substrates. They are not isolated examples. Meanwhile, a number of other mammalian sulphatases have been discovered, all of which appear to be concerned with glycosaminoglycan turnover. Their discovery has stemmed from the study of inherited metabolic diseases of connective tissues and, indeed, most work on mammalian sulphatases is now directed towards their relationship to enzyme deficiency diseases.

BIOLOGICAL ROLE OF APS SULPHATASES

Ten years ago we first noted the extraordinary variety of animal organisms capable of degrading APS and PAPS and, since then, we have attempted to understand the significance of the enzyme systems involved. There are at least two distinct routes for the release of sulphate from PAPS. Route 1 involves the sequential action of a PAPS sulphatase (3′-phosphoadenylylsulphate sulphohydrolase, EC 3.6.2.2) and a PAP phosphatase (adenosine-3′,5′-biphosphate 3′-phosphohydrolase, EC 3.1.3.7) and Route 2 the sequential action of a 3′-nucleotidase (3′-phosphoadenylylsulphate 3′-phosphohydrolase, EC 3.1.3.30) and an APS sulphatase (adenylylsulphate sulphohydrolase, EC 3.6.2.1). Both routes apparently operate in the mammalian liver cell and PAPS can be degraded by the lysosomes, cytosol and the endoplasmic reticulum of that cell. Both routes are certainly present in the cytosol and we have separated the APS and PAPS sulphatases and the cobalt-activated 3′-nucleotidase for further study. Absolute proof is difficult to obtain but we gained the impression that Route 2 was the more active of the two cytosol pathways. Moreover, we were doubtful about the significance of Route 1 because of the exceptionally high optimum pH (9.4) of the PAPS sulphatase (see Dodgson & Rose 1975). We therefore concentrated on APS sulphatase and soon realized that the enzyme enjoyed a bimodal distribution in the liver cell, with apparently distinct enzymes in cytosol and lysosomes.

Concentrating first on cytosol, we were aware that this cell fraction also contains the APS–PAPS biosynthetic system. However, we purified the ox liver enzyme to homogeneity and examined its properties (Table 2). The enzyme appeared to be specific for APS and was strongly inhibited (competitively) by ATP, suggesting that it would be inoperative under conditions favourable to PAPS biosynthesis. The enzyme was only feebly inhibited (non-competitively) by AMP and we were led to the somewhat unsatisfactory conclusion that it was perhaps concerned with the regeneration of ATP (via AMP) from unwanted PAPS and/or APS.

Our attention then turned to the lysosomal APS sulphatase, an enzyme that seemed to have no reason to be present in lysosomes, and we purified it to homogeneity (Rogers et al 1978). Somewhat to our relief it turns out to be a relatively unspecific enzyme which is also active towards ATP, ADP, FAD and pyrophosphate. These compounds (including APS) are all, strictly speaking, acid anhydrides and the bonds cleaved by the enzyme are anhydride links between two phosphoric acid residues or between sulphuric acid and a phosphoric acid residue. The enzyme was active towards bis(4-nitrophenyl) phosphate, 4-nitrophenyl 5′-phosphothymidine and 4-

TABLE 2

Some general properties of the purified APS sulphatases of bovine liver and *Comamonas terrigena*

	Bovine liver enzymes		*C. terrigena*
	Supernatant	*Lysosomal*	
Mol.wt. (Gel filtration)	69 000	56 000	27 000
(SDS)	68 000	52 000	25 000
Optimum pH	5.4	6.2	8.5
K_m(mM)	0.95	0.04	1.0
Effect of ATP	Competitive inhibition ($K_i = 7.5x10^{-6}M$)	Enzyme substrate ($K_m = 1.4x10^{-4}M$)	Relieves inhibition by AMP
Effect of AMP	Non-competitive inhibition	No effect	Competitive inhibition ($K_i = 3x10^{-4}M$)
Specificity of APS	Specific	Not specific	Specific
Products of APS hydrolysis	AMP, SO_4^{2-}	AMP, SO_4^{2-}	AMP, SO_4^{2-}

nitrophenyl phosphate, but this is perhaps not unexpected as these compounds could be considered as mixed anhydrides of a phosphoric acid with the weak acid, 4-nitrophenol. The pattern emerges of an enzyme hydrolysing acid anhydrides containing at least one phosphoric acid residue. These findings bring some relief to us in our search for the function of a specific APS sulphatase. Meanwhile, this work again illustrates the specificity pitfalls that face the unwary worker in the sulphatase field.

A third line of approach to the problem is now under way, based on the somewhat dubious grounds that microbial systems might prove to be more amenable to attack than those of mammalian origins. A source of the enzyme has been found in the soil bacterium *Comamonas terrigena* (White et al 1979). Enzyme activity appears during the exponential phase of growth in nutrient broth and reaches maximum levels at mid-log phase, with levels being maintained into the stationary phase (Fig. 1). Growth on defined media with a variety of different but relevant sulphur sources gave a similar pattern of enzyme production, although with some indication of two peaks of maximum activity. However, the specific enzyme activities attained varied appreciably in relation to the type of compound selected to supply sulphur for growth (Fig. 2). Maximum activities were obtained with compounds at the beginning and end of the biosynthetic pathway from SO_4^{2-} to methionine, with significantly lower levels obtained with pathway intermediates. The enzyme thus

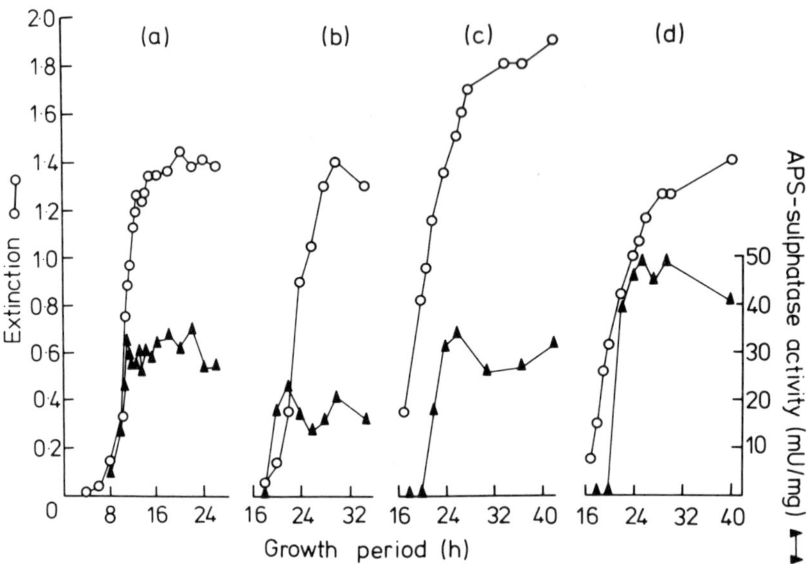

FIG. 1. Appearance of APS sulphatase during batch culture of *Comamonas terrigena* in various media. (a) Nutrient broth; (b) basal salts/pyruvate/0.5mM-APS; (c) basal salts/pyruvate/2mM-cysteine; (d) basal salts/pyruvate/2mM-methionine. A unit (U) of enzyme activity is the amount of enzyme liberating 1 μmole of sulphate per minute.

seems to be constitutive but subject to partial control. The enzyme remains with the bacterium after osmotic shock or during spheroplast formation, indicating an intracellular localization. There is no obvious significance in the two peaks of enzyme activity and separate experiments have established that one and the same enzyme is present at these points.

How do we explain these findings? High enzyme levels when the organism is grown on sulphate seem illogical, since the production (not destruction) of APS is vital for growth. Does the enzyme have a scavenging role to play? This also seems illogical in the light of its early appearance during growth. Could the enzyme be other than a specific APS sulphatase, analogous perhaps to the liver lysosomal enzyme? Could it be an ATP sulphurylase (EC 2.7.7.4) exhibiting hydrolytic activity towards APS in a manner similar to that described for the *Penicillium chrysogenum* enzyme (Farley et al 1978), or could it be an APS cyclase with sulphatase activity analogous to that found in *Chlorella pyrenoidosa* (Tsang & Schiff 1976)?

In attempts to resolve these problems we have now purified the enzyme to virtual homogeneity and have shown it to be a specific hydrolase liberating AMP and SO_4^{2-} from APS. It is not ATP sulphurylase, arylsulphatase, an APS cyclase or a non-specific enzyme attacking anhydride links similar to the liver lysosomal APS

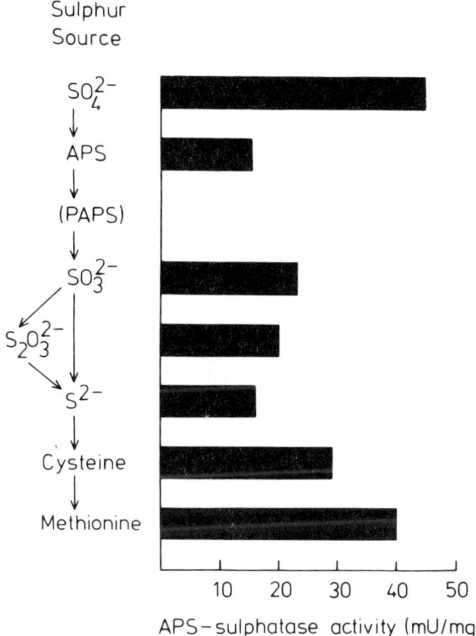

FIG. 2. Effect of sulphur source on APS sulphatase activity in extracts of *Comamonas terrigena*. The results are the average values for samples taken at two separate times during the late-log/stationary phase period. The effect of PAPS is not yet known.

sulphatase. One of the intriguing features of the enzyme, which establishes a clear difference from the specific liver cytosol enzyme, is the failure of ATP to inhibit the enzyme and the effectiveness of AMP as a competitive inhibitor (Table 2). This points to some essential functional difference between the bacterial and liver cytosol enzymes.

The question, therefore, remains as to the function of the bacterial enzyme. Here we are struck by the interesting observations made by Fitzgerald's group that *C. terrigena*, in common with *Pseudomonas* C12B and *E. coli*, is able to produce and secrete into the medium large quantities of choline sulphate when provided with choline. Furthermore, for *Pseudomonas* C12B Fitzgerald has established that peripheral enzymes, readily liberated from the cell by osmotic shock, are responsible for the synthesis of choline sulphate from choline and SO$_4^{2-}$. In short, the ability to synthesize APS and PAPS outside the plasma membrane seems to be established for this particular organism and it seems likely that *C. terrigena* and *E. coli* might also possess this ability. If this is so, perhaps the ability to produce APS outside the plasma membrane and the ability to liberate sulphate from APS intracellularly may take on a particular significance in relation to sulphate transport in these microorganisms. If this were so, some control over the system would be possible because of the competitive inhibition of the APS sulphatase by AMP. Accumulation of sulphate inside the cell as a result of APS sulphatase action would eventually be halted because of a parallel accumulation of AMP. It would then follow that ATP should be without direct effect on the APS sulphatase, and this proves to be the case. Moreover, ATP relieves the inhibitory effect of AMP, a finding that perhaps lends further support to the possibility that APS is involved in sulphate transport. Further studies on these various possibilities are now in progress.

Clearly, the APS sulphatases form a fascinating group of enzymes that might yet be found to play vital roles in the metabolism of microorganisms, plants and animals. In any event they present challenges to the investigator that we found difficult to resist.

ACKNOWLEDGEMENTS

We wish to thank the Science Research Council and the Wellcome Trust for support given to our work on APS sulphatases and to acknowledge the part played by colleagues in our studies, particularly Dr. G.F. White of our department and Drs. J.W. Fitzgerald and W.J. Payne of the Department of Microbiology of the University of Georgia, USA.

References

Adachi T, Murooka Y, Harada T 1975 Regulation of arylsulfatase synthesis by sulfur compounds in *Klebsiella aerogenes*. J Bacteriol 121:29-35

Bartholomew B, Dodgson KS, Matcham GWJ, Duncan JS, White GF 1977 A novel mechanism of enzymic ester hydrolysis. Biochem J 165:575-580

Burns GRJ, Wynn CH 1977 Differential repression of arylsulphatase synthesis in *Aspergillus oryzae*. Biochem J 166:415-420

Burns GRJ, Galanopoulou E, Wynn CH 1977 Kinetic studies of the phenol sulphate-phenol sulphotransferase of *Aspergillus oryzae*. Biochem J 167:223-227

Chandramohan D, Devendran K, Natarajan R 1974 Arylsulphatase activity in marine sediments. Marine Biol (NY) 27:89-92

Cloves JM, Dodgson KS, Games DE, Shaw DJ, White GF 1977 The mechanism of action of primary alkylsulphohydrolase and arylsulphohydrolase from a detergent degrading micro-organism. Biochem J 167:843-846

Dodgson KS, Rose FA 1975 Sulfohydrolases. In: Greenberg DM (ed) Metabolic pathways, 3rd edn. Academic Press, New York, vol 7: Metabolism of sulfur compounds, p 359-431

Farley JR, Nakayama G, Cryns D, Segel IH 1978 Adenosine triphosphate sulfurylase from *Penicillium chrysogenum*. Equilibrium binding, substrate hydrolysis and isotope exchange studies. Arch Biochem Biophys 185:376-390

Fitzgerald JW 1973 The formation of choline *O*-sulphate by *Pseudomonas* C12B and other *Pseudomonas* species. Biochem J 136:361-369

Fitzgerald JW 1976 Sulfate ester formation and hydrolysis: a particularly important yet often ignored aspect of the sulfur cycle of aerobic soils. Bacteriol Rev 40:698-721

Fitzgerald JW, Payne WJ 1972 The regulation of arylsulphatase formation in *Pseudomonas* C12B. Microbios 6:147-156

Fitzgerald JW, Scott CL 1974 Utilization of choline-*O*-sulphate as a sulphur source for growth by a *Pseudomonas* isolate. Microbios 10:121-131
Fitzgerald JW, Cline ME, Rose FA 1977 Sulphate esters as de-repressors of arylsulphatase formation and the problem of sulphate ester utilization by *Pseudomonas* C12B. FEMS (Fed Eur Microbiol Soc) Microbiol Lett 2:217-220
Gorham SD, Cantz M 1978 Arylsulphatase B, an exo-sulphatase for chondroitin 4-sulphate tetrasaccharide. Hoppe-Seyler's Z Physiol Chem 359:1811-1814
Houghton C, Rose FA 1976 Liberation of sulfate from sulfate esters by soils. Appl Environ Microbiol 31:967-976
Meh E, Jatzkewitz H 1968 Cerebroside 3-sulfate as a physiological substrate for arylsulfatase A. Biochim Biophys Acta 151:619-627
Murooka Y, Higashiura T, Harada T 1978 Genetic mapping of tyramine oxidase and arylsulphatase genes and their regulation in intergeneric hybrids of enteric bacteria. J Bacteriol 136:714-722
Rogers KM, White GF, Dodgson KS 1978 Purification and properties of bovine liver lysosomal adenosine 5′-phosphosulphate sulphohydrolase. Biochim Biophys Acta 527:70-85
Roy AB 1976 Sulphatases, lysosomes and disease. Aust J Exp Biol Med Sci 54:111-135
Scott JM, Spencer B 1968 Regulation of choline sulphatase synthesis and activity in *Aspergillus nidulans*. Biochem J 106:471-477
Spencer B 1958 Enzymic cleavage of aryl hydrogen sulphates in the presence of $H_2{}^{18}O$. Biochem J 69:155-159
Suzuki S, Takahashi N, Egami F 1957 Enzymic transsulfation from a phenol to carbohydrates. Biochim Biophys Acta 24:444-445
Tsang ML-K, Schiff JA 1976 Properties of enzyme fraction A from *Chlorella* and co-purification of 3′(2),5′-biphosphonucleoside 3′(2)-phosphohydrolase, adenosine 5′-phosphosulfate sulfohydrolase and adenosine 5′-phosphosulfate cyclase activities. Eur J Biochem 65:113-121
White GF, Rowlands MG, Dodgson KS, Payne WJ 1979 Adenosine 5′-phosphosulphate sulphohydrolase activity in *Comamonas terrigena*. FEMS (Fed Eur Microbiol Soc) Microbiol Lett 5:267-271

Discussion

Segel: Are you suggesting that a covalent modification of sulphate is involved in its transport, similar to Dr Meister's system for amino acids?

Dodgson: Sulphate transport in bacteria is an energy-dependent process; a system in which APS was synthesized outside the plasma membrane and hydrolysed within the cell proper would provide the necessary energy gradient and would allow some degree of control inside the cell.

Roy: You said that the lysosomal APS sulphatase does not hydrolyse *p*-nitrophenyl sulphate. Did you try the bis(*p*-nitrophenyl) sulphate?

Dodgson: No, but that's a good point, in view of the enzyme's ability to hydrolyse bis(nitrophenyl) phosphate.

Segel: The K_m of 1 mM for APS (for the *Comamonas terrigena* enzyme) sounds very high, doesn't it?

Dodgson: Yes. However, the K_m may well turn out to be much lower, because it has since become clear that AMP (a competitive inhibitor) was also present in the enzyme preparation that we used. This determination has to be repeated in the absence of AMP.

Rassin: I am intrigued by the possibility that cysteine and methioinine may be sources of sulphate, because in the whole animal one of the best inhibitors of sulphate incorporation into sulphatides has been shown to be methionine (Chase & O'Brien 1970). Methionine reduces radio-labelled sulphate incorporation by about 90%. I have often wondered whether this inhibition was competitive incorporation rather than simply inhibition of transport of sulphate into the brain.

Kredich: Since this was done by radio-labelling, the result could have been due to a dilution effect of unlabelled sulphate derived from methionine.

Segel: Has anyone prepared, or been able to buy, phosphosulphate, the inorganic sulphur analogue of pyrophosphate?

Kelly: I don't think it is available commercially. The closest thing is probably sodium thiophosphate, but that contains rather reduced sulphur.

Hamilton: Have you found the sulphate transport system in *Escherichia coli,* Professor Dodgson?

Dodgson: Fitzgerald has looked at *E. coli* (personal communication) and indicated that it can produce choline sulphate outside the plasma membrane. We studied nutrient broth-grown *E. coli* and found very weak APS sulphatase activity. However, on defined media we have indications of much greater activity.

Hamilton: Thinking back to Dr Kredich's paper, I wonder whether the *cysA* gene product in *Salmonella typhimurium* could be one of these enzyme systems?

Kredich: I don't know of any evidence to that effect. One could compare a strain with a large deletion in *cysA* to the parental wild-type strain grown under similar conditions and ask whether the activity is there. Such strains are readily available.

Jones-Mortimer: The system is presumably one of the ATP-dependent transport systems, so one might expect ATPase activity.

Hamilton: Why do you say that it's an ATP-dependent system?

Jones-Mortimer: Because of the probable correlation between binding-protein transport systems and energization of transport by a product of glycolysis (Berger 1973).

Hamilton: ATP or some similar compound (acetyl phosphate has been suggested) is implicated in energy coupling with certain transport systems which often also contain binding-protein components. The entire sulphate transport system is so unknown, however, that it might not be a simple binding-protein system requiring ATP, although a binding protein certainly has been found.

Dodgson: Another point is that almost all animals can degrade PAPS by removing the 3′-phosphate group to yield APS. I am astonished at the wide distribution of this capacity in the invertebrate and vertebrate world. There is also some indication that *Comamonas terrigena* can degrade PAPS, possibly to APS, so that the system present in the soluble fraction of the liver cell may also have a parallel in microorganisms. I asked Dr Segel earlier whether he had found PAPS sulphatase in his *Penicillium* preparations (p 68). We have tended to neglect the mammalian PAPS sulphatase because its optimum pH is so high that it would seem unlikely to be an efficient enzyme *in vivo*. We have separated the mammalian enzyme and purified it to some extent, not to homogeneity. It appeared to be a simple hydrolytic enzyme. We didn't establish whether it was specific for the phosphosulphate linkage.

Segel: There is so little APS and PAPS in the cell at any time that they are not even a good source to scavenge for left-over nucleotides.

Dodgson: I agree that the concept of a scavenging role for APS sulphatase is not very convincing. However, in spite of the difficulty of assigning a physiological function we still feel that the mammalian cytosol APS sulphatase is an enzyme of high specificity.

Roy: The animal APS and PAPS sulphatases seem to be very different enzymes from the other sulphatases. In multiple sulphatase deficiency, for instance, where every other sulphatase is lacking, APS and PAPS sulphatases are apparently normal (Austin 1973).

Segel: But they are anhydrases; all the others are sulphate ester hydrolases.

Dodgson: Thinking about the magnesium-stimulated 3′-phosphatase enzyme of plants which Dr Schiff suggested may have a role to play in facilitating APS formation (see p 49, 62), I believe that we never tested magnesium on our liver cytosol 3′-nucleotidase. This probably reflected our preoccupation with cobalt because of the work of others who believed that PAPS degradation was initiated by a cobalt-stimulated PAPS sulphatase. In fact, we were able to show that it was the 3′-nucleotidase which was cobalt activated. We didn't study that enzyme in detail and have no idea what inhibits it or whether it has anything to do with facilitating APS formation.

Segel: What is known about ascorbate 2-sulphate?

Roy: Mohamram et al (1976) showed that preparations from rat liver and colon could use PAPS for the synthesis of ascorbate 2-sulphate. After that the story becomes complicated because, although non-radioactive ascorbate 2-sulphate is stable, ascorbate 2-[^{35}S]sulphate is not. It rapidly decomposes to $^{35}SO_4^{2-}$ (Shapiro & Poon 1975) or more complex, but unidentified, products (Powell et al 1978).

Dodgson: There must be some intramolecular transformation with the radio-labelled form. Unless there is some form of in-built radiation sink, such as an aromatic ring, radioactive sulphate esters cannot be kept in the solid state. Unless you keep them in frozen aqueous solution they undergo decomposition by self-irradiation, even when the specific activity is quite low.

References

Austin JH 1973 Studies in metachromatic leukodystrophy. Multiple sulfatase deficiency. Arch Neurol 28:258-264

Berger EA 1973 Different mechanisms of energy coupling for the active transport of proline and glutamine in *Escherichia coli*. Proc Natl Acad Sci USA 70:1514-1518

Chase HP, O'Brien D 1970 Effect of excess phenylalanine and of other amino acids on brain development in the infant rat. Pediatr Res 4:96-102

Mohamram M, Rucker RB, Hodges RE 1976 Formation *in vitro* of ascorbic acid 2-sulfate. Biochim Biophys Acta 437:305-310

Powell GM, Parry TJ, Curtis CG 1978 An enterogastric circulation. Biochem Soc Trans 6:141-3

Shapiro SS, Poon JS 1975 Apparent sulfation of glycosaminoglycans by ascorbic acid 2-[^{35}S]sulfate: an explanation. Biochim Biophys Acta 385:221-231

Sulphatase A: an arylsulphatase and a glycosulphatase

A.B. ROY

Department of Physical Biochemistry, John Curtin School of Medical Research, Australian National University, P.O. Box 334, Canberra City, A.C.T. 2601, Australia

Abstract Sulphatase A was first described as an arylsulphatase but was subsequently shown to have cerebroside sulphatase activity, hydrolysing galactose 3-sulphate residues in certain lipids. A characteristic feature of the arylsulphatase activity is the substrate-induced inactivation of the enzyme which occurs during the catalytic reaction. Present views of the course of this modification are considered. A similar modification of the enzyme does not occur during the cerebroside sulphatase reaction and the fall in velocity noted during most such assays can be explained by disappearance of substrate and accumulation of sulphate. Possible reasons for the difference between the two types of activity are considered. Sulphatase A also hydrolyses hexose sulphates at rates comparable to those of aryl sulphates. The specificity of sulphatase A towards its natural substrates, sulpholipids, is considered in the light of these findings.

The first mammalian sulphatase to be described was an arylsulphatase (aryl-sulphate sulphohydrolase, EC 3.1.6.1) found by Neuberg & Simon (1925) in rabbit tissues, but despite the rapid demonstration of the quite widespread distribution of these enzymes little detailed work was possible with the limited methods then available. With the resurgence of interest in the sulphatases in the 1950s it was again the arylsulphatases which attracted attention and it soon became apparent that most mammals contained three sulphatases which showed arylsulphatase activity and were non-committally termed sulphatases A, B and C. The first two were lysosomal enzymes and the last was a microsomal enzyme. Within the past fifteen years several sulphatases A have been obtained as homogeneous proteins, rather fewer sulphatases B have been so obtained, while the sulphatases C remain available only in the form in which they were originally described, essentially as washed microsomal preparations.

In the early 1960s interest began to develop in an inborn error of metabolism, metachromatic leucodystrophy, which is characterized by a defective catabolism of cerebroside sulphate and an accumulation of that lipid in various tissues, implying

Sulphur in biology
(Ciba Foundation Symposium 72) p 177-190

that normal tissues contain a cerebroside sulphatase capable of hydrolysing the galactose 3-sulphate residues of the sulpholipid. Great was the surprise when this enzyme was demonstrated (Mehl & Jatzkewitz 1968) to be identical to sulphatase A, so making it a glycosulphatase, a type of enzyme first found by Soda & Hattori (1931) in molluscs and believed to be quite distinct from the arylsulphatases. There were, however, differences between the arylsulphatase and cerebroside sulphatase activities of sulphatase A: the most important of these was that the latter activity required the presence of an accessory factor and was non-competitively inhibited by SO_4^{2-} whereas the arylsulphatase required no accessory factor and was competitively inhibited by SO_4^{2-}. The role of the accessory factor has been extensively studied by Jatzkewitz: essentially it is a lysosomal protein (Mraz et al 1976) which binds one molecule of cerebroside sulphate (Fischer & Jatzkewitz 1977) to give what is presumed to be the physiological substrate (Fischer & Jatzkewitz 1978) of sulphatase A. Very recently another inborn error of metabolism, a variant of metachromatic leucodystrophy, has been described (Stevens et al 1979): in this condition the defect is the absence of the activator protein.

The sulphatase A of ox liver, and sulphatases A in general, are self-interacting systems and this is very pertinent to the determination of the various activities. At concentrations of the order of mg ml^{-1} the sulphatase A of ox liver exists as a monomer of molecular weight 107 000 at pH 6.5 and above, as a tetramer at pH 5.5 and below, and as an equilibrium mixture of these and intermediate polymers between pH 5.5 and 6.5. At lower, or 'enzymic', concentrations of the order of μg ml^{-1} the tetramer dissociates, even at pH 4.5 (Jerfy et al 1976). This is shown in Fig. 1. In most assays of the arylsulphatase activity of sulphatase A the protein concentration is of the order of 0.5 μg ml^{-1} and the pH about 5.6: as is clear from Fig. 1, the monomer is the dominant species under these conditions. On the other hand, in the cerebroside sulphatase determinations made in this laboratory the protein concentration is about 10 μg ml^{-1} and the pH 4.5: under these conditions the tetramer is dominant. In effect, different enzyme species are present in these two types of assay. There have been no studies of the arylsulphatase activity of sulphatase A under conditions where the tetramer is the dominant species at pH 5.6 because of practical problems inherent in studying the rapid reactions catalysed by such high enzyme concentrations (0.1–1 mg ml^{-1}) but at pH 4.5, where the tetramer is important at lower concentrations, there are indications that there are negatively cooperative interactions between the monomer units in the tetramer. On the other hand, there are no obvious differences in the specific activities of the monomer and tetramer.

It should be noted that these association–dissociation reactions are too rapid to be followed by conventional light-scattering techniques, so that one can be confident that, whatever be the polymeric form of the enzyme existing in a concentrated

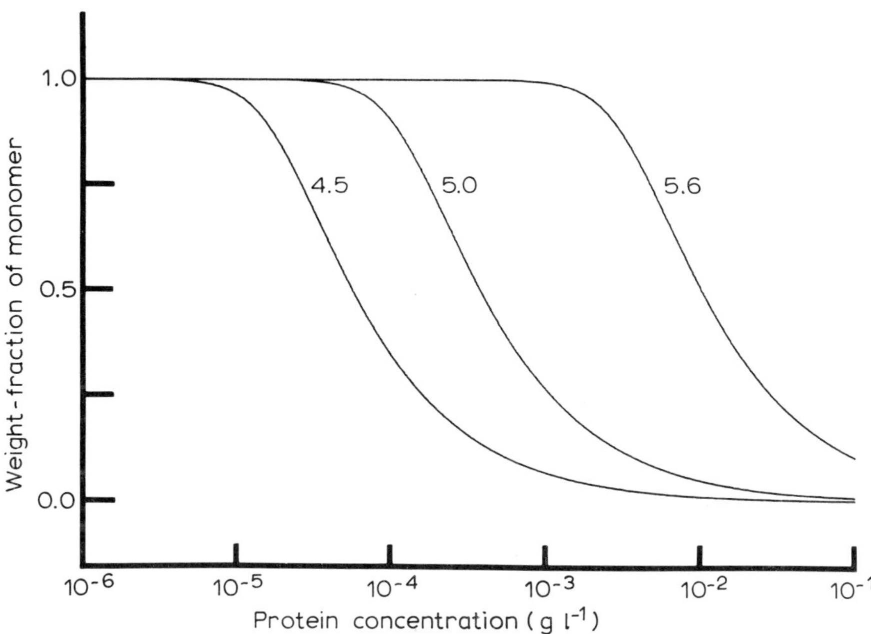

FIG. 1. The weight-fraction of sulphatase A occurring as monomer at different total concentrations and pH. The curves are computed for a temperature of 20 °C and $\mu = 0.1$: the values of K^*_{ass}, the apparent association constant for tetramer formation, were 4.1×10^{13}, 1.4×10^{11} and 7.4×10^{6} l^3 g^{-3} at pH 4.5, 5.0 and 5.6 respectively.

solution of sulphatase A added to a reaction mixture, the appropriate equilibrium will be rapidly attained, certainly in much less than a minute, on dilution.

The kinetics of the arylsulphatase activity of sulphatase A have attracted much attention over the past 25 years because the enzyme undergoes an inactivation during its catalytic cycle. Although the general pattern of the behaviour was worked out by Baum & Dodgson in 1958 the details still remain obscure. Briefly, during the catalytic cycle sulphatase A rapidly attains an equilibrium which lies much in favour of a modified form of the enzyme which is devoid of enzyme activity. The inactivation occurs with all the aryl sulphates so far used as substrates for the enzyme, although at somewhat different rates: with nitrocatechol sulphate the half-time for the inactivation is about 3 min at 37°C. This modified enzyme can be isolated and is quite stable in the cold although it reverts to the native enzyme, with a half-time of about 3 h, at 37 °C. Further, the arylsulphatase activity of the modified enzyme is restored by the addition of SO_4^{2-}, or any of several other anions, together with an aryl sulphate substrate. This activation by SO_4^{2-} and substrate does not reform native sulphatase A nor cause the formation of yet a third form of the enzyme — or at least not a form stable in the absence of substrate and SO_4^{2-} — because the original modified sulphatase A can be re-isolated from the reaction mixture (C.

Prosser, unpublished work 1979). The nature of the change which leads to the formation of the modified enzyme has not been clarified but it is presumed to be conformational and it must be slight, because there is no difference between the circular dichroic spectra of the native and modified enzymes. There is no change in molecular weight or in the ability to form a tetramer, nor is there any spectroscopic evidence for the occurrence of nitrocatechol or nitrocatechol sulphate (the latter being the usual substrate) in the modified enzyme. On the other hand, recent experiments with nitrocatechol [^{35}S]sulphate suggest that the modified enzyme contains 1 mole SO_4^{2-} per mole of enzyme (C. Prosser, unpublished work 1979).

Recently we have developed methods for the quantitative characterization of the inactivation. As this occurs only in the presence of substrate it can be followed only by measuring the decrease in the rate of the catalysed reaction. Theoretical treatment leads to Equation 1 for the description of the initial stages of the progress curve for the arylsulphatase reaction

$$u = \frac{v_o}{k^*}(1 - e^{-k^*t}) \tag{1}$$

of sulphatase A: u is the amount of product in time t, v_o is the initial velocity of the hydrolytic reaction and k^* is an apparent velocity constant for the substrate-induced modification, which is related to the velocity constant, k, by Equation 2. In the

$$k^* = \frac{ks}{K + s} \tag{2}$$

latter expression, s is the substrate concentration and K a constant analogous to, but not identical with, K_m for the catalytic reaction (Roy 1978). Some pertinent values are summarized in Table 1, from which it can be seen that despite considerable variations in K_m, V_0 and K, there is little variation in k which has a value of about 0.23 min^{-1}. It therefore seems unlikely that the inactivation occurs, as had been believed to be the case, through an enzyme–substrate complex because the chemical nature of this is likely to vary considerably from substrate to substrate. It is likely that a common intermediate is involved in the modification, and a reasonable possibility is the enzyme–SO_4^{2-} complex which may be the last intermediate in the catalytic cycle, because SO_4^{2-} is a competitive inhibitor of the arylsulphatase activity of sulphatase A. The picture which then emerges is the following (Roy 1978), where F is the substrate-modified enzyme:

$$\begin{array}{l}
E + S \rightleftarrows ES \rightleftarrows E.\Phi OH.SO_4^{2-} \longrightarrow E.SO_4^{2-} \longrightarrow E + SO_4^{2-} \\
\qquad\qquad\qquad\qquad\qquad \Phi.OH \qquad\qquad \uparrow\downarrow \\
E + S \rightleftarrows E'S \rightleftarrows E!\Phi OH.SO_4^{2-} \longrightarrow F.SO_4^{2-} \longrightarrow F + (SO_4^{2-}\ ?)
\end{array}$$

With a number of substrates V_0 is essentially constant (Table 1) and this is consistent with the last step, the breakdown of $E.SO_4^{2-}$ being rate-limiting. For reasons discussed elsewhere (Roy 1978) the equilibrium between $E.SO_4^{2-}$ and $F.SO_4^{2-}$ appears to be attained only after several minutes.

TABLE 1

Kinetic constants for the hydrolysis of aryl sulphates by sulphatase A, and for the simultaneous substrate-induced modification of the enzyme, at pH 5.6, μ = 0.1

	K_m (mM)	v_0 (μmol mg^{-1} min^{-1})	K (mM)	k (min^{-1})
Nitrocatechol sulphate	1.0	234	0.34	0.22
Nitroquinol sulphate	5.92	243	2.56	0.23
2-Nitropyridyl 3-sulphate	5.88	252	1.29	0.22
2-Nitrophenyl sulphate	12.3	270	2.52	0.23
4-Methylumbelliferone sulphate	41.0	48	9.56	0.28
2-Naphthyl sulphate	40.6	37	10.6	0.26
3-Nitrophenyl sulphate	104	225	25.9	0.28
4-Nitrophenyl sulphate	223	277	52.8	0.30

It is interesting to speculate that $E.SO_4^{2-}$ may itself be a sulphate ester and that the transition state for its hydrolysis might, like that in the acid-catalysed hydrolysis of sulphate esters (Kice & Anderson 1966), have a considerable SO_3-like character. This could explain the powerful inhibition ($K_i << 1$ μM) of the arylsulphatase reaction by SO_3^{2-}, which has a similar shape to SO_3 and so might function as a transition-state inhibitor. This could account for the K_i for SO_3^{2-} being several orders of magnitude less than that for SO_4^{2-}. Attempts to detect and characterize the $E.SO_4^{2-}$ complex must be undertaken. It is tempting to suggest that it might contain a sulphated imidazole group: this seems possible both because of the participation of a histidyl residue in the sulphatase A reaction (Jerfy & Roy 1969, 1974) and because of the role of imidazole groups in the intramolecular catalysis of the hydrolysis of sulphate esters (Benkovic & Dunikoski 1970).

Before going on to the situation with cerebroside sulphate it is necessary to consider the formation of substrate-modified sulphatase A functioning as an arylsulphatase at pH 4.5. The situation is similar to, but not identical with, that at pH 5.6 because the modified enzyme is not activated by SO_4^{2-} at pH 4.5: it is in fact inhibited, as shown in Fig. 2. The change in structure of the enzyme is probably similar, if not identical, at both pH values because the modified enzyme formed at pH 5.6 is inhibited by SO_4^{2-} at pH 4.5 while, conversely, the enzyme formed at pH 4.5 is activated by SO_4^{2-} at pH 5.6 although inhibited at pH 4.5 (Fig. 2). The effect of pH on the behaviour of the different forms of sulphatase A with different substrates certainly requires further examination because most studies have been made with nitrocatechol sulphate which may not be typical in so far as the reactivation of the modified enzyme is concerned.

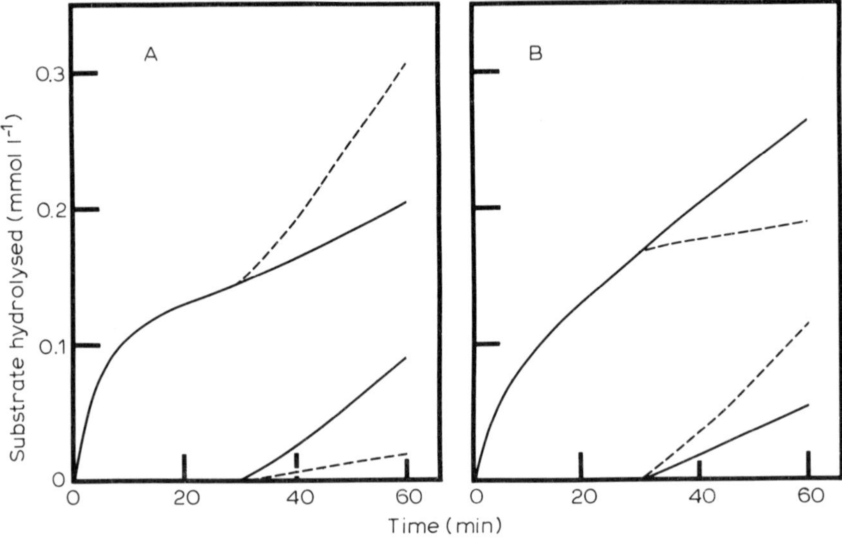

FIG. 2. The effect of SO_4^{2-} on nitrocatechol sulphate-modified sulphatase A at pH 4.5 and 5.6. In A, modification was done at pH 5.6 and after 30 min SO_4^{2-} was added, with or without a change in pH. In B the modification was done at pH 5.6. - - - - , SO_4^{2-} added to the reaction mixture, ———, no SO_4^{2-} added.

With the cerebroside sulphatase activity of sulphatase A the situation is intrinsically more complex, first because the pH optimum is at 4.5 with virtually no enzyme activity being detectable at pH 5.6 under the usual conditions of assay and, second, because of the requirement for the activator protein or for highly artificial conditions such as a very low ionic strength (Stinshoff & Jatzkewitz 1975) or the presence of taurodeoxycholate and $MnCl_2$ (Porter et al 1972). In the former system the substrate is a sonicated suspension of cerebroside sulphate which is grossly heterogeneous and contains micelles, or aggregates, with particle weights ranging from below 180 000 to above 500 000 (Jeffrey & Roy 1977). In reaction mixtures containing taurodeoxycholate and $MnCl_2$ (Porter et al 1972, Jerfy & Roy 1973, Roy 1979) the substrate is cerebroside sulphate incorporated into a bile salt micelle with a particle weight of about 24 000 and containing about four molecules of cerebroside sulphate per micelle (Roy 1977). It may be pertinent that the particle weight of this micelle must be very similar to that — 21 000 — of the activator protein–cerebroside sulphate complex (Fischer & Jatzkewitz 1975).

In our initial studies of the cerebroside sulphatase activity of sulphatase A (Jerfy & Roy 1973) we could find no evidence for the formation of a substrate-modified enzyme nor, conversely, for the hydrolysis of cerebroside sulphate by nitrocatechol sulphate-modified sulphatase A. The latter finding has been confirmed using a preparation containing about 10% of native sulphatase A (C. Prosser, private communication 1979): the specific activity of this preparation towards cerebroside sulphate was only about 10% of that of the native enzyme. It was not possible to show any activation by SO_4^{2-} because, as noted above, the modified enzyme is inhibited by SO_4^{2-} at pH 4.5, and at pH 5.6, where SO_4^{2-} does activate, there is no detectable cerebroside sulphatase activity. The view that a cerebroside sulphate-modified form of sulphatase A is not formed is contrary to the findings of Stinshoff & Jatzkewitz (1975) who claim that such a form of human sulphatase A is produced, although they did not isolate it nor show a reactivation by SO_4^{2-}. On the other hand, Stevens et al (1975), while not commenting directly on the formation of a cerebroside sulphate-modified sulphatase A, did state that the purified human enzyme displayed 'linear kinetics with time', a finding which is clearly incompatible

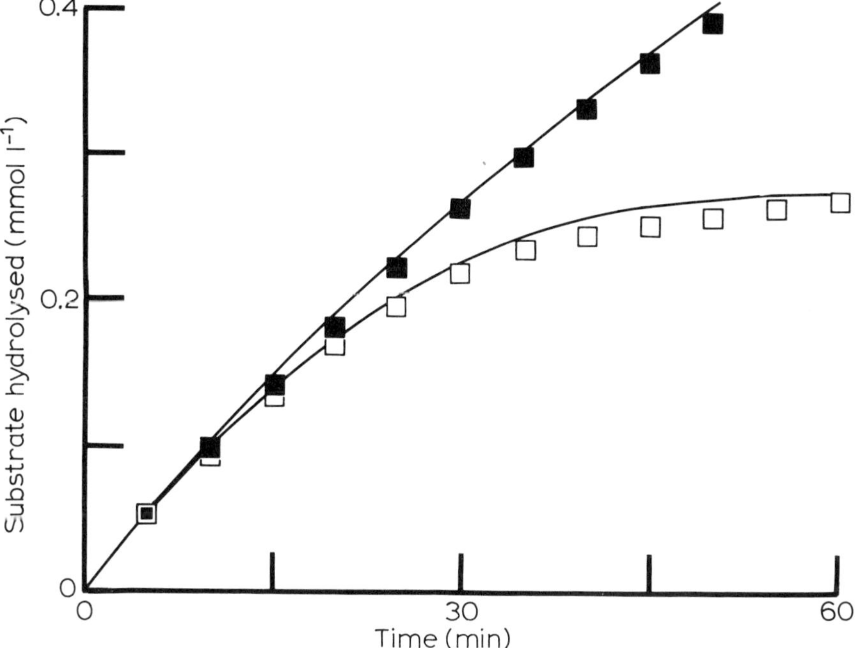

FIG. 3. Progress curves for the hydrolysis of 0.28 mM-cerebroside sulphate by sulphatase A at pH 4.5 in 0.035 M-$MnCl_2$ and 2 mM-taurodeoxycholate. In the lower curve (□) the substrate concentration was allowed to decrease as the reaction proceeded while in the upper curve (■) the substrate concentration was held constant by the addition of cerebroside sulphate at suitable intervals. The symbols show points taken from pH-stat recordings and the lines are theoretical ones computed using the appropriate values of the kinetic constants.

with a substrate-induced inactivation. In view of this uncertainty we have re-examined the behaviour of sulphatase A as a cerebroside sulphatase and have confirmed (Roy 1979) our previous observation that there is no detectable substrate-induced inactivation of sulphatase A functioning as a cerebroside sulphatase in the presence of taurodeoxycholate and $MnCl_2$. This is clearly shown by the results in Fig. 3, which demonstrates that the fall in the velocity during the reaction can be completely accounted for by the decreasing substrate concentration and increasing concentration of SO_4^{2-}. The decreasing velocity noted by Stinshoff & Jatzkewitz (1975) seems, at first sight, to be too great to be explained in this way but their results are complicated by the fact that the substrate is not homogeneous with respect to particle size, as has already been mentioned, and only part of the total concentration of cerebroside sulphate might be available as substrate. The net effect would be that the concentration of cerebroside sulphate was less than thought and, if this were so, their results are explicable without invoking a substrate-induced inactivation. Undoubtedly further work is required in this area and it is unfortunate that it has not been possible to detect the hydrolysis of cerebroside sulphate by the sulphatase A of ox liver under the conditions used by Stinshoff & Jatzkewitz for the enzyme from human tissues.

If it is correct that there is no cerebroside sulphate-induced inactivation of sulphatase A, the reason for this behaviour must be sought. An explanation can be proposed, based on the accepted fact that the hydrolysis of cerebroside sulphate by sulphatase A is non-competitively inhibited by SO_4^{2-}. Taken at its simplest, this must mean that SO_4^{2-} cannot be the last-released product in the catalytic cycle, so that the cerebroside sulphatase reaction can be represented as follows:

$$\mathrm{E} + \mathrm{S} \rightleftharpoons \mathrm{ES} \rightleftharpoons \mathrm{E.Cer.SO_4^{2-}} \xrightarrow[\searrow \mathrm{SO_4^{2-}}]{} \mathrm{E.\ Cer} \longrightarrow \mathrm{E} + \mathrm{Cer}$$

If the arguments put forward about the arylsulphatase activity are correct, then a substrate-induced inactivation would not be expected in this case because an enzyme.SO_4^{2-} complex is not formed. There is nothing inherently impossible in the above scheme because all that need be postulated is a rather firm binding of cerebroside to the enzyme. This could arise by non-polar interactions and so account for the different order of release of products in the arylsulphatase and cerebroside sulphatase reactions.

Further information on this point might be obtained by investigating artificial substrates which are closer models of the physiological substrate, cerebroside sulphate, than the usual aryl sulphates. The first of these are the simple monosaccharide sulphates which obviously cannot bind to the enzyme by non-polar interactions. All the monosaccharide sulphates so far tested have proved to be substrates for sulphatase A: these are hydrolysed both at pH 4.5 and 5.6 but the optimum pH is the latter one and the reaction therefore apparently more closely resembles the arylsulphatase than the cerebroside sulphatase reaction. Some data obtained at pH 4.5 are given in Table 2: these were obtained by measuring the amount of SO_4^{2-} produced during an incubation time of 5 min so the kinetic constants are only apparent ones. Detailed studies in the pH-stat, to give v_0, have so far been made only with glucose 3-sulphate: at pH 5.6 this had a K_m of 220 ± 20 mM and a V_0 of 97 μmol mg^{-1} min^{-1}.

Substrate-induced inactivation occurred and for this reaction K and k were 51 ± 6 mM and 0.21 min^{-1} respectively, the latter value being very close to those listed in Table 1 for aryl sulphates. The hydrolysis of glucose 3-sulphate by sulphatase A was competitively inhibited by SO_4^{2-} with a K_i of approximately 1.5 mM, again in agreement with values obtained with aryl sulphates (Roy 1978). Investigation of the hydrolysis at pH 4.5 gave values not inconsistent with those in Table 2 but, as previously found at this pH with aryl sulphates, the kinetic behaviour was not strictly Michaelis in type so that values of K_m and V_0 are not available: apparent values were 17 ± 2 mM and 9.2 μmol mg^{-1} min^{-1} respectively, and k^* did not vary with substrate concentration, once again as found with aryl sulphates (Roy 1978).

TABLE 2

Kinetic constants for the hydrolysis of monosaccharide sulphates by sulphatase A at pH 4.5 in 0.1 M-pyridine–acetic acid buffer at 37°C. The reactions were followed by measuring the amount of SO_4^{2-} liberated over a 5 min period

Substrate	K_m *(mM)*	*V* *(μmol mg^{-1} min^{-1})*
Glucose 3-sulphate	14	4.5
Glucose 6-sulphate	20	1.0
Galactose 2-sulphate	110	0.28
Galactose 6-sulphate	36	0.43

The kinetic parameters of the reaction with glucose 3-sulphate make it difficult to achieve a high degree of conversion of sulphatase A to the modified form. There can be no doubt, however, that it is formed, because the reaction velocity falls too rapidly to be caused by the disappearance of substrate and appearance of SO_4^{2-}: indeed, it still occurs in the presence of Ba^{2+}, added to remove the latter. In general, therefore, monosaccharide sulphates behave in a similar way to aryl sulphates.

The second group of model compounds comprises aryl sulphates having long alkyl or acyl substituents, such as 4-hexadecanoylphenyl sulphate. The synthesis of these compounds is being undertaken and it is predicted that they will more closely resemble cerebroside sulphate in their behaviour: in particular, their hydrolysis by sulphatase A will be non-competitively inhibited by SO_4^{2-} and will not be accompanied by a substrate-induced inactivation.

It is therefore now clear that the specificity of sulphatase A is quite different from what was assumed to be the case when it was isolated as an arylsulphatase, and was believed to be the case only a few years ago. It will hydrolyse many aryl sulphates — essentially all that have been examined, although at very different rates — ascorbate 2–sulphate, cerebroside sulphate and related glycolipids, and all monosaccharide sulphates so far tested, either glucose or galactose derivatives. Some data are summarized in Table 3. Recently the situation has become even more complex because F. Egami (personal communication 1979) has shown that sulphatase A will hydrolyse adenosine 2′,3′-monophosphate to adenosine 3′-phosphate. The rate of the reaction is low, only about 0.03 μmol mg^{-1} min^{-1}, and although it has been shown by all the samples of sulphatase A so far investigated its significance must remain in doubt.

TABLE 3

The rates of hydrolysis of some sulphate esters by sulphatase A at pH 5.6, 37°C, $\mu = 0.1$

	V_0 (μmol mg^{-1} min^{-1})
Nitrocatechol sulphate	240
2-Nitrophenyl sulphate	270
4-Nitrophenyl sulphate	277
4-Methylumbelliferone sulphate	48
Oestrone sulphate	1
Ascorbate 2-sulphate	90
Cerebroside sulphate	10[a]
Glucose 3-sulphate	97

[a]At pH 4.5 in 35 mM-$MnCl_2$ and 2 mM-taurodeoxycholate, V_1 in place of V_0.

This situation is not a characteristic of sulphatase A because it also exists in other so-called arylsulphatases. Sulphatase B, like sulphatase A, was isolated and characterized as an arylsulphatase but is now known to hydrolyse the *N*-acetylgalactosamine 4-sulphate residues in oligosaccharides derived from glycosaminoglycans (Gorham & Cantz 1978) as well as in simpler compounds such as UDP-*N*-acetylgalactosamine 4-sulphate (Fluharty et al 1975). A similar low specificity may exist in the arylsulphatases of the invertebrates. The arylsulphatase of the marine gastropod *Charonia lampas* has been much studied and it seems certain that it also will hydrolyse not only aryl sulphates but also ascorbate 2-sulphate, monosaccharide sulphates (Hatanaka et al 1976), and adenosine 2′,3′-diphosphate (Egami & Uchida 1978).

If, as now seems certain, arylsulphatases are not the specific enzymes they were thought to be, whether with the 'unphysiological' arylsulphatase activity or the 'physiological' glycosulphatase activity, how can the specificity of the effects of their deficiency be explained? Why, for example, does the deficiency of sulphatase A (metachromatic leucodystrophy) differ from the deficiency of sulphatase B (Maroteaux-Lamy syndrome)? The answer must lie in the nature of their physiological substrates and the specificity must reside not only in the nature of the residue to which the sulphate group is attached but also in the nature of the group to which the entire monosaccharide sulphate is itself linked — an *N*-acylsphingosine (ceramide) or other lipid in the case of the substrate for sulphatase A and a disaccharide, or similar simple oligosaccharide, for the substrate for sulphatase B, and presumably also for the other types of glycosulphatase now known to occur in mammalian tissues (Roy 1977).

ACKNOWLEDGEMENT

I am indebted to Mrs P.J. Archbald for the synthesis and characterization of the hexose sulphates used in this work.

References

Baum H, Dodgson KS 1958 Studies on sulphatases. 22. The anomalous kinetics of arylsulphatase A of human tissues: interpretation of the anomalies. Biochem J 69:573-582

Benkovic SJ, Dunikoski LK 1970 Intramolecular catalysis of sulfate ester hydrolysis. A model for aryl sulfate sulfohydrolase. Biochemistry 9:1390-1397

Egami F, Uchida T 1978 May certain sulphatases have any specific phosphodiesterase activity? Purified arylsulphatase from a marine snail, *Charonia lampas*, hydrolysed nucleoside 3′,5′-cyclic phosphates. Seikagaku 50:1297-1303

Fischer G, Jatzkewitz H 1975 The activator of cerebroside sulphatase. Purification from human liver and identification as a protein. Hoppe-Seyler's Z Physiol Chem 356:605-613

Fischer G, Jatzkewitz H 1977 The activator of cerebroside sulphatase. Binding studies with enzyme and substrate demonstrating the detergent function of the activator protein. Biochim Biophys Acta 481:561-572

Fischer G, Jatzkewitz H 1978 The activator of cerebroside sulphatase. A model of the activation. Biochim Biophys Acta 528:69-76

Fluharty AL, Stevens RL, Fung D, Peak S, Kihara H 1975 Uridine diphospho-*N*-acetylgalactosamine 4-sulfate sulfohydrolase activity of human arylsulfatase B and its deficiency in the Maroteaux-Lamy syndrome. Biochem Biophys Res Commun 64:955-961

Gorham ST, Cantz M 1978 Arylsulphatase B, an exosulphatase for chondroitin 4-sulphate tetrasaccharide. Hoppe-Seyler's Z Physiol Chem 359:1811-1814

Hatanaka H, Ogawa Y, Egami F 1976 Arylsulphatase and glycosulphatase of *Charonia lampas*. Substrate specificity towards sugar sulphate derivatives. Biochem J 159:445-448

Jeffrey HJ, Roy AB 1977 Micelles of cerebroside sulphate. Aust J Exp Biol Med Sci 55:339-346
Jerfy A, Roy AB 1969 The sulphatase of ox liver. XII. The effect of tyrosine and histidine reagents on the activity of sulphatase A. Biochim Biophys Acta 175:355-364
Jerfy A, Roy AB 1973 The sulphatase of ox liver. XVI. A comparison of the arylsulphatase and cerebroside sulphatase activities of sulphatase A. Biochim Biophys Acta 293:178-190
Jerfy A, Roy AB 1974 The sulphatase of ox liver. XVIII. An essential histidyl residue in sulphatase A. Biochim Biophys Acta 371:76-88
Jerfy A, Roy AB, Tomkins HJ 1976 The sulphatase of ox liver. XIX. On the nature of the polymeric forms of sulphatase A present in dilute solutions. Biochim Biophys Acta 422:335-348
Kice JL, Anderson JM 1966 The mechanism of the acid hydrolysis of sodium aryl sulfates. J Am Chem Soc 88:5242-5245
Mehl E, Jatzkewitz H 1968 Cerebroside 3-sulfate as a physiological substrate of arylsulfatase A. Biochim Biophys Acta 151:619-627
Mraz W, Fischer G, Jatzkewitz H 1976 The activator of cerebroside sulfatase. Lysosomal localisation. Hoppe-Seyler's Z Physiol Chem 357:1181-1191
Neuberg C, Simon E 1925 On sulfatase. V. Animal sulfatase. Biochem Z 156:365-373
Porter MT, Fluharty AL, de la Flor SD, Kihara H 1972 Cerebroside sulfatase determination in cultured human fibroblasts. Biochim Biophys Acta 258:769-778
Roy AB 1977 Sulphatase deficiencies. In: Harkness RA, Cockburn F (eds) The cultured cell and inherited metabolic disease. MTP Press, Lancaster, p 120-138
Roy AB 1978 The sulphatase of ox liver. XXI. Kinetic studies of the substrate-induced inactivation of sulphatase A. Biochim Biophys Acta 526:489-506
Roy AB 1979 The sulphatase of ox liver. XXII. Further observations on the cerebroside sulphatase activity of sulphatase A. Biochim Biophys Acta 568: 103-110
Soda T, Hattori C 1931 An enzyme which hydrolyses glucose monosulphate:glucosulphatase. Bull Chem Soc Jpn 6:258-264
Stevens RL, Fluharty AL, Skokut MH, Kihara H 1975 Purification and properties of arylsulphatase A from human urine. J Biol Chem 250:2495-2501
Stevens RL, Fluharty AL, Kihara H, Kaback MM, Shapiro LJ, Sandhoff K, Fischer G 1979 Metachromatic leucodystrophy: a variant with apparent cerebroside sulfatase activator deficiency. Clin Res 27:104A
Stinshoff K, Jatzkewitz H 1975 Comparison of the cerebroside sulphatase and arylsulphatase activity of human arylsulphatase A in the absence of activators. Biochim Biophys Acta 377:126-138

Discussion

Kredich: With respect to the phosphodiesterase activity of sulphatase A, I don't recall whether methylated xanthines inhibit highly purified cyclic AMP phosphodiesterase, but assuming that they do, have you tried such inhibitors with sulphatase A?

Roy: The phosphodiesterase activity is inhibited by papaverine (T. Uchida, personal communication 1979) but I have no information about the effects of methyl xanthines.

Kredich: It would be interesting to determine the effect of theophylline on arylsulphatase activity.

Roy: Papaverine did not inhibit the arylsulphatase activity (T. Uchida, personal communication 1979). Incidentally, both adenosine and guanosine 2′,3′-phosphates are hydrolysed by sulphatase A, but not the corresponding pyrimidine nucleotides.

Segel: Is this a typical ping-pong type of hydrolase where one might expect, for example, the enzyme–sulphate complex to be a seryl sulphate?

Roy: I would like to think of the modified enzyme as an imidazole sulphate, but I have no evidence for this. The detailed reaction mechanism has been difficult to study kinetically because the activity falls off so rapidly that it has been hard to measure initial velocities. With Equation 1 (p 180) we can now do this. Sulphate is a competitive inhibitor, as I said, which is consistent with it being the last-released product. Unfortunately, the phenolic product is such a poor inhibitor that we can't say anything about its inhibition pattern, so the detailed mechanism is not known.

Segel: For those sulphatases where sulphate is not competitive there is another explanation, namely an 'iso' type of mechanism where the enzyme, after release of sulphate, has to isomerize to a new form before picking up the substrate. Here, sulphate would not be competitive.

Roy: When we first looked at the non-competitive inhibition by SO_4^{2-} (Jerfy & Roy 1973) we explained it by a complicated mechanism which took into account the existence of the substrate as mixed micelles. However, Stinshoff & Jatzkewitz (1975) also found non-competitive inhibition in their system where the substrate was simply sonicated cerebroside sulphate. This seemed to rule out the explanation in terms of micelles — unless it also applied to micelles of cerebroside sulphate itself.

Swaisgood: Are the self-association properties of the 'F' form of the enzyme the same as those of the native enzyme? If not, studies of such properties could provide a method for further characterizing the structural changes involved.

Roy: As far as we can tell, they seem to be the same.

Whatley: When the sulphatase enzyme is reactivating 'naturally' and has recovered from the inhibition, what is the reactivating species — is it the sulphate?

Roy: Primarily the sulphate, and to a much lesser extent the phenol.

Whatley: I don't understand how it recovers its activity after a while.

Roy: Neither do we! The fact that you can get back the same modified enzyme from the reaction mixture after activation by sulphate doesn't help our understanding.

Bright: You say that deficiencies in sulphatase A are rapidly lethal. Does this mean that cerebrosides are continuously turning over in mammals?

Roy: Yes. The children die about four years after onset, apparently because of the accumulation of cerebroside sulphates causing degeneration of the myelin sheath of the nerves.

Dodgson: Dr Colin Wynn, in Manchester, was looking at arylsulphatase enzymes from *Aspergillus oryzae,* and found several isoenzymes, one of them having the capacity to take a substrate like nitrocatechol sulphate and transfer the sulphate group to tyramine (Burns et al 1977). This seems hard to explain, but the experiments have not been faulted.

Roy: It seems surprising, but Suzuki et al (1959) claimed that molluscan sulphatases transfer sulphate from *p*-nitrophenyl sulphate to certain sulphur-poor fractions of charonin sulphate.

Dodgson: Yes, but Suzuki later indicated to me that PAP was probably present and this would allow the transfer to take place.

Kredich: Is the tyramine–sulphate adduct a sulphamate?

Dodgson: No; the sulphate was going on the OH group, not on the amino group. What worries me is that J.W. Fitzgerald when studying choline sulphate in microorganisms found a pseudomonad that grows quite well on choline sulphate but has never found any sulphatase activity that will free that sulphate. Fitzgerald claims to have found choline but not free sulphate.

Rose: I have the uneasy feeling that so many conditions have to be right for sulphatase A to be active towards cerebroside sulphate that its sole function as the regulator of the amount of this material in various tissues seems almost unlikely. It is difficult to imagine, although admittedly we see the devastating effect of its absence at an early age in metachromatic leucodystrophy. On this basis the feeble activity sulphatase A shows towards other sulphate esters *in vitro,* such as the compounds Dr Roy tested — catecholamine sulphates (Jenner & Rose 1978), etc. — might be equally important *in vivo.*

Roy: That argument is not entirely valid. For cerebroside sulphatase activity you need the low pH (about 4.5), which you can get in lysosomes. *In vivo* the activator protein (accessory protein) acts as a transporter of cerebroside sulphate to the enzyme. And Stevens et al (1979) have now found a variant of metachromatic leucodystrophy in which sulphatase A activity is normal but the activator protein is lacking.

References

Burns GRJ, Galanopoulou E, Wynn CH 1977 Kinetic studies of the phenol sulphate-phenol sulphotransferase of *Aspergillus oryzae*. Biochem J 167:223-227

Jenner WN, Rose FA 1978 Dopamine 3- and 4-*O*[^{35}S]sulphate as substrates for arylsulphatases *in vitro* and their metabolism by the rat *in vivo*. In: Aito A (ed) Conjugation reactions in drug biotransformation. Elsevier/North-Holland Biomedical Press, Amsterdam/New York/Oxford, p 501

Jerfy A, Roy AB 1973 The sulphatase of ox liver. XVI. A comparison of the arylsulphatase and cerebroside sulphatase activities of sulphatase A. Biochim Biophys Acta 293:178-190

Stevens RL, Fluharty AL, Kihara H, Kaback MM, Shapiro LJ, Sandhoff K, Fischer G 1979 Metachromatic leucodystrophy: a variant with apparent cerebroside sulfatase activator deficiency. Clin Res 27: 104A

Stinshoff K, Jatzkewitz H 1975 Comparison of the cerebroside sulphatase and arylsulphatase activity of human sulphatase A in the absence of activators. Biochim Biophys Acta 377: 126-138

Suzuki S, Takahashi N, Egami F 1959 Para nitrophenyl sulfate hydrolysis and transsulfation to charonin sulfuric acid by cell free extracts of *Charonia lampas*. J Biochem (Tokyo) 46:1-10

Studies on the nature and regulation of the cellular thiol:disulphide potential

D.M. ZIEGLER, M.W. DUFFEL and L.L. POULSEN

Clayton Foundation Biochemical Institute, and Department of Chemistry, The University of Texas at Austin, Austin, Texas 78712, USA

Abstract Microsomal fractions separated from homogenates of liver, kidney and corpora lutea contain a monooxygenase (dimethylaniline monooxygenase [*N*-oxide forming], EC 1.14.13.8) that catalyses NADPH- and oxygen-dependent oxidation of cysteamine to cystamine. The monooxygenase purified to homogeneity from hog liver also catalyses oxygenations of diverse xenobiotics, but it does not catalyse oxidation of any other physiological sulphur- or nitrogen-containing compounds. All the available evidence indicates that cysteamine is the physiological substrate for the monooxygenase, and the oxidation of this thiol to the disulphide may be a significant source of disulphide maintaining the cellular thiol:disulphide potential. The concentration of protein–low molecular weight mixed disulphide is a function of this potential. Changes in concentration of this protein-mixed disulphide reflect changes in thiol:disulphide balance. At constant substrate concentrations the potential would depend primarily on activity of the cytosol glutathione reductase (NAD(P)H: oxidized-glutathione oxidoreductase, EC 1.6.4.2) relative to that of the membrane-bound monooxygenase. In hepatic tissue from adult mice and hamsters there is a correlation between the concentration of protein-mixed disulphide and the activity of the monooxygenase relative to the reductase. Hepatic glutathione reductase is relatively constant in mice, but the monooxygenase is much higher in the female than in the male. After gonadectomy monooxygenase activity decreases in the female and increases in the male. Activities are restored to control levels by treating males with testosterone and females with progesterone. Testosterone decreases and progesterone increases activity. These two hormones apparently regulate the level of this enzyme in hepatic tissue.

Alterations of cellular thiol to disulphide ratios, usually reflected as changes in the concentration of protein-glutathione mixed disulphides, are known to accompany changes in a variety of complex biological processes. For example, cyclic variations in glutathione (GSH) and glutathione disulphide (GSSG) affect protein synthesis in sea urchin embryos (Mano 1977).

Also, synthesis of peptide disulphide bonds *in vitro* requires media containing both thiol and disulphide (Saxena & Wetlaufer 1970). Polymerization and

Sulphur in biology
(Ciba Foundation Symposium 72) p 191-204

depolymerization of tubulin is also influenced by thiol–disulphide exchange (Kimura 1973). In addition to these *in vitro* effects, the disulphide:thiol content of rat liver proteins varies diurnally. The ratio is highest in the early evening and lowest in the morning (Isaacs & Binkley 1977a). For a more complete list of physiological events associated with changes in cellular thiol:disulphide potentials, see the recent review by Kosower & Kosower (1978).

The possible relationship between cellular thiol:disulphide ratios and the regulation of metabolic reactions by disulphide exchange between enzymes and small disulphides has been recognized for some time (Barron 1951). This view is reinforced by observations that activities of several enzymes are regulated *in vitro* by thiol–disulphide exchange. At pH 7.4 fructose 1,6-diphosphatase (D-fructose-1,6-diphosphate 1-phosphohydrolase, EC 3.1.3.11) is activated by the disulphide cystamine, and inactivated upon incubation with thiols (Pontremoli et al 1967). Glycogen synthase (UDPglucose: glycogen 4-α-glucosyltransferase, EC 2.4.1.11), on the other hand, is reversibly inactivated by GSSG (Ernest & Kim 1973) and reversible inactivation of enzymes by low molecular weight disulphides is a fairly common phenomenon (cf. Kosower & Kosower 1978).

While the experimental evidence for changes in cellular thiol:disulphide in different metabolic states and possible effects on enzymic activities is substantial, mechanisms regulating this potential are not clear. Effective regulation of metabolic reactions through modification of enzymic activities by thiol–disulphide exchange will be a function of the thiol–disulphide ratio and precise control of this ratio for normal metabolism is essential. In turn, one or more of the components of such an enzymic redox system controlling the ratio must be closely regulated. Glutathione and glutathione reductase provide an adequate source for reducing equivalents, but a satisfactory molecular mechanism for the *in vivo* generation of disulphides is not as well defined.

CELLULAR THIOL OXIDANTS

While several metabolic reactions involving oxidation of a thiol to a disulphide are known, those capable of net generation of disulphides are quite limited. *Escherichia coli* contains an enzyme catalysing the reduction of nucleotides by GSH (Holmgren 1976). The GSSG formed by this route may be a significant source of low molecular weight disulphides during rapid cell division of microorganisms, but whether this reaction could maintain significant disulphide concentrations in the presence of glutathione reductase has not been explored. In mammalian tissues GSSG can be generated by H_2O_2-dependent oxidation of GSH catalysed by the glutathione peroxidase (glutathione: H_2O_2 oxidoreductase, EC 1.11.1.9). Isaacs & Binkley (1977b) have collected an impressive amount of information suggesting that this reaction may account for diurnal variations in rat liver peptide-mixed disulphide. They suggest that a decrease in catalase directs more peroxide into the glutathione peroxidase pathway, increasing cellular disulphide, and changes in protein-mixed disulphide correlate with changes in catalase (H_2O_2:H_2O_2 oxidoreductase, EC 1.11.1.6) activity. While this mechanism is potentially possible, it presents some serious problems. Glutathione peroxidase and catalase, the two principal enzymes competing for peroxide, are both present in the cytosol and even in animals treated with dibutyryl cyclic AMP to depress catalase the activity of this enzyme is still 1000 times greater than that of the glutathione peroxidase (Isaacs & Binkley 1977b). Net generation of GSSG under these conditions probably does not occur. However, if some glutathione peroxidase is present in a cell compartment containing lesser amounts of catalase it could be a significant source of cellular disulphide. Since this is extremely difficult to demonstrate experimentally, the peroxidation of glutathione as a major source of cellular disulphides remains an attractive but unresolved possibility.

Two years ago we proposed that the oxidation of cysteamine to cystamine catalysed by a membrane-bound monooxygenase could be a significant source of cellular disulphides (Ziegler & Poulsen 1977). The essential features of this system for maintaining thiol:disulphide potentials are illustrated in Fig. 1. Disulphide generated by the monooxygenase is coupled to the glutathione reductase through thioltransferases (thiol:disulphide oxidoreductases).

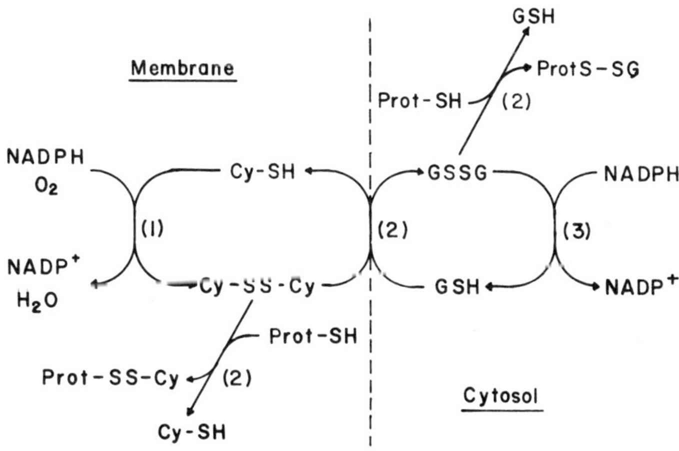

FIG. 1. Possible reactions maintaining thiol:disulphide balance. Reaction 1, oxidation of cysteamine to cystamine catalysed by the membrane-bound monooxygenase. Reaction 2, thiol-disulphide exchange catalysed by membrane-bound and cytosolic thioltransferases. Reaction 3, reduction of GSSG catalysed by cytosolic glutathione reductase.

Neither the number nor substrate specificities of thioltransferases have been totally defined, but in a recent review Freedman (1979) suggests that there are several different enzymes catalysing thiol–disulphide exchange. Highly specific membrane-bound thioltransferases would enhance regulatory capabilities, but this would not solve the basic problem of the cellular source of disulphides.

The system shown in Fig. 1 can provide disulphides required for peptide disulphide synthesis *in vitro* (Poulsen & Ziegler 1977), but little is known about its possible operation or regulation *in vivo*. On the basis of this model, at constant substrate concentrations, thiol to disulphide ratios would depend primarily on levels of glutathione reductase relative to the monooxygenase. Changes in the amount of either of these enzymes should be reflected in an altered tissue thiol:disulphide balance. Although glutathione reductase levels vary in different tissues and organisms, there is no evidence for large variations in activity under normal conditions in a given tissue. The membrane-bound monooxygenase, however, has been reported to increase during maturation of corpora lutea from hogs (Heinze et al 1970). We have confirmed this observation. The monooxygenase, very low in activity at ovulation, increases about 45-fold by Day 13–15 of the cycle and then decreases sharply on Days 16–20 (Fig. 2). The glutathione reductase changes less

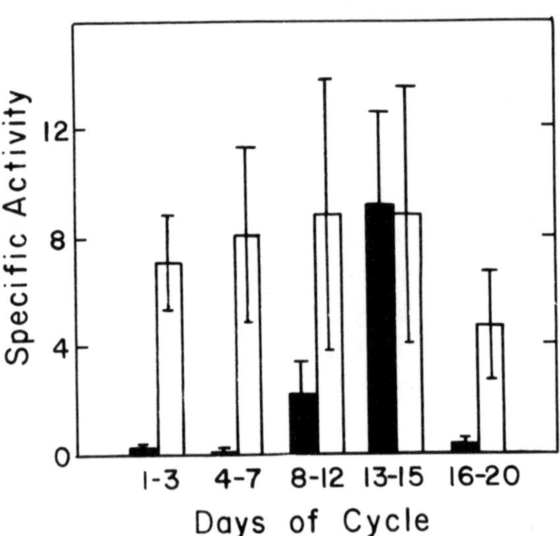

FIG. 2. Activities of dimethylaniline (cysteamine) monooxygenase and glutathione reductase in corpora lutea from hogs. The day of the cycle for each corpus luteum was determined by comparison with the photographs by Akins & Morrissette (1968). Monooxygenase (■) and reductase (□) specific activities were measured as described in Table 2 and are expressed as nmoles product formed min^{-1} mg^{-1} homogenate protein.

than two-fold during this period. Unfortunately, this is not a suitable tissue for measuring changes in protein-mixed disulphides because of rather high concentrations of unknown reducing agents that interfere with these measurements.

Protein-mixed disulphides can be accurately measured in hepatic tissue, and age- (Uehleke et al 1971) and sex- (Wirth & Thorgeirsson 1978) dependent changes in hepatic monooxygenase activity in rodents have been described. Most of our studies on changes in protein-mixed disulphide as a function of glutathione reductase and microsomal monooxygenase activities have been confined to liver from mice and hamsters.

HEPATIC GLUTATHIONE REDUCTASE AND (CYSTEAMINE) MONOOXYGENASE ACTIVITIES

Glutathione reductase activity was determined by following GSSG-dependent NADPH oxidation (Carlberg & Mannervik 1975). Activity of the monooxygenase was assessed by following *N*-oxidation of dimethylaniline (Ziegler & Pettit 1964), since the oxidation of cysteamine to cystamine is difficult to measure quantitatively in whole homogenates. However, this method accurately measures potential cysteamine monooxygenase activity of tissue homogenates (Poulsen & Ziegler 1977) since *N*-oxidation of the amine and oxidation of cysteamine to cystamine are catalysed by the same monooxygenase. The close correlation of these two activities is also apparent from rates listed in Table 1.

TABLE 1

Dimethylaniline and cysteamine monooxygenase activities of microsomes

	Specific activity[a]	
Source	*Dimethylaniline monooxygenase*	*Cysteamine monooxygenase*
Pig liver (female)	41	42
Hamster liver (female)	12	11
Rat liver (male)	7	5
Mouse liver (female)	12	12
Mouse liver (male)	4	3
Pig corpus luteum	21	19

[a]Specific activities are expressed as nmoles of substrate-dependent oxygen uptake per min per mg protein, measured by the method described earlier (Poulsen & Ziegler 1977) at pH 8.4 with 3 mM-octylamine present to suppress other microsomal oxidases. Concentration of substrates: dimethylaniline, 1 mM; cysteamine, 2.5 mM.

Furthermore, during isolation of the monooxygenase from hepatic microsomes cysteamine monooxygenase/tertiary amine *N*-oxygenase activities are constant in all fractions including the final homogeneous preparations of the monooxygenase (unpublished experiments).

We have not detected significant sex-related differences in hepatic monooxygenase activity in hamsters from measurements made on several hundred animals over a three-year period. Measurements of hepatic glutathione reductase, although not as extensive, suggest that this activity is higher in females than in males (Table 2) and for the group of hamsters used in these experiments the sex-dependent differences in the ratio of monooxygenase to reductase activities are due primarily to increased glutathione reductase in the female. There is a greater sex difference in activities of these two enzymes in the mouse, but in this species glutathione reductase is constant and the monooxygenase activity is almost three times greater in female than male liver.

TABLE 2

Protein-mixed disulphide as a function of the ratio of dimethylaniline (cysteamine) monooxygenase to glutathione reductase activities in rodent hepatic tissue

		Enzyme activities[b]			*Protein-mixed disulphide*[c]
Species[a]	*Sex*	*Dimethylaniline monooxygenase*	*Glutathione reductase*	*Monooxygenase / Reductase*	*(nmoles/mg protein)*
Mouse	Male	0.94 ± 0.08	54.2 ± 3.2	0.017	1.2 ± 0.5
Mouse	Female	2.46 ± 0.29	49.7 ± 3.3	0.049	2.4 ± 1.0
Hamster	Male	2.77 ± 0.28	51.1 ± 5.2	0.055	2.6 ± 0.4
Hamster	Female	3.07 ± 0.41	93.0 ± 9.6	0.033	0.4 ± 0.3

[a]White mice were 90–120 days old and of the CF1 strain from the same colony. The females had been isolated from males 14–20 days before sacrifice. Golden hamsters, males 140–150 and females 90–100 days of age, were obtained from different local colonies and maintained on a 14/10 hour light/dark diurnal cycle.

[b]Monooxygenase activities were measured by the method of Ziegler & Pettit (1964) and glutathione reductase as described by Carlberg & Mannervik (1975). Monooxygenase was assayed at pH 8.4 and 38 °C; reductase was assayed at pH 7.4 and 38 °C. Both activities are expressed as nmoles product formed min^{-1} mg^{-1} homogenate protein.

[c]Tissue samples were homogenized in 0.3 M-trichloroacetic acid (TCA) within a few seconds after the animal was decapitated. The TCA-insoluble fraction, separated by centrifugation, was washed three times with TCA and once with methanol: chloroform (1:9). The protein was then resuspended in 5 mM-SDS, reduced with borohydride and TCA-soluble thiol measured by the method of Habeeb (1972). All values are means ± S.D. for at least three animals.

The greater concentration of monooxygenase in female mouse hepatic tissue is reflected in an increased protein disulphide:thiol ratio, largely due to an increase in protein-low molecular weight disulphide (Table 2). In hamsters the decrease in reductase in males results in a higher monooxygenase-to-reductase ratio in males than in females and this is reflected in higher protein-mixed disulphide in the male of this species. Though the correlations are not exact, protein-mixed disulphides do change with changes in the ratio of reductase to monooxygenase activities in hepatic tissue from mice and hamsters and the data in Table 2 are consistent with the proposed mechanism (Fig. 1) for maintaining thiol:disulphide potential. The ratio of reductase to monooxygenase in liver also changes during the first few weeks after birth. This is due largely to the more rapid increase of the monooxygenase than the reductase in neonatal rodents (Uehleke et al 1971). Preliminary experiments suggest that during the first four weeks changes in hepatic protein-mixed disulphide also correlate with changes in the ratio of reductase to monooxygenase, and neonatal rodent tissue may be a valuable model for further studies on the enzymic regulation of thiol-disulphide balance *in vivo*.

STEROID SEX HORMONE-DEPENDENT CHANGES

The sex difference in hepatic monooxygenase activity in mice suggests that this enzyme is hormone regulated. The activities of this enzyme and of glutathione

TABLE 3

Effect of gonadectomy on dimethylaniline (cysteamine) monooxygenase and glutathione reductase activities[a] of mouse tissues

Sex	*Treatment*[b]	*Liver*		*Kidney*		*Lung*	
		Monooxygenase	*Reductase*	*Monooxygenase*	*Reductase*	*Monooxygenase*	*Reductase*
Female	Control	2.56±0.35	43.4±12.9	0.71±0.06	97.1±15.3	0.79±0.12	29.4± 6.4
Female	Gonadectomized	2.05±0.34	41.7±11.6	0.78±0.06	93.9±10.4	0.83±0.21	36.2±13.9
Male	Control	0.94±0.07	42.2±12.5	1.08±0.06	111.0±12	0.83±0.10	24.6±0.9
Male	Gonadectomized	1.65±0.06	53.5±13.7	1.25±0.20	132.0±22	0.68±0.11	21.9± 7.0

[a]Enzyme activities were measured by methods described in Table 2 and are expressed as nmoles of dimethylaniline *N*-oxide formed or GSSG reduced min^{-1} mg^{-1} homogenate protein. All activities are the means ± S.D. of at least three animals.
[b]The mice used were described in Table 2. They were killed 14 days after gonadectomy.

reductase in liver, kidney and lung from control and gonadectomized mice are listed in Table 3. The specific activity of glutathione reductase is not affected by gonadectomy in any of these tissues, but significant changes in hepatic monooxygenase are apparent. After gonadectomy monooxygenase activity decreases in the female and increases in the male. Hepatic monooxygenase activity in gonadectomized animals of both sexes is depressed by testosterone and oestradiol (Fig. 3). The effect of testosterone is more pronounced than that of oestradiol. Progesterone, on the other hand, increases the specific activity of the monooxygenase in females to levels not significantly different from those of control females.

Changes in specific activity in lung and kidney are not significant, but the decrease in kidney size of males after gonadectomy reduces the total amount of this enzyme present in this organ.

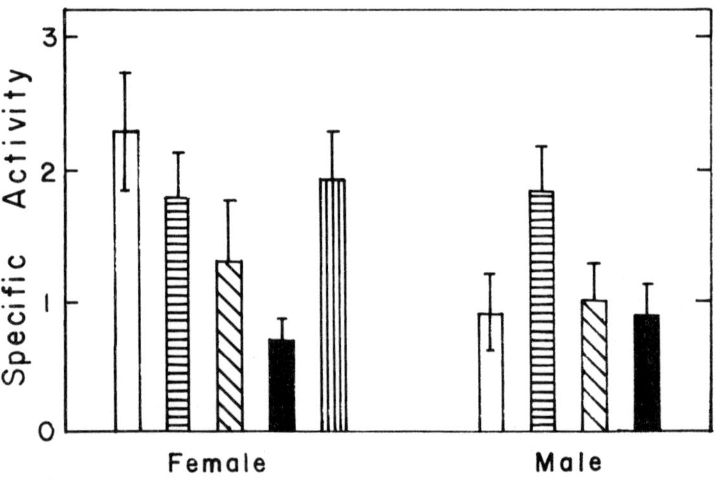

FIG. 3. Effect of steroid sex hormones on mouse hepatic dimethylaniline (cysteamine) monooxygenase activity. The mice used in these experiments were the same as those described in Table 2. Activities of control, □; and after gonadectomy, ▤. Fourteen days after gonadectomy, steroid hormones were administered by subdermal capsule implantation (Stratton et al 1973) and capsules were left in place for 14 days. 17β-Oestradiol, ▧ (approximately 42 μg/day per animal); testosterone, ■ (approximately 230 μg/day per animal); progesterone, ▥ (approximately 250 μg/day per animal). Monooxygenase specific activity measured as described in Table 2 is expressed as nmoles dimethylaniline *N*-oxide formed min^{-1} mg^{-1} homogenate protein.

GENERAL DISCUSSION

Molecular mechanisms regulating cellular thiol:disulphide balance are undoubtedly complex and the system illustrated in Fig. 1 may be only one of several operating in the intact organism. Though all components of this system are present in most mammalian tissues, the monooxygenase is not detectable in skeletal muscle and any peptide-mixed disulphide in muscle must be generated by some other mechanism. However, protein thiol-mixed disulphides are highest in hepatic tissue and in this organ changes in amounts of protein mixed-disulphide do correlate with changes in enzyme activities (Table 2). It is generally accepted that generation of disulphide is the rate-limiting reaction maintaining the thiol:disulphide potential and effects of changes in monooxygenase concentration (activity) on protein-mixed disulphide are consistent with this conclusion. If this is the case, steroid hormone-induced changes in monooxygenase concentration could affect the hepatic thiol:disulphide balance and this may account for some metabolic effects of progesterone and testosterone. However, the evidence for this hypothesis is quite circumstantial and more extensive studies are necessary before it can be accepted or rejected.

At constant enzyme concentration the kinetic mechanisms regulating the cycle shown in Fig. 1 are not known. Such a control mechanism must exist, since all the components are present in tissues and unregulated operation would quickly deplete tissue of glycogen through regeneration of NADPH by the phosphogluconate pathway. Since generation of disulphide is apparently the limiting reaction, regulation by availability of cysteamine appears most likely for the following reasons. The work of Huxtable & Bressler (1976) has definitely established that cysteamine is a normal metabolite. Although tissue concentrations are not accurately known, their studies suggest that free cysteamine probably does not exceed 10–20 μM. These concentrations are below the K_m (110 μM) of cysteamine for the monooxygenase. This is frequently true for regulated metabolic reactions and small changes in the concentration of free cysteamine in this range markedly affect thiol:disulphide potential *in vitro* (Poulsen & Ziegler 1977). Furthermore, *in vivo* separation of the enzymes catalysing thiol oxidation and disulphide reduction by different substrate specificities and by localization in different cell compartments would also retard operation of the cycle. Free cysteamine available to the membrane-bound monooxygenase is probably rather small and would depend on rates of cysteamine synthesis and degradation. Oxidation to hypotaurine is the major route for degradation (Huxtable & Bressler 1976) but regulated disposition by this route has not been demonstrated. The extensive work of Cavallini and his associates (Cavallini et al 1976) has established a pathway for synthesis of cysteamine through phosphopantenyl cysteine and there is some evidence that cysteamine synthesis responds to thiol:disulphide balance. Two of thc enzymes in this pathway require very high concentrations of thiol for activity (Cavallini et al 1976) and this could be a significant regulatory mechanism. Increased synthesis of cysteamine under strong reducing conditions would increase concentrations of cysteamine available for the generation of disulphide, thereby restoring the thiol: disulphide potential. However, there is no direct *in vivo* evidence supporting regulated cysteamine synthesis by this or any other mechanism. A definite answer to this question will require better analytical methods than those currently available for accurately measuring fluctuations in cysteamine at the low amounts present in tissues.

References

Akins EL, Morrissette MC 1968 Gross ovarian changes during estrous cycle of swine. Am J Vet Res 29:1953-1957

Barron ES 1951 Thiol groups of biological importance. Adv Enzymol Relat Areas Mol Biol 11:201-266

Carlberg I, Mannervik B 1975 Purification and characterization of the flavoenzyme glutathione reductase from rat liver. J Biol Chem 250:5475-5480

Cavallini D, Scandurra R, Dupre S, Santoro L, Barra D 1976 A new pathway of taurine biosynthesis. Physiol Chem Phys 8:157-160

Ernest MJ, Kim Ki-Han 1973 Regulation of rat liver glycogen synthetase: reversible inactivation of glycogen synthetase D by sulfhydryl-disulfide exchange. J Biol Chem 248:1550-1555

Freedman B 1979 How many distinct enzymes are responsible for the several cellular processes involving thiol:protein-disulphide interchange? FEBS (Fed Eur Biochem Soc) Lett 97:201-210

Habeeb AFSA 1972 Reaction of protein sulfhydryl groups with Ellman's reagent. In: Hirs CH, Timasheff SN (eds) Methods in enzymology. Academic Press, New York, vol 25:457-464

Heinze E, Hlavica P, Kiese M, Lipowsky G 1970 *N*-Oxygenation of arylamines in microsomes prepared from corpora lutea of the cycle and other tissues of the pig. Biochem Pharmacol 19:641-649

Holmgren A 1976 Hydrogen donor system for *Escherichia coli* ribonucleoside-diphosphate reductase dependent upon glutathione. Proc Natl Acad Sci USA 73:2275-2279

Huxtable R, Bressler R 1976 The metabolism of cysteamine to taurine. In: Huxtable R, Barbeau A (eds) Taurine. Raven Press, New York, p 45-57

Isaacs J, Binkley F 1977a Glutathione dependent control of protein disulfide-sulfhydryl content by subcellular fractions of hepatic tissue. Biochim Biophys Acta 497:192-204

Isaacs J, Binkley F 1977b Cyclic AMP-dependent control of the rat hepatic glutathione disulfide-sulfhydryl ratio. Biochim Biophys Acta 498:29-38

Kimura I 1973 Evidence of the similarity of microtubule protein from mitotic apparatus and sperm tail of the sea urchin, as a substrate in thiol-disulfide exchange reaction. Exp Cell Res 79:445-446

Kosower NS, Kosower EM 1978 The glutathione status of cells. Int Rev Cytol 54:109-160

Mano Y 1977 Interaction between glutathione and the endoplasmic reticulum in cyclic protein synthesis in sea urchin embryos. Dev Biol 61:273-286

Pontremoli S,. Traniello S, Enser M, Shapiro S, Horecker BL 1967 Regulation of fructose diphosphatase activity by disulfide exchange. Proc Natl Acad Sci USA 58:286-293

Poulsen LL, Ziegler DM 1977 Microsomal mixed-function oxidase-dependent renaturation of reduced ribonuclease. Arch Biochem Biophys 183:563-570

Saxena VP, Wetlaufer DB 1970 Formation of three-dimensional structure in proteins. I. Rapid nonenzymic reactivation of reduced lysozyme. Biochemistry 9:5015-5022

Stratton LG, Ewing LL, Desjardins C 1973 Efficacy of testosterone-filled polydimethylsiloxane implants in maintaining plasma testosterone in rabbits. J Reprod Fertil 35:235-244

Uehleke H, Reiner O, Helmer KH 1971 Perinatal development of tertiary amine N-oxidation and NADPH cytochrome *c* reduction in rat liver microsomes. Res Commun Chem Pathol Pharmacol 2:793-805

Wirth PJ, Thorgeirsson SS 1978 Amine oxidase in mice—sex differences and developmental aspects. Biochem Pharmacol 27:601-603

Ziegler DM, Pettit GH 1964 Formation of an intermediate *N*-oxide in the oxidative demethylation of *N,N*-dimethylaniline catalyzed by liver microsomes. Biochem Biophys Res Commun 15:188-193

Ziegler DM, Poulsen LL 1977 Protein disulfide bond synthesis: a possible intracellular mechanism. Trends Biochem Sci 2:79-80

Discussion

Roy: Is there no direct decarboxylation of cysteine?

Ziegler: No. Investigators in the taurine field have failed to find a direct cysteine decarboxylase in any mammalian tissue.

Rassin: One point of concern about the cysteamine pathway is that the associated enzymes have all been measured *in vitro* but *in vivo* when we give labelled cysteine we almost never see these intermediates. Experiments done *in vivo* with labelled cysteine have only indicated the presence of the 'classical' pathways through cysteine sulphinic acid in the rat, cat and monkey (Sturman et al 1970, 1973, Knopf et al 1978).

Ziegler: Cysteamine is a reactive metabolite and could not be present in the cell in concentrations higher than catalytic. If it were there in higher than catalytic amounts, the high level of cysteamine monooxygenase would increase the disulphide potential to such an extent that destruction of the tissue would occur.

Rassin: How should we then demonstrate that this is an important pathway *in vivo*?

Ziegler: This would be difficult to demonstrate directly, since we still do not have inhibitors specific for either the monooxygenase or glutathione reductase. However, indirect evidence obtained with alternative (xenobiotic) substrates for the monooxygenase indicates that *in vivo* the thiol:disulphide ratio depends on activity of the monooxygenase relative to activity of the glutathione reductase. For example, a number of thioureas and 2-mercaptoimidazoles are specifically *S*-oxygenated by the monooxygenase to sulphenates. The sulphenates are rapidly reduced by glutathione, regenerating the parent thiocarbamate and glutathione disulphide.

The sequence of reactions is illustrated in Fig. 1 with the antithyroid drug, methimazole. Reaction 1 is catalysed by the membrane-bound monooxygenase and reaction 4 by the glutathione reductase present in the cytosol. If the rate of reaction 1 exceeds that of reaction 4, the excess GSSG diffuses from the tissue, leading to a decrease in GSH concentration. Our preliminary studies on this drug

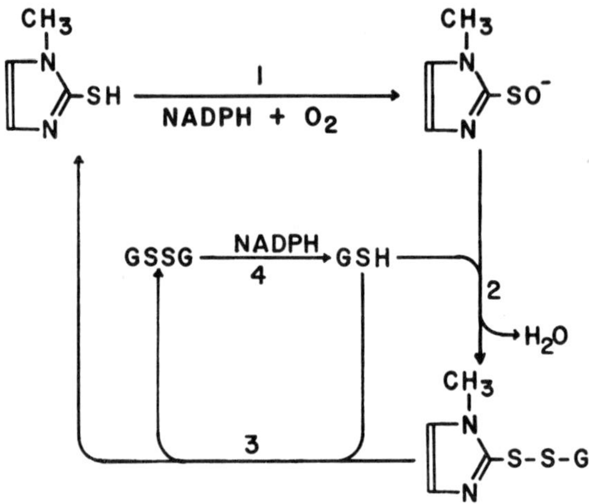

FIG. 1 (Ziegler). *S*-Oxygenation of methimazole (see text).

with adult hamsters demonstrate that this does occur *in vivo* and that methimazole-dependent GSH depletion in liver, lung and bone marrow is a function of tissue monooxygenase and glutathione reductase activities. High concentrations of exogenous cysteamine would probably also reduce tissue GSH by the same mechanism, but we have not tested it experimentally.

Rassin: You used an antithyroid agent; the cystathionase enzyme which catalyses the synthesis of cysteine has been proposed to be under control of the thyroid (Heinonen 1977). This might affect the interpretation of your results.

Ziegler: It's possible. All antithyroid compounds that inhibit the incorporation of iodine into thyroglobulin are substrates for the monooxygenase and the K_m for each is roughly proportional to antithyroid activity.

Kredich: You suggest that a lot of cysteamine or perhaps cystamine would destroy the cell. We examined the toxicity of cystamine in cultured human fibroblasts. At 200 μM-cystamine, the cells were extruding blebs of cytoplasm within three hours and were dead in six hours.

Ziegler: Did you look at their GSH:GSSG ratio?

Kredich: We did not; this would be interesting to do.

Ziegler: We have not measured the amount of monooxygenase in fibroblasts. If you added the disulphide, and I believe disulphide does get into these tissues, it would be interesting to see what happens to GSH and to the protein-mixed disulphide level of the membrane proteins. Excess disulphide would certainly destroy the integrity of the membrane, since it could change the properties of the membrane proteins by forming mixed sulphides.

Bright: The examples you gave of cellular reactions which are affected by the thiol:disulphide ratio all change over a much shorter time-scale than the changes in enzyme amounts you discussed. Do you think the amount of monooxygenase or the cysteamine concentration fluctuates diurnally, or sufficiently rapidly to regulate gluconeogenesis and such things?

Ziegler: Yes, there would be kinetic regulation of cysteamine synthesis, undoubtedly by some type of feedback inhibition. To demonstrate this we shall have to develop a method of measuring cysteamine.

Kägi: How much mixed disulphide exists in the cell normally?

Ziegler: The concentrations that we find are considerably lower than some of the literature values, and range from 0.5 nmol/mg protein to a high value in neonatal hamster liver of 8 nmol of peptide-mixed disulphide per mg protein.

Segel: A value of 2 nmol/mg means that about 10% of the total cell protein is present as a mixed disulphide. Not all proteins have SH groups, so it would seem that a significant proportion of all the SH enzymes (20–30%) is present as mixed disulphides at any time.

Ziegler: Yes. This is correct.

Dodgson: Is this monooxygenase restricted to the parenchymal cells of the liver?

Ziegler: It is present in the parenchymal cells and is distributed in the endoplasmic reticulum and possibly in the nuclear membrane, but microsomal contamination of the nuclear fraction has not been excluded. It is also present in high concentration in the bile ducts.

Rose: Is the enzyme also present in the mitochondrial outer membrane?

Ziegler: The enzyme is totally absent, so far as we can tell.

Rose: That is surprising. So many enzymes which are located primarily in the endoplasmic reticulum now seem to be present in the outer mitochondrial membrane fraction. There is always the problem of contamination to contend with, of course, but in certain cases the distribution is real and has even been taken to indicate that the endoplasmic reticulum and the outer mitochondrial membrane are continuous for functional reasons (Mier et al 1978).

Ziegler: There is a little activity in isolated mitochondria. Residual activity in this fraction could be due to microsomal contamination.

References

Heinonen K 1977 Effects of hypo- and hyperthyroidism on the activity of cystathionase in mammalian parenchymatous organs during early development. Experientia (Basel) 33:1427-1428

Knopf K, Sturman JA, Armstrong M, Hayes KC 1978 Taurine: an essential nutrient for the cat. J Nutr 108:773-778

Mier J, Spycher MA, Meyer UA 1978 Isolation of a subfraction of rough endoplasmic reticulum closely associated with mitochondria. Exp Cell Res 111:479-483

Sturman JA, Cohen PA, Gaull GE 1970 Metabolism of L-^{35}S-methionine in vitamin B_6 deficiency: observations on cystathioninuria. Biochem Med 3:510-523

Sturman JA, Niemann WH, Gaull GE 1973 Metabolism of ^{35}S-methionine and ^{35}S-cystine in the pregnant rhesus monkey. Biol Neonate 22:16-37

Sulphydryl oxidase: oxidation of sulphydryl groups and the formation of three-dimensional structure in proteins

HAROLD E. SWAISGOOD and H. ROBERT HORTON

Departments of Food Science and Biochemistry, North Carolina State University, Raleigh, North Carolina 27650, USA

Abstract Sulphydryl oxidase, an enzyme isolated from milk, catalyses the *de novo* synthesis of disulphide bonds. Thiol groups in amino acids or their derivatives, peptides, and proteins are oxidized; molecular oxygen serves as the electron acceptor and undergoes a two-electron reduction to hydrogen peroxide. Michaelis constants vary considerably amongst various substrates; glutathione is a particularly good substrate. Inhibition studies and oxidation of 1,3-diphenylisobenzofuran suggest a mechanism involving an electron transfer to singlet O_2 forming an enzyme-bound hydroperoxy group. Evidence for a direct interaction of the enzyme with horseradish peroxidase was also obtained. Although protein-folding appears to be thermodynamically favoured, rates of spontaneous acquisition of functional three-dimensional structures in disulphide-containing proteins have appeared disturbingly slow. In the presence of sulphydryl oxidase, functional structure is rapidly acquired by both reductively unfolded ribonuclease A and reductively denatured immobilized chymotrypsinogen A as judged by restoration of native fluorescence characteristics and biological activity. Preliminary data suggest that unlike thiol: protein-disulphide oxidoreductase, protein-disulphide isomerase, or GSSG/GSH redox systems, sulphydryl oxidase does not permit a 'reshuffling' of disulphide bonds.

The oxidation of thiols represents an important aspect of sulphur metabolism. Oxidoreductions of thiol/disulphide compounds appear to be involved in many cellular functions, including the reductive degradation of polypeptide hormones and proteins, regulation of protein synthesis, maintenance of the intracellular redox potential, and protection of the cell from oxidative damage (Freedman 1979, Kosower & Kosower 1976, Flohé et al 1976).

A group of enzymes catalysing thiol:protein-disulphide interchange has been identified in recent years (for a review, see Freedman 1979). Two of these enzymes are: glutathione:protein-disulphide (insulin) oxidoreductase (EC 1.8.4.2) and protein-disulphide isomerase (EC 5.3.4.1), the former catalysing reduction of insulin and the latter catalysing interchange of disulphide bonds in randomly oxidized ribonuclease (Goldberger et al 1963, Anfinsen 1973). Although some

Sulphur in biology
(Ciba Foundation Symposium 72) p 205-222

evidence has indicated that these activities result from two distinct enzymes, Morin et al (1978) have purified a thiol:protein-disulphide oxidoreductase which apparently exhibits both activities.

None of these enzymes produces a net increase in disulphide bonds. It is apparent that an oxidative step is necessary at some point for the synthesis of protein disulphides. Recently, Ziegler & Poulsen (1977) have characterized a microsomal mixed-function oxidase catalysing specifically the oxidation of cysteamine to cystamine. They further proposed that cystamine, through the subsequent formation of mixed disulphides, catalysed the synthesis of protein disulphides. Another oxidative enzyme using molecular oxygen as the electron acceptor was identified in bovine milk (Kiermeier & Petz 1967 a, b, c). Subsequently, we have purified and characterized this enzyme (Janolino & Swaisgood 1975) and have shown that it is capable of catalysing the net synthesis of disulphide bonds both in low molecular weight substrates and in proteins. An enzyme with similar activity has also been reported to be present in epididymal fluid, although it appears to differ from the milk enzyme in several respects (Chang & Morton 1975).

PHYSICOCHEMICAL CHARACTERISTICS OF SULPHYDRYL OXIDASE

This enzyme exists physically in a highly associated or aggregated form (Janolino & Swaisgood 1975). Furthermore, the size of the associated enzyme particle is concentration dependent, although the rate of re-equilibration may not be very rapid. This observation led to the development of the first means for enzyme purification; namely, starting with the precipitate obtained from whey (by rennin treatment of skim milk) which was half-saturated with ammonium sulphate (Fraction C), the concentration-dependent aggregation was exploited to yield the major portion of the activity, first, in the supernatant and, finally, in the pellet (Fraction H) upon low-speed centrifugation. Fraction H appeared reasonably homogeneous as judged by gel electrophoresis in the Ornstein-Davis system, sodium dodecyl sulphate (SDS)-gel electrophoresis, and by lack of reactivity with antibodies prepared against bovine xanthine oxidase. In the absence of SDS, the preparation did not penetrate the 7.5% acrylamide separation gel upon electrophoresis; however, both activity and protein staining were noted at the interface of the separation and the stacking gels and at the top of the stacking gel.

The unusually large size of the enzyme aggregate led to yet another means of isolating this enzyme involving chromatography on 3000 Å pore-diameter controlled-pore glass (CPG) beads. Chromatography of Fraction H on this material indicated that nearly all of the activity was excluded from the pore volume (Janolino & Swaisgood 1978). Furthermore, such chromatographic studies showed that the aggregate was not an artifact of the isolation procedure and further substantiated the concentration dependence of the size. Accordingly, slightly more than half the activity in untreated whey was excluded from the pore volume, whereas the excluded activity increased to nearly 100% when the whey was first concentrated. Preparations of the enzyme representing the fraction excluded from the pore volume of 3000 Å pore diameter CPG (Fraction V) can thus be obtained either from whey or from Fraction C. More recently, similar preparations have been obtained using Bio-Gel A-150m. These preparations, unlike Fraction H, contain non-proteinaceous material, most likely lipid, the amount depending on the efficiency of removal of fat globules in the skimming operation. Although such fractions show greater variation in homogeneity than Fraction H, careful technique has yielded preparations with similar specific activity which were homogeneous as judged by gel electrophoresis in the presence and absence of SDS and which did not react with antibodies against xanthine oxidase.

Properties of the enzyme in the presence of various dissociating agents have been investigated in an effort to identify smaller soluble forms which possess enzymic activity. Size, activity, and tryptophyl fluorescence were examined in a series of guanidinium chloride solutions of increasing concentration (V.G. Janolino & H.E. Swaisgood, unpublished paper, Am. Dairy Sci. Assoc. Meetings, July 1978). Extensive unfolding occurred between 1.5 and 2.0 M-guanidinium chloride as evidenced by loss of activity and red-shift in the fluorescence emission spectra. However, extensive dissociation of the aggregate apparently did not occur until higher concentrations of the denaturant were reached. In 5 M-guanidinium chloride, complete dissociation occurred, yielding a subunit molecular weight of roughly 85 000. Interestingly, after exposure to 7 M-guanidinium chloride, purified Fraction V remained clear upon removal of the denaturant and migrated as a single band in the separation gel upon electrophoresis; however, enzymic activity was not restored. Although the enzyme was not extensively dissociated in 1 M-guanidinium chloride (solutions were still turbid), apparently some reduction in aggregate size had taken place as reflected by an increase in specific activity. Earlier studies had shown the specific activity to be dependent upon the concentration of the enzyme stock solution (Janolino & Swaisgood 1975). Similar observations were made for the enzyme in 0.1% Triton X-100 and, moreover, increased penetration of the pore volume was noted when the detergent-treated enzyme was chromatographed.

Determination of the subunit molecular weight by gel electrophoresis in the presence of SDS using a series of acrylamide concentrations gave a value of 89 000. The validity of this value was supported by light scattering measurements in 5 M-guanidinium chloride. These and other physical characteristics of the enzyme are summarized in Table 1.

Perhaps the most significant features to be noted from chemical analyses of

TABLE 1

Some physical characteristics of sulphydryl oxidase

Subunit size	
Method	*Molecular weight*
SDS-gel electrophoresis	89 000 ± 900
Light scattering in 5M-guanidinium chloride	85 000 ± 5000
Size of active enzyme	
Method	*Observations*
Controlled-pore glass chromatography (3000 Å diameter)	More than 50% of the activity does not penetrate the pore volume
Scanning electron microscopy	Visible particles larger[a] than *c*.200 nm in diameter
Centrifugation (2000 x ***g*** for 30 min at 4 °C)	
A 0.15% protein solution	Enzyme activity remains in the supernatant
A 3% protein solution	More than 50% of the enzyme activity is pelleted
Gel electrophoresis	Activity does not penetrate the separation gel

[a]C.S. Barnes, H.R. Horton & H.E. Swaisgood, unpublished paper. Am. Soc. Biol. Chem. Meetings, June 1978 (Abstr) Fed Proc 37: 1780

purified sulphydryl oxidase can be summarized by indicating that the enzyme is a metallo-glycoprotein. Values for the various constituents based on the proposed subunit weight are listed in Table 2. Iron seems to be essential for enzymic activity; its removal appears to be largely reversible (Janolino & Swaisgood 1975). Two sulphydryl groups are reactive with 5,5′-dithiobis(2-nitrobenzoic acid); they appear to be required for enzymic activity. An additional reactive thiol, titratable only in the apoenzyme, is thought to be liganded to iron. The carbohydrate moiety does not appear to be essential for activity, although the enzyme became increasingly insoluble as its removal approached completion.

KINETIC CHARACTERISTICS OF SULPHYDRYL OXIDASE

Aside from differences in physicochemical characteristics in comparison with other known enzymes which are involved in thiol metabolism, sulphydryl oxidase appears to be unique in that it catalyses the aerobic oxidation of sulphydryl groups in

TABLE 2

Results of chemical analyses of sulphydryl oxidase preparations

Amino acid composition[a]		*Other chemical analyses*	
Residue	*Mol/89 000*	*Constituent*	*Mol/89 000*
Lys	45	Fucose[b]	4
His	11	Mannose[b]	11
Arg	30	Galactose[b]	19
Asp	62	Glucose[b]	10
Thr	42	*N*-Acetylgalactosamine[b]	20
Ser	46	*N*-Acetylglucosamine[b]	12
Glu	72	*N*-Acetylneuraminic acid[b]	23
Pro	37		
Gly	75	Iron	0.5[a]
Ala	86		
Half-cys	5	Sulphydryl[c]:	
Val	56	Active enzyme	2
Met	7	Inactive enzyme (iron removed)	3
Ile	32		
Leu	61		
Tyr	18		
Phe	29		
Trp	5		

[a]Values obtained for Fraction H (Janolino & Swaisgood 1975).
[b]Values obtained for Fraction V by chromatography of Fraction C.
[c]Values obtained for Fraction H (P. Abraham & H.E. Swaisgood, unpublished observation, Am. Dairy Sci. Assoc. Meetings, 20-23 June 1976).

proteins resulting in the generation of disulphide bonds *de novo*. The stoichiometry of the reaction has been established (Janolino & Swaisgood 1975) as:

$$2RSH + O_2 \longrightarrow RSSR + H_2O_2$$

A rather general substrate specificity is exhibited in that both small thiols and protein sulphydryl groups may be oxidized. However, selectivity was noted among various small thiols. Results listed in Table 3 indicate that a number of these compounds are very poor substrates if they are oxidized at all.

Judging from the Michaelis constants, glutathione is the best substrate followed by L- and D-cysteine. Thus, amino acids or peptides appear to be the best substrates. With glutathione, substantial substrate inhibition was observed at concentrations above 1 mM so that the activity was maximal at about 0.8 mM (Janolino & Swaisgood 1975), suggesting possible binding to a site other than the catalytic site.

TABLE 3

Some kinetic characteristics of sulphydryl oxidase

Potential thiol substrates			*Potential inhibitors*		
Compound	K_m *(mM)*	$\frac{V_{0.8mM}}{V_{max}}$[a]	*Compound*	*Concn (mM)*	*Percentage inhibition*[b]
Glutathione	0.30	0.65	*o*-Phenylenediamine	0.55	36
L-Cysteine	0.80	0.50	*o*-Dianisidine	0.71	55
D-Cysteine	1.33	0.38	Diazabicyclooctane	0.82	18
N-Acetyl-L-cysteine			Guanine	0.58	30
	3.85	0.17	Dioxane	3.30	0
Cysteamine	30.0	0.03			
2-Nitro-5-thiobenzoic acid	100	0.008	Catechol	0.92	0
Dithiothreitol		0	3,4 Dihydroxybenzoic acid	0.82	0
Dithioerythritol		0	Mannitol	0.80	0
Mercaptoethanol		0	Ethanol	1.50	0
Mercaptoacetic acid		0			
Mercaptopropionic acid		0			
Lipoic acid		0			

[a]Ratio of the activity at a substrate concentration of 0.8 mM to the maximum velocity.
[b]For each case the glutathione concentration was 0.8 mM. P. Abraham, H.E. Swaisgood & H.R. Horton, unpublished observation. Am. Soc. Biol. Chem. Meeting, June 1978 (Abstr) Fed Proc 37:1341

Interestingly, such substrate inhibition was not observed in the presence of 0.8 mM-GSSG, and furthermore, increasing the GSSG concentration led to decreases in the intensity of tryptophyl fluorescence of the enzyme. Another intriguing observation was that oxytocin also enhanced the enzyme's activity toward 0.8 mM-glutathione and affected the fluorescence intensity at even lower concentrations than GSSG.

Lack of inhibition of glutathione oxidation in the presence of equivalent concentrations of catechol or 3,4-dihydroxybenzoic acid indicates that, unlike the enzyme known as thiol oxidase (EC 1.8.3.2), sulphydryl oxidase does not oxidize these compounds (Table 3). The effects of a number of reagents on enzymic activity were examined in an effort to identify oxygen intermediates in the two-electron reduction of molecular oxygen to hydrogen peroxide. No inhibition was observed in the presence of mannitol or ethanol; hence the intermediate formation of hydroxyl radicals would seem to be unlikely. Similarly, intermediate formation of superoxide radicals does not appear to occur, as judged by (1) lack of oxidation of nitrobluetetrazolium chloride, and (2) a lack of inhibition of glutathione oxidation upon addition of superoxide dismutase.

However, significant inhibition was observed with singlet O_2 trappers such as guanine and diazabicyclooctane, even at a glutathione concentration corresponding to the maximum velocity. Enzymic formation of singlet O_2 was further implicated by the sulphydryl oxidase-catalysed oxidation of 1,3-diphenylisobenzofuran, a compound which is known to be oxidized specifically by singlet O_2 (Singh et al 1978). The rate at which this compound was oxidized by the enzyme (Fig. 1) was markedly reduced by the substrate glutathione. Oxidation of diphenylisobenzofuran was also inhibited by known sulphydryl oxidase inhibitors, KCN (Kiermeier & Petz 1967a) and *o*-dianisidine (see below), and was completely eliminated by heat inactivation of the sulphydryl oxidase activity.

Sulphydryl oxidase activity toward glutathione was inhibited by two substrates for horseradish peroxidase, *o*-phenylenediamine and *o*-dianisidine. Gunsalus et al (1965) have suggested that these compounds bind to an enzyme–hydroperoxy group complex, and may thereby inhibit oxidative enzymes which form such complexes. The suggested formation of a hydroperoxy group–sulphydryl oxidase complex during the catalysed oxidation of glutathione was further supported by

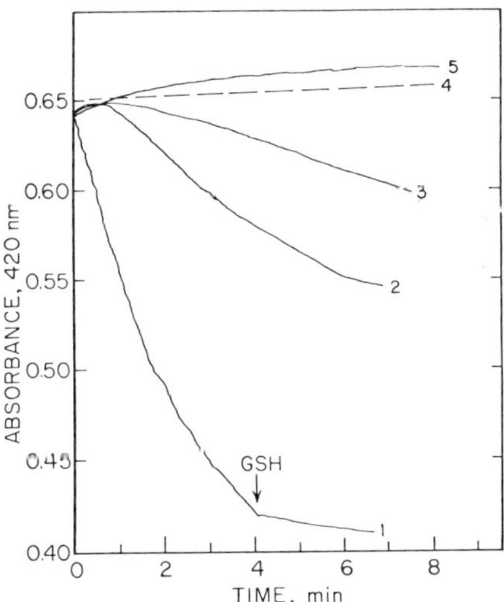

FIG. 1. Oxidation of 1,3-diphenylisobenzofuran (DPBF) by sulphydryl oxidase. A 3.0-ml reaction mixture contained 0.13 mg enzyme (Fraction V) and 0.097 μmol DPBF. *Curve 1:* DPBF; the solution was made to 0.8 mM in glutathione (GSH) at the point indicated. *Curve 2:* DPBF plus 0.21 mM-glutathione. *Curve 3:* DPBF plus 0.23 mM-KCN. *Curve 4:* DPBF with heat-inactivated enzyme. *Curve 5:* DPBF plus 0.2 mM- *o*-dianisidine. The absorbance scale for heat-inactivated enzyme was adjusted to compensate for the turbidity.

direct observations of interaction with horseradish peroxidase. The rate of oxidation of glutathione was enhanced three- to five-fold in the presence of equal concentrations of horseradish peroxidase (Fig. 2). The enhancement depended upon the presence of active peroxidase but was not affected by addition of an equivalent amount of catalase. Furthermore, the characteristic spectrum of horseradish peroxidase (II) (Hasinoff & Dunford 1970), with an absorption maximum centred at 417 nm, was generated in the presence of sulphydryl oxidase upon the addition of glutathione. These observations may point to a direct interaction between these two enzymes in which an intermediate reduced form of oxygen is transferred to the peroxidase.

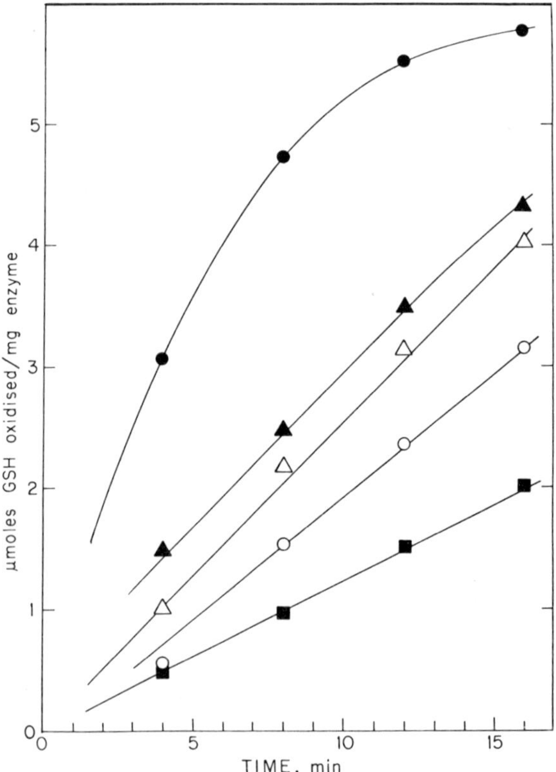

FIG. 2. Effect of horseradish peroxidase (HRP) on the oxidation of glutathione by sulphydryl oxidase. All assay mixtures contained 0.8 mM-glutathione and 0.22 mg sulphydryl oxidase (Fraction V) in 0.05 M-sodium phosphate, pH 7.0. ●, 0.34 mg HRP; ○, 0.34 mg partially inactivated (78% inactivation) HRP; ▲, 0.14 mg HRP; Δ, 0.14 mg HRP and 0.11 mg catalase; ■, control (sulphydryl oxidase without HRP). Sulphydryl concentrations were determined by reaction with 5,5′-dithiobis(2-nitrobenzoic acid). In the absence of sulphydryl oxidase, controls containing HRP and glutathione did not show any change.

Evidence for formation of an enzyme–substrate covalent intermediate during the sulphydryl oxidase-catalysed oxidation of glutathione has also been obtained. Sulphydryl oxidase was bound to immobilized glutathione and was not released by washing, even in the presence of the protein denaturant, 4M-urea. However, rinsing with 0.8 mM-glutathione solution released the active enzyme. Starting with Fraction C, an enzyme preparation with a purity similar to that of Fraction H was obtained in a single step by such chromatography on an immobilized glutathione column.

Consideration of all of the available data leads us to suggest a tentative reaction scheme whereby sulphydryl oxidase may catalyse the oxidation of thiols (Fig. 3). According to this scheme, the first molecule of thiol forms a mixed disulphide with the enzyme while transferring two electrons to a singlet O_2–enzyme complex, thereby forming an enzyme-bound hydroperoxy group. Subsequently, this group could bind a proton forming H_2O_2 which dissociates while a second thiol interchanges with the enzyme-thiol disulphide yielding the corresponding disulphide product and the free enzyme.

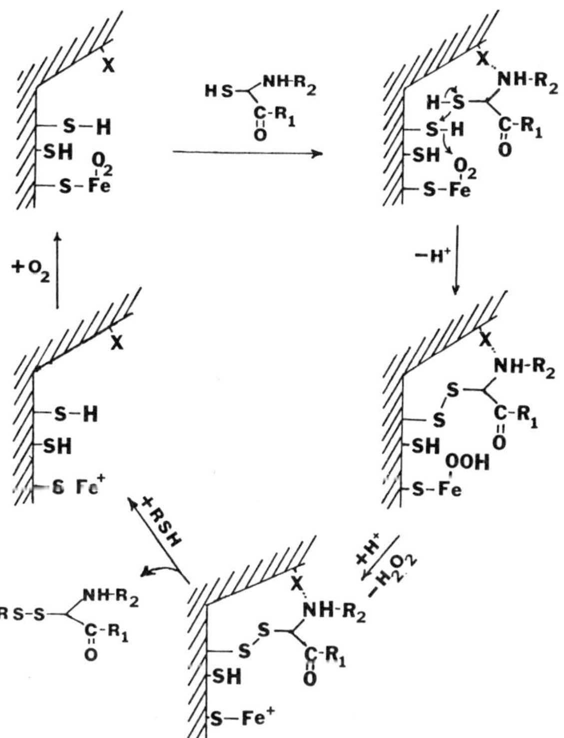

FIG. 3. Postulated reaction scheme for sulphydryl oxidase-catalysed thiol oxidation.

CATALYSIS OF DISULPHIDE FORMATION AND RESULTING THREE-DIMENSIONAL STRUCTURE IN PROTEINS

The acquisition of biological function by protein molecules depends upon the formation of three-dimensional structures from linear sequences of amino acid residues. In spite of a large number of studies on protein folding, the molecular details of the folding process are not yet completely understood, although evidence has accumulated supporting the concept that such folding is a thermodynamically controlled process (Wetlaufer & Ristow 1973, Anfinsen 1973, Brown et al 1972). In the case of disulphide-containing proteins, the formation of biologically functional molecules involves the oxidative coupling of sulphydryl groups. Although protein folding appears to be thermodynamically favoured, rates of spontaneous acquisition of functional three-dimensional structures in these proteins have appeared disturbingly slow (Anfinsen 1973, Anfinsen et al 1961). Such considerations led to the discovery of various enzyme systems and a non-enzymic GSH/GSSG system which were capable of catalysing disulphide formation. However, as noted previously, none of these systems allow *de novo* synthesis of disulphides. Hence the observed aerobic, sulphydryl oxidase-catalysed formation of disulphide bonds led us to examine the effect of this enzyme on the formation of three-dimensional structure in proteins. The enzyme proved capable of catalysing restoration of activity and 'native' structure for each of the proteins examined, namely chymotrypsinogen A (Janolino et al 1978), ribonuclease A (Janolino & Swaisgood 1975, Janolino et al 1978) and soybean trypsin inhibitor.

The rate of oxidation of sulphydryl groups in reductively denatured ribonuclease A and chymotrypsinogen A was increased more than ten-fold by the addition of sulphydryl oxidase to the solution (Fig. 4). Accordingly, the half-time of the oxidation of these reduced proteins in the presence of sulphydryl oxidase was about 16 min, whereas nearly three hours was required in its absence. In order to prevent intermolecular interactions which lead to non-native states in the case of chymotrypsinogen, this protein was immobilized to derivatized porous glass beads for these studies (Janolino et al 1978), providing a rough analogy to polypeptide chain folding while attached to a ribosome. Unlike the sulphydryl-disulphide interchange systems (Anfinsen 1973, Wetlaufer & Ristow 1973) sulphydryl oxidase did not restore enzymic activity when incubated with 'scrambled' ribonuclease, even in the presence of small amounts of 2-mercaptoethanol (Janolino et al 1978). Thus, catalytically formed disulphides must be correctly paired, at least prior to the formation of the last disulphide, for these proteins. These results differ from those obtained by Hantgan et al (1974) for anaerobic reoxidation of ribonuclease which, in the presence of a GSH/GSSG system, appeared to follow a 'random search mechanism'.

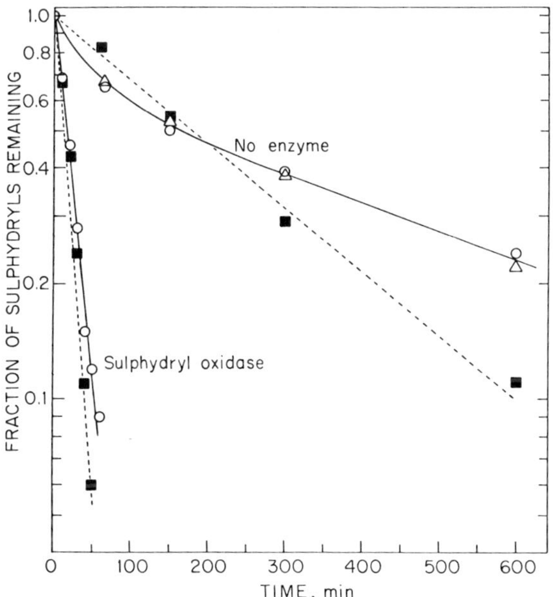

FIG. 4. First-order kinetic plots for oxidation of sulphydryl groups in reductively denatured soluble ribonuclease A and immobilized chymotrypsinogen A in the presence and absence of sulphydryl oxidase at pH 7.0. ■, ribonuclease A (8 mols SH/mol); ○, chymotrypsinogen A (10 mols SH/mol) first reoxidation; Δ, second reoxidation.

Kinetics of the refolding of immobilized chymotrypsinogen were monitored using both tryptophyl fluorescence and restoration of trypsin-activatable structure (Janolino et al 1978). The results, illustrated in Fig. 5, show that sulphydryl oxidase catalysed the formation of functional biological structure as indicated by restoration of both 'native' fluorescence properties and trypsin-activatable chymotryptic activity. Although the sulphydryl oxidase-catalysed oxidation of sulphydryl groups followed first-order kinetics, the rates of restoration of 'native' fluorescence and trypsin-activatable structure required two exponentials to fit the experimental data. Furthermore, the fluorescence data correlate with activity data and the results obtained for two cycles of reductive denaturation and reoxidation appear identical, thus eliminating the possibility that two first-order reactions with different rates led to active and inactive species, respectively. Rather, it appears that an intermediate with a structure close to that of the native zymogen may be formed. Other than an increase in the rate, the kinetics of restoration of 'native' fluorescence and trypsin-activatable structure are similar for aerobic autoxidation and sulphydryl oxidase-catalysed oxidation, suggesting that the pattern of folding is not altered by the enzyme. The results are consistent with a mechanism involving rapid refolding bringing sulphydryl groups into juxtaposition for oxidation, followed

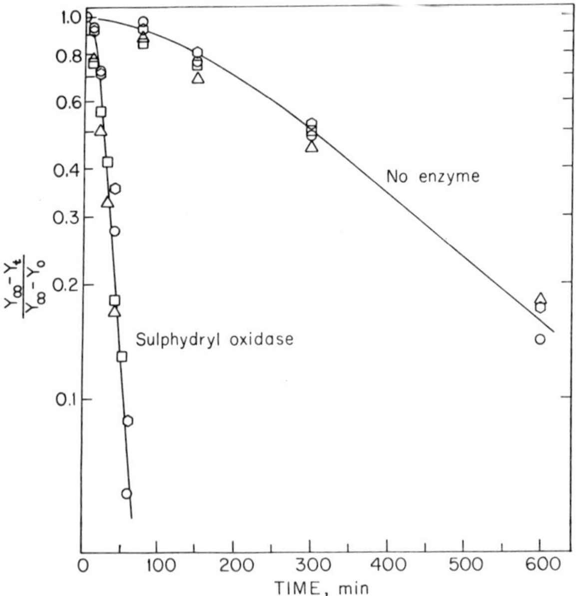

FIG. 5. Semilogarithmic kinetic plots for appearance of trypsin-activatable chymotryptic activity and for fluorescence changes during refolding and reoxidation of immobilized chymotrypsinogen A at pH 7.0 in the presence and absence of sulphydryl oxidase. The left-hand ordinate is $[Y(\infty) - Y(t)]/[Y(\infty) - Y(0)]$ where Y is the enzymic activity, first reoxidation (○), second reoxidation (⬡); the wavelength of maximum fluorescence (□); and the fluorescence emission intensity (Δ).

by more subtle conformational changes, leading to trypsin-activatable structure and 'native' fluorescence.

The mechanism by which an enzyme can catalyse the formation of a disulphide bond which resides in the internal three-dimensional structure of a protein remains obscure. However, two possible mechanisms can be suggested. These are illustrated for a hypothetical protein with two disulphide bonds, one (between S_1 and S_3) which is within the internal structure and the other (between S_2 and S_4) which is on the surface. In mechanism I, a non-native disulphide is catalytically formed on the surface which subsequently rearranges through an intramolecular sulphydryl-disulphide interchange to yield the internalized disulphide bond, followed by catalytic oxidation of the remaining sulphydryls:

$$\begin{matrix} HS_1 & & S_4H \\ & P & \\ HS_3 & & S_2H \end{matrix} \xrightarrow[O_2]{E} \begin{matrix} S_1 & — & S_4 \\ & P & \\ HS_3 & & S_2H \end{matrix} \rightleftharpoons \begin{matrix} S_1 & & S_4H \\ | & P & \\ S_3 & & S_2H \end{matrix} \xrightarrow[O_2]{E} \begin{matrix} S_1 & & S_4 \\ | & P & | \\ S_3 & & S_2 \end{matrix} \qquad \mathrm{I}$$

In mechanism II, the conformational motility of protein molecules is invoked. Indeed the concentration of unfolded species should be greater for reduced proteins since the disulphide bonds contribute to the free energy of stabilization of the native structure. Thus, partially unfolded forms with sulphydryl groups juxtaposed on the surface may exist which are subsequently oxidized, followed by refolding according to the conformational equilibrium:

$$\begin{matrix} HS_1 & & SH_4 \\ & P & \\ HS_3 & & SH_2 \end{matrix} \rightleftharpoons \begin{matrix} HS_1 & & SH_4 \\ & >P< & \\ HS_3 & & SH_2 \end{matrix} \xrightarrow[O_2]{E} \xrightarrow[O_2]{E} \begin{matrix} S_1 & & S_4 \\ & >P< & \\ S_3 & & S_2 \end{matrix} \rightleftharpoons \begin{matrix} S_1 & & S_4 \\ | & P & | \\ S_3 & & S_2 \end{matrix} \qquad \text{II}$$

This latter pathway would obviate the formation of non-native disulphides.

In the case of bovine pancreatic trypsin inhibitor, the elegant studies of Creighton (1975, Creighton et al 1978) have revealed a pathway of the type given as 'mechanism I' for the 'anaerobic' GSH/GSSG catalysed oxidative folding; whether the same pathway occurs in the aerobic folding catalysed by sulphydryl oxidase remains to be established.

ACKNOWLEDGEMENTS

This investigation was supported in part by the National Science Foundation under grant AER77-12724 and by the North Carolina Agricultural Research Service. Any opinions, findings, and conclusions or recommendations expressed in this publication are those of the authors and do not necessarily reflect the views of the National Science Foundation.

References

Anfinsen CB 1973 Principles that govern the folding of protein chains. Science (Wash DC) 181:223-230

Anfinsen CB, Haber E, Sela M, White FH 1961 The kinetics of formation of native ribonuclease during oxidation of the reduced polypeptide chain. Proc Natl Acad Sci USA 47:1309-1314

Brown JC, Swaisgood HE, Horton HR 1972 Air-reoxidation of reduced, denatured chymotrypsinogen A covalently attached to a solid matrix. Biochem Biophys Res Commun 48:1068-1073

Chang TSK, Morton B 1975 Epididymal sulfhydryl oxidase: a sperm-protective enzyme from the male reproductive tract. Biochem Biophys Res Commun 66:309-315

Creighton TE 1975 The two-disulphide intermediates and the folding pathway of reduced pancreatic trypsin inhibitor. J Mol Biol 95:167-199

Creighton TE, Kalef E, Arnon R 1978 Immunochemical analysis of the conformational properties of intermediates trapped in the folding and unfolding of bovine pancreatic trypsin inhibitor. J Mol Biol 123:129-147

Flohé L, Günzler WA, Ladenstein R 1976 Glutathione peroxidase. In: Arias IM, Jakoby WB (eds) Glutathione: metabolism and function. Raven Press, New York (Kroc Found Ser vol 6) p 115-138

Freedman RB 1979 How many distinct enzymes are responsible for the several cellular processes involving thiol: protein-disulphide interchange? FEBS (Fed Eur Biochem Soc) Lett 97:201-210

Goldberger RF, Epstein CJ, Anfinsen CB 1963 Acceleration of reactivation of reduced bovine pancreatic ribonuclease by a microsomal system from rat liver. J Biol Chem 238:628-635

Gunsalus IC, Conrad HE, Trudgill PW 1965 Generation of active oxygen for mixed-function oxidation. In: King TE, Mason HS, Morrison M (eds) Oxidases and related redox systems. Wiley, New York, p 417-447

Hantgan RR, Hammes GG, Scheraga HA 1974 Pathways of folding of reduced bovine pancreatic ribonuclease. Biochemistry 13:3421-3431

Hasinoff BB, Dunford HB 1970 Kinetics of the oxidation of ferrocyanide by horseradish peroxidase compounds I and II. Biochemistry 9:4930-4939

Janolino VG, Swaisgood HE 1975 Isolation and characterization of sulfhydryl oxidase from bovine milk. J Biol Chem 250:2532-2538

Janolino VG, Swaisgood HE 1978 Effect of support pore size on activity of immobilized sulfhydryl oxidase. J Dairy Sci 61:393-399

Janolino VG, Sliwkowski MX, Swaisgood HE, Horton HR 1978 Catalytic effect of sulfhydryl oxidase on the formation of three-dimensional structure in chymotrypsinogen A. Arch Biochem Biophys 191:269-277

Kiermeier F, Petz E 1967a Über ein sulfhydrylgruppenoxydierendes Enzym in der Milch I. Isolierung und Charakterisierung des Enzyms. Z Lebensm-Unters-Forsch 132:342-351

Kiermeier F, Petz E 1967b Über ein sulfhydrylgruppenoxidierendes Enzym in der Milch II. Einfluss auf die Sulfhydrylgruppen erhitzter Milch und Molke. Z Lebensm-Unters-Forsch 134:97-102

Kiermeier F, Petz E 1967c Über ein sulfhydrylgruppenoxydierendes Enzym in der Milch III. Einfluss von Erhitzungstemperatur und Lagerzeit. Z Lebensm-Unters-Forsch 134:149-156

Kosower NS, Kosower EM 1976 Functional aspects of glutathione disulfide and hidden forms of glutathione. In: Arias IM, Jakoby WB (eds) Glutathione: metabolism and function. Raven Press, New York (Kroc Found Ser vol 6) p 159-174

Morin JE, Carmichael DF, Dixon JE 1978 Characterization, kinetics and comparative properties of thiol: protein disulfide oxidoreductase. Arch Biochem Biophys 189:354-363

Singh A, Mcintyre NR, Koroll GW 1978 Photochemical formation of metastable species from 1,3-diphenylisobenzofuran. Photochem Photobiol 28:595-601

Wetlaufer DB, Ristow S 1973 Acquisition of three-dimensional structure of proteins. Annu Rev Biochem 42:135-158

Ziegler DM, Poulsen LL 1977 Protein disulfide bond synthesis: a possible intracellular mechanism. Trends Biochem Sci 2:79-81

Discussion

Postgate: On some of your plots (Figs. 4 and 5) you had curves without added enzyme. This is presumably spontaneous renaturation that occurs anyway.

Ziegler: This occurs only if there is a thiol oxidant, probably copper plus oxygen in these experiments. In the presence of GSH and glutathione reductase there is no spontaneous renaturation.

Swaisgood: These are classical systems for refolding and autoxidation such as those first used by Anfinsen and Haber (1961, Haber & Anfinsen 1962). Although de-ionized water, which was subsequently glass-distilled, was used for preparing these solutions, no other special precautions were taken; hence, traces of metals such as copper may be present to catalyse such autoxidations. It should be noted, however, that the autoxidation rates that we observed are very similar to those previously reported.

Postgate: Is there not a third possibility? You raised the dilemma of how do large enzyme proteins get access to buried SH groups in small proteins like ribonuclease A and bound chymotrypsin which, being bound, is probably only accessible from one side anyway. It would all be much easier to envisage if there were a small carrier molecule, capable of disulphide: SH exchange, present in your system.

Swaisgood: I think the key point to be made here is that we believe a net synthesis of disulphide is occurring rather than thiol:disulphide interchange. Certainly this is the case for catalysed oxidation of glutathione and other small molecular substances. For the case of oxidation of sulphydryls in proteins, I would emphasize that no thiol or disulphide was added to the assay system; the enzyme solution was simply added to the reduced protein which had been treated to remove all the reductant.

One can also raise the question as to whether or not the observed catalysis is truly enzymic or if the enzyme produces a small molecular oxidant which diffuses into the solution. First of all we feel that it is definitely enzymic, as evidenced by: (1) loss of catalysis by heat inactivation; (2) loss of activity with increasing denaturant (e.g. guanidinium chloride) concentration which correlates precisely with structural transitions indicated by fluorescence measurements; (3) coincidence of activity with specific protein bands observed on gel electrophoresis; (4) substrate specificity; and (5) lack of dialysable catalytic activity. We have also tried to find evidence for a small molecular oxidant produced and released by the enzyme. However, evidence for such a substance has not turned up; for example, when the reduced protein and the enzyme are separated by a dialysis membrane, catalytic oxidation and formation of structure does not occur.

Postgate: From what Dr Ziegler says, you have evidence from the existence of spontaneous oxidative renaturation that there must be something catalytic present.

Ziegler: You are speaking of two very different time scales. Dr Swaisgood's scale was very long, up to more than an hour and a half.

Swaisgood: Spontaneously, complete oxidation and full regain of structure and activity for the reductively denatured proteins require a time period of 20–24 hours. Catalytically, this reaction occurs in a matter of minutes.

Postgate: I am thinking of the situation where a reagent like 5,5′-dithiobis (2-nitrobenzoate) (DTNB) is used to investigate the accessibility of SH groups within the protein. Thorneley & Eady (1973) studied a nitrogenase protein which has 14 SH groups: the first two react in 50 seconds, the others react more and more slowly. It is a long time before all 14 groups react. By inducing a small conformational change in the protein they caused the SH groups to react much faster. This is simply a matter of DTNB being able to get into the protein more rapidly in the slightly different conformation. I am suggesting that you may need small amounts of a carrier to get in and out of your renaturable protein.

Swaisgood: Again I would have to say that we have not found any evidence for such a carrier molecule. Activity is definitely associated with the protein and it is rather easily denatured. If such a carrier molecule is produced enzymically, at least its life-time is so short that it does not diffuse through a dialysis membrane.

Ziegler: If you completely denature ribonuclease, can it be reactivated by sulphydryl oxidase alone?

Swaisgood: Yes.

Ziegler: This is somewhat unexpected, since you apparently do not need a low molecular weight thiol to facilitate the rearrangement of peptide disulphides.

Swaisgood: We do not start with enzyme which has been treated with reductant, nor is it necessary for us to add a disulphide or a thiol to the reactivation assay solution. All our evidence indicates that the denatured protein can be oxidized directly from the completely reduced form with molecular oxygen serving as the electron acceptor.

Meister: Is it possible that one molecule of glutathione is attached to the enzyme?

Swaisgood: We do not have any direct analytic method which could rule out that possibility, if it is firmly attached to the enzyme.

Meister: Could there be a mixed disulphide?

Swaisgood: Perhaps I have misled you by noting that we can purify the enzyme by affinity chromatography on a column of immobilized GSH. This is not our standard method for preparation. Normally, as was the case for the enzyme preparation used for these studies, the enzyme is never exposed to either GSH or GSSG; thus any mixed disulphides would have to be formed *in vivo*. Also I should point out that the enzyme can be inactivated by treatment with iodoacetate or iodoacetamide.

Furthermore, in studies of an immobilized form of sulphydryl oxidase (Janolino & Swaisgood 1978), activity can be maintained for relatively long periods of continuous oxidation of substrate molecules flowing through immobilized enzyme reactors. Hence thiol–disulphide interchange *per se* does not account for the extent of substrate oxidation observed.

Dodgson: In your spontaneous re-oxidation experiments, what pH do you use?

Swaisgood: We have measured it at both pH 7 and 8.6. It takes about 20 hours or so for complete spontaneous reoxidation. We didn't examine the effect of pH on rate very closely, because we didn't try to control variables like how much copper ion might be contaminating the system. There did not appear to be much difference between the rates at pH 8.6 and pH 7, although refolding was a little more rapid at 8.6 ($t_{1/2}$, pH 8.6 was 160 min; $t_{1/2}$, pH 7 was 180 min).

Kägi: How do you think peroxidase interacts with sulphydryl oxidase?

Swaisgood: At present we can only speculate. As I noted, the stimulation of GSH oxidative activity by peroxidase was not affected by adding excess catalase, and an effect on the absorption spectrum of peroxidase was observed, suggesting the formation of the HRP (II) form of this enzyme. Furthermore, it is known that GSH *per se* is not a good substrate for peroxidase, as we have confirmed by control experiments. Hence we propose a direct interaction between the two enzymes. Perhaps peroxidase directly abstracts the hydroperoxy group from sulphydryl oxidase and if, in fact, the protonation of this group were the rate-limiting step in sulphydryl oxidase catalysis, such abstraction would enhance the activity.

Kägi: Are peroxide derivatives of thiols known?

Ziegler: Thiol peroxides are extremely unstable. I would suspect them to have a half-life of nanoseconds.

Swaisgood: We are not proposing a thiol peroxide as an intermediate. However, we looked for the possible oxidation of thiol by peroxide formed during the reaction. Controls, to which was added the amount of peroxide expected from complete oxidation of the thiol compound, did not show measurable rates of oxidation within the time scale of the enzymic assays.

Ziegler: You isolate sulphydryl oxidase from milk. Have you any information on its cellular origin?

Swaisgood: Not yet. We have produced antibodies to the enzyme which we shall use to try to localize it. We find it in milk of a number of species, including human milk.

Ziegler: Milk is designed to be consumed by the neonate. Could it be involved in a transport function in the neonatal intestine?

Swaisgood: That is one possibility. I haven't completely ruled out the possibility that it is involved in the formation of disulphides in proteins during their synthesis, but we have no evidence for that other than its activity.

It would be interesting to know more about its cellular location. The fact that it is in milk is interesting, because we know that intact proteins are transported across the gut wall, up to a certain age, and these proteins, especially the immunoglobulins, contain disulphides. We normally obtain the enzyme from bovine milk because that is its most abundant source, but we have also measured activity in goat's milk and in human milk.

Rassin: Do you see a difference between human and cow's milk? Cow's milk is very poor in cysteine; it contains predominantly casein protein which has very little cysteine, whereas human milk contains mostly whey proteins, relatively rich in cysteine.

Swaisgood: Human milk contains a lot of α-lactalbumin, which contains four SS bonds.

Rassin: The amount of cysteine in 100 ml of cow's milk is about 110 μmol, whereas human milk contains almost 200 μmol cysteine per 100 ml.

Swaisgood: Although we have measured enzyme activity in human milk, these were preliminary experiments and the amounts available did not allow us to calculate the quantities present. Hence the relative amounts in human milk and cow's milk cannot be given yet.

Rassin: Human milk has evolved to supply the needs of man. It would be interesting to know whether the immune functions proposed for human milk are specifically related to the enzymes discussed here and to the transport of immunoglobulin.

Ziegler: Have you looked at the distribution of sulphydryl oxidase in tissues other than the mammary gland?

Swaisgood: We haven't examined any other tissue yet. Sulphydryl oxidase activity has been reported in epididymal tissue. It may not be the same enzyme, however.

Kelly: In cell-free extracts of thiobacilli, which obtain energy by oxidizing reduced sulphur compounds to sulphate, phosphorylation was shown to be coupled to the oxidation of mercaptoethanol or reduced glutathione (Hempfling & Vishniac 1965). Is there any similarity with the system you have been looking at? The beginning and the end-product is the same in the two systems; one wonders what goes on in the middle.

Swaisgood: I have no experimental evidence that would shed light on this question.

Le Gall: The iron binding to sulphydryl oxidase is rather unusual; do you know any other examples where iron is bound in this way to protein?

Swaisgood: We don't know how it's bound.

References

Anfinsen CB, Haber E 1961 Studies on the reduction and re-formation of protein disulfide bonds. J Biol Chem 236:1361-1363

Haber E, Anfinsen CB 1962 Side-chain interactions governing the pairing of half-cystine residues in ribonuclease. J Biol Chem 237:1839-1844

Hempfling WP, Vishniac W 1965 Oxidative phosphorylation in extracts of *Thiobacillus* X. Biochem Z 342:272-287

Janolino, VG, Swaisgood HE 1978 Effect of support pore size on activity of immobilized sulfhydryl oxidase. J Dairy Sci 61:393-399

Thorneley RNF, Eady RR 1973 Nitrogenase of *Klebsiella pneumoniae:* evidence for an adenosine triphosphate-induced association of the iron-sulphur protein. Biochem J 133:405-408

Metallothionein: an exceptional metal thiolate protein

JEREMIAS H.R. KÄGI, YUTAKA KOJIMA, MARGRIT M. KISSLING and KONRAD LERCH

Biochemisches Institut der Universität Zürich, Zürichbergstrasse 4, CH-8028 Zürich, Switzerland

Abstract Metallothioneins are unusual, low molecular weight proteins of extremely high sulphur and metal content. They occur in substantial quantity and in multiple variant forms in parenchymatous tissues (liver, kidney, intestines) of vertebrates and certain microorganisms *(Neurospora crassa,* yeast). They are thought to play a central role in the cellular metabolism of metals such as zinc, copper and cadmium.

All mammalian forms studied are single chains with 20 cysteinyl residues among a total of 61 amino acid residues and highly characteristic amino acid sequences. Their most conspicuous common features are seven -Cys-X-Cys- sequences where X stands for an aliphatic residue other than Cys. Together with additional cysteinyl residues located elsewhere in the chain and brought into juxtaposition by appropriate chain folding, these dithiol sequences are believed to form the basis of the trithiolate chelating structures typical of most of the six or seven metal-binding sites of the mammalian cadmium- and/or zinc-containing metallothioneins.The positions of the cysteinyl residues are preserved in evolution: the copper-containing metallothionein from *Neurospora crassa,* containing only 25 amino acid residues, has a distribution of metal-binding cysteinyl residues identical to that of the N-terminal portion of the mammalian chains.

The detailed physiological role of metallothionein remains to be clarified but its biosynthesis is known to be modulated by nutritional and endocrine factors. Recent evidence suggests that metallothionein is a critical determinant in the homeostasis of zinc.

One of the specific roles of sulphur in proteins is to serve as a ligand for metal binding. The recognition of non-haem iron proteins as ubiquitous components of electron transport systems has directed attention especially to the participation of cysteinyl side-chains as powerful ligands in the formation of the iron-sulphur clusters (Palmer 1975). More recently, crystallographic and spectroscopic studies have uncovered an analogous metal-binding function of the thiolate group of cysteinyl residues in certain zinc enzymes (Eklund et al 1976, Nelbach et al 1972). However, the most conspicuous examples of metal thiolate proteins are found among a group of metalloproteins which seem to play a role in the metabolism of heavy metals

Sulphur in biology
(Ciba Foundation Symposium 72) p 223-237

and in the defence mechanism of living organisms against reactive metal ions. Their best-studied representative is metallothionein, a protein of highly unusual compositional and structural features.

Metallothionein is a low molecular weight protein that was discovered in 1957 by Margoshes and Vallee in equine renal cortex and has received its designation on the basis of its extraordinarily high metal and sulphur content (Kägi & Vallee 1960). The protein was initially found in the search for a tissue constituent responsible for the natural accumulation of cadmium in animal tissues. It is still the only defined macromolecular biological compound known to contain cadmium. However, and more importantly, it is also the protein with the highest zinc content known and it also contains substantial amounts of copper (Table 1). As first noted by Piscator (1964) and confirmed since in numerous experimental studies, the biosynthesis of metallothionein is elicited in response to the administration of cadmium and zinc salts (for references see Nordberg & Kojima 1979). Mainly as a consequence of this observation, the protein has become a subject of intensive study in the fields of metal toxicology and trace metal nutrition (for references see Nordberg & Kojima 1979).

Metallothionein is widely distributed and occurs primarily in the supernatant fractions of the parenchymatous tissues of liver, kidney and gut. Table 1 lists some of the species in which the protein has been reported to occur 'naturally'; that is,

TABLE 1

Occurrence of metallothionein (without induction)

Species or class	*Organ*	*Isoproteins*	*Metal composition*		
			Zn	*Cd*	*Cu*
Man	Liver	2	+ + + + +	±	±
	Kidney	2	+ + +	+ + +	+
Horse	Liver	2	+ + + + +	+	±
	Kidney	2	+ + +	+ + +	±
	Intestine		+ + +	+ + +	±
	Spleen				
Calf	Liver	3	+ + + + +		±
Pig	Liver	3	+ + + +		+ +
Rat	Liver	2	+ + + +	−	+ +
Seal	Liver	3			
Chicken	Liver		+ + + + +	−	−
Fish	Liver				

without experimental pretreatment of the animal with metal salts. Besides vertebrates it has now also been identified in *Neurospora crassa* (see p 231) and in yeast (Prinz & Weser 1975) and there are reports that similar proteins are also present in invertebrates (Olafson et al 1979). An important outcome of these comparative studies (Table 1) is also that all mammalian tissues contain at least two isoproteins of metallothionein (isometallothioneins) and that, unrelated to this heterogeneity, there is a substantial variation in metal composition depending on the origin of the protein. Thus, while most forms contain zinc, cadmium and copper in amounts adding up to the same total molar sum, the relative proportions of the three metals vary appreciably (Kägi & Vallee 1961). Zinc is generally the predominant component but even within a single species there are quite large individual differences. Especially noteworthy is the organ-specific disparity in composition between the kidney and liver metallothioneins. Thus, while the renal forms often contain very appreciable amounts of cadmium, this metal is practically absent in the hepatic metallothioneins isolated from the same organism (Kägi et al 1974).

Table 2 gives a summary of the principal physico-chemical and compositional features of a typical metallothionein. The molecular weight and the hydrodynamic properties measured by ultracentrifugation and gel filtration are very similar for all forms and are consistent with an elongated shape of the molecules with an axial ratio of 6 to 1. All mammalian metallothioneins characterized thus far are single-chain proteins with a chain length of 61 residues and containing six to seven group 2b metal ions. Besides the high metal content, there are 20 thiolate groups that are titratable with silver ions and all participate in metal binding with a sulphur-to-metal ratio of nearly 3:1. There are no disulphide bonds, no free thiol groups and no acid-labile sulphide ions in the protein (Kägi & Vallee 1961, Kägi et al 1974, Kojima et al 1976).

TABLE 2

Properties of metallothionein

Molecular weight	6100 + metal
Molecular shape	prolate ellipsoid (axial ratio 6:1)
Chemical composition:	
Amino acids	61 residues/mole
Thiolate sulphur	20 g-atoms/mole
Thiol sulphur	None
Disulphide sulphur	None
Sum of metals (Zn + Cd + Cu)	6-7 g-atoms/mole
Thiolate sulphur/sum of metals	3

The amino acid composition (Table 3) shows that this high thiolate content is accounted for by the presence of 20 cysteinyl residues per chain, which is the highest cysteine content of any protein known. There is also a relatively large number of basic amino acid and of seryl residues but there are no aromatics and no histidine.

In view of its extraordinary composition and its high metal-binding capacity, it was of special interest to explore the primary structure of this protein (Kissling & Kägi 1977, Kojima et al 1976). A typical sequence is shown in Fig. 1. In all mammalian metallothioneins sequenced so far, the amino- and carboxyl-terminal residues are acetylmethionine and alanine, respectively (Kojima & Kägi 1978). The cysteinyl residues are distributed fairly uniformly over the entire chain. While there are no extended repetitive sequences, there is considerable regularity in their distribution. The most striking feature is the abundance of -Cys-X-Cys- sequences, where X stands for an amino acid residue other than cysteine (Cys). This arrangement occurs seven times along the chain and thus corresponds to the number of bivalent metal ions bound to these proteins. Besides this peculiar distribution of the cysteinyl residues it should be noted that most of the seryl and the basic amino acid residues tend to cluster with the cysteinyl residues. A 20-residue sequence in the centre of the polypeptide chain is composed almost exclusively of these amino acid residues.

TABLE 3

Amino acid composition of equine renal metallothionein-1B

Residue	*Number/molecule*
Cys	20
Ser	8
Lys	7
Arg	1
Ala	7
Gly	5
Val	3
Asp	2
Asn	1
Glu	1
Gln	2
Pro	2
Thr	1
Met	1
Total residues	61
Chain weight	6114

(From Kojima et al 1976.)

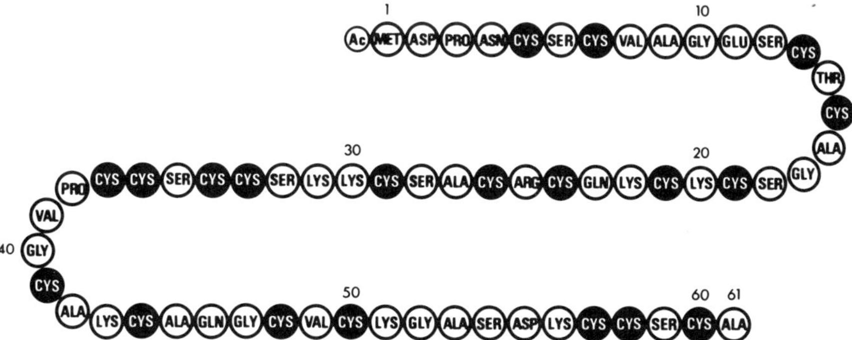

FIG. 1. Amino acid sequence of equine renal metallothionein-1B (Kojima et al 1976).

The peculiarities of the amino acid sequence of metallothionein led us to suggest that each of the seven -Cys-X-Cys- oligopeptide sequences forms part of a separate metal-binding site (Kojima et al 1976). As shown schematically in Fig. 2, the binding of the metal ion could be initiated by chelation to these sites. Since in the final complex most cadmium and/or zinc ions are bound to three cysteinyl residues *(vide supra)*, it is thought that after their formation these primary dithiolate complexes interact with a third cysteinyl residue (and perhaps also with other ligands) located elsewhere in the chain and brought into juxtaposition by appropriate chain folding. On binding of the metal ion, a proton is displaced from each cysteinyl side-chain yielding the negatively charged trithiolate complexes that determine the overall electrostatic features of most metallothioneins (Kojima et al 1976).

The formation of the metal thiolate complexes is also signalled by spectroscopic changes (Fig. 3). The spectrum of the metal-free protein, thionein, is remarkably plain, owing to the total absence of aromatic amino acids in the chain. The only absorption band is that attributable to the secondary amide and thiol transitions near 190 nm (Bühler & Kägi 1979). The metal-containing forms exhibit, superimposed upon the polypeptide absorption spectrum, both thiolate ligand absorption bands and metal-specific sulphur–metal charge-transfer absorption bands (Kägi & Vallee 1961, Bühler & Kägi 1979).

On the basis of spectrophotometric titration data and the distribution of the thiolate ligands along the polypeptide chain, we have proposed that the metal-binding sites are separated in the molecule and that they are largely independent of one another (Kägi & Vallee 1961, Kojima et al 1976). This view is now also supported by studies of metallothionein using the technique of dark-field electron microscopy (Fiskin et al 1977). The structure of the molecule deduced from the electron

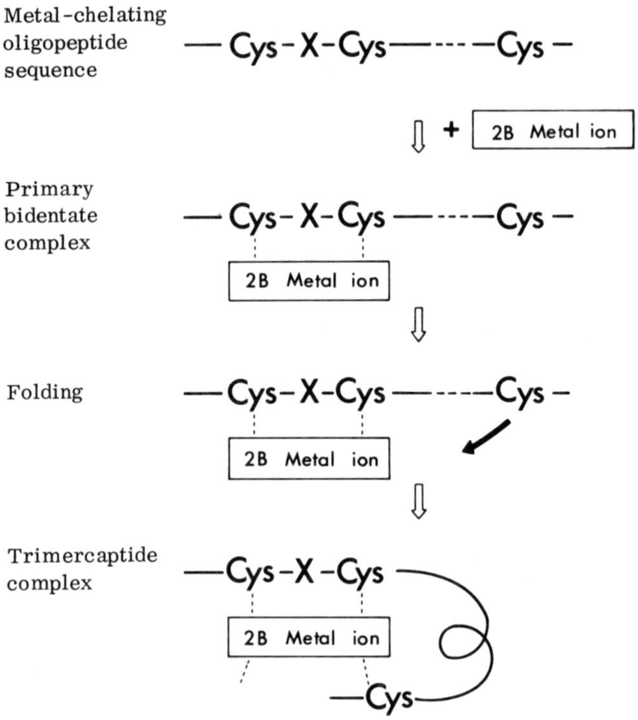

Metal-chelating oligopeptide sequence — Cys-X-Cys——--- —Cys –

\+ 2B Metal ion

Primary bidentate complex — Cys-X-Cys ——---—Cys –

2B Metal ion

Folding —Cys-X-Cys ——----Cys –

2B Metal ion

Trimercaptide complex —Cys-X-Cys

2B Metal ion

—Cys

FIG. 2. Metal-binding in metallothionein.

micrographs using image-processing methods is in agreement with the elongated shape deduced from hydrodynamic studies (Kägi et al 1974) and indicates the existence of well-separated strongly electron-scattering metal–sulphur domains.

Our sequencing studies have also given insight into the nature and significance of the various polymorphic forms of metallothioneins (Kojima & Kägi 1978, Kojima et al 1979). Fig. 4 lists the sequences of different molecular species that have been isolated from the kidney and the liver of the horse. Both tissues contain two major variants, designated MT-1B and MT-1A, which vary in a total of seven amino acids and which thus indicate the existence of two different cistrons for metallothionein. Of MT-1A there are in addition two minor variants that differ only in a single residue (in Leu 54 and Arg 39, respectively) and are thought to represent allelic subforms. The most important result is that all isometallothioneins are closely similar. The positions of the cysteinyl residues and the surrounding seryl and basic amino acid residues are especially highly conserved. With one exception, all substitutions have occurred outside these residues. As a consequence, there is also no appreciable difference in their metal-binding properties and,

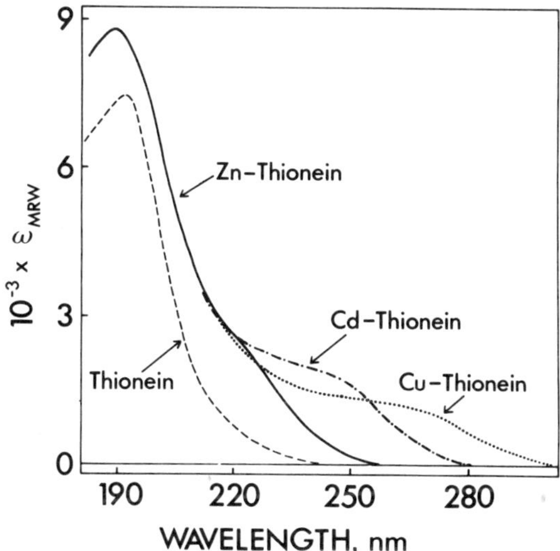

FIG. 3. Absorption spectra of thionein and of metallothioneins containing zinc (Zn-Thionein), cadmium (Cd-Thionein) or copper (Cu-Thionein). The spectra were obtained at pH 2 for thioneins and at pH 7 for the metallothioneins. ϵ_{MRW} is based on a mean residue weight (MRW) of 100.

```
                                                      1
Metallothionein-1B            N-Acetyl-Met-Asp-Pro-Asn-Cys-Ser-Cys-Val-
Metallothionein-1A                                                 -Pro-
   Variant Leu-54                                                  - " -
   Variant Arg-39                                                  - " -

   10                                      20
-Ala-Gly-Glu-Ser-Cys-Thr-Cys-Ala-Gly-Ser-Cys-Lys-Cys-Lys-Gln-Cys-Arg-Cys-
-Thr-    -Gly-                                          -Glu-
- " -    - " -                                          - " -
- " -    - " -                                          - " -

           30                                39  40
-Ala-Ser-Cys Lys Lys-Ser-Cys-Cys-Ser-Cys-Cys-Pro-Val-Gly-Cys-Ala-Lys-Cys-
-Thr-                                            -Gly-           -Arg-
- " -                                            - " -           - " -
- " -                                            -Arg-           - " -

                   50             54                          60  61
-Ala-Gln-Gly-Cys-Val-Cys-Lys-Gly-Ala-Ser-Asp-Lys-Cys-Cys-Ser-Cys-Ala-OH
                                    - " -
                                    -Leu-
                                    -Ser-
```

FIG. 4. Polymorphism of equine metallothionein.

since kidney and liver contain the same set of isometallothioneins, there appears to be no relationship between the occurrence of sequence polymorphism and the organ-specific variation in metal composition *(vide supra)*. The generally higher cadmium content in the kidney proteins is not the result of a difference in primary structure but must be attributed to a difference in chemical environment in the two organs.

The highly conservative character of the primary structure is also revealed by the similarity in the sequences of metallothionein from different species (Fig. 5). Thus, the recently established structures of the proteins from equine, human and mouse liver are in complete agreement with respect to the position of nearly all cysteinyl and most seryl and basic amino acid residues. By these criteria, the metallothioneins are rather slowly evolving proteins (Kissling & Kägi 1979, Nordberg & Kojima 1979).

This evolutionary conservatism implies that similar forms of metallothionein may

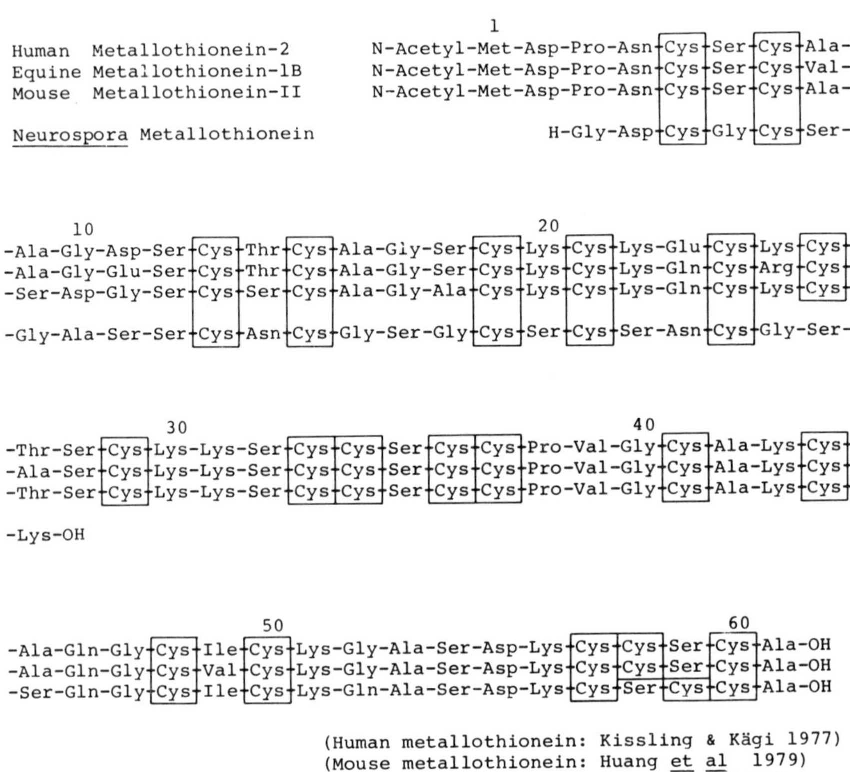

FIG. 5. Evolutionary relationship of mammalian and *Neurospora* metallothioneins.

also occur in quite distant classes of organisms. This is borne out by our recent identification and characterization of a copper-containing metallothionein in *Neurospora crassa* (Ammer et al 1978, Lerch 1979). This protein, which on copper supplementation accumulates in large quantity in the stationary growth phase of the mould, is much smaller than the mammalian metallothioneins (25 amino acid residues) and has an even more rudimentary amino acid composition, with cysteine, serine and glycine accounting for 80% of all residues. The amino acid sequence also listed in Fig. 5 indicates complete agreement in the position of its seven cysteinyl residues and of one of its seryl residues with the amino-terminal half of the mammalian chains, thus establishing an unambiguous evolutionary link among these metal-binding proteins.

The wide occurrence of the metallothioneins in the biosphere and the remarkable preservation in evolution of the metal-binding cysteinyl residues in the sequence implies their involvement in some important role in metal biochemistry and physiology. However, no single biological function has as yet been assigned to them with certainty. Of the early suggestions made (Table 4), some now seem rather unlikely. Thus, judged from their general structure and from the variability in metal composition, an involvement in immune phenomena, in enzymic functions or in redox processes seems improbable. On the other hand, a participation of metallothioneins in processes of metal metabolism such as metal storage, metal detoxification and metal homeostasis is receiving increasing experimental support (for references see Nordberg & Kojima 1979). There is thus little doubt left that the protein plays some role in the response of the organism to certain toxic metals by altering cellular tolerance (Webb 1979) and there is also good evidence that the protein can serve in short-term intracellular storage of zinc (Chen et al 1977, Cherian 1977) and thus may modulate the supply of this metal within the cell (Richards & Cousins 1975). Recent experimental studies on the influence of stress factors such as strenuous exercise, cold environment and infectious illness (Oh et al 1978, Sobocinski et al 1978), on the formation of metallothionein in the liver and on plasma zinc concentration (Table 5) suggest furthermore that this protein may be a key factor

TABLE 4

Suggested functions for metallothionein

1. Catalysis
2. Storage
3. Detoxification
4. Transport
5. Immune phenomena
6. Homeostatic mechanisms

(From Kägi & Vallee 1960.)

TABLE 5

Factors reported to stimulate accumulation of metallothionein in rat liver

Factors
Metals (Cd, Zn, Cu, Hg)
Bacterial endotoxin
CCl_4 poisoning
Strenuous exercise
Cold exposure
Starvation

in the homeostatic regulation of zinc metabolism in higher organisms. The participation of metallothionein in these various functions does not seem to be mutually exclusive. Recent reports suggest in fact that its biosynthesis is regulated both by the supply of metal salts and by certain hormones (Failla & Cousins 1978, Karin & Herschman 1979). However, additional and more extensive studies on factors regulating the genetic expression of this unusual protein are needed before its detailed role in metal metabolism can be fully unravelled.

References

Ammer D, Budry R, Lerch K 1978 Structure and function of a copper-metallothionein from *Neurospora crassa*. Experientia (Basel) 34:905

Bühler RHO, Kägi JHR 1979 Spectroscopic properties of zinc metallothionein. In: Kägi JHR, Nordberg M (eds) Metallothionein. Birkhäuser, Basel, p 211-220

Chen RW, Vasey EJ, Whanger PP 1977 Accumulation and depletion of zinc in rat liver and kidney metallothioneins. J Nutr 107:805-813

Cherian MG 1977 Studies on the synthesis and metabolism of zinc-thionein in rats. J Nutr 107:965-972

Eklund H, Nordström B, Zeppezauer E, Söderlund G, Ohlsson I, Boiwe T et al 1976 Three-dimensional structure of horse liver alcohol dehydrogenase at 2.4 Å resolution. J Mol Biol 102:27-59

Failla ML, Cousins RJ 1978 Zinc accumulation and metabolism in primary cultures of adult rat liver cells. Regulation by glucocorticoids. Biochim Biophys Acta 543:293-304

Fiskin AM, Peterson G, Brady FO 1977 Cd-metallothionein-A test object for dark-field studies of protein structure. Ultramicroscopy 2:389-395

Huang I-Y, Tsunoo H, Kimura M, Nakashima H, Yoshida A 1979 Primary structure of mouse liver metallothionein-I and -II. In: Kägi JHR, Nordberg M (eds) Metallothionein. Birkhäuser, Basel, p 169-179

Kägi JHR, Vallee BL 1960 Metallothionein, a cadmium- and zinc-containing protein from equine renal cortex. J Biol Chem 235:3460-3465

Kägi JHR, Vallee BL 1961 Metallothionein, a cadmium- and zinc-containing protein from equine renal cortex. J Biol Chem 236:2435-2442

Kägi JHR, Himmelhoch SR, Whanger PD, Bethune JL, Vallee BL 1974 Equine hepatic and renal metallothioneins. Purification, molecular weight, amino acid composition, and metal content. J Biol Chem 249:3537-3542

Karin M, Herschman HR 1979 Dexamethasone stimulation of metallothionein synthesis in HeLa cell cultures. Science (Wash DC) 204:176-177

Kissling MM, Kägi JHR 1977 Primary structure of human hepatic metallothionein. FEBS (Fed Eur Biochem Soc) Lett 82:247-250

Kissling MM, Kägi JHR 1979 Amino acid sequence of human hepatic metallothionein. In: Kägi JHR, Nordberg M (eds) Metallothionein. Birkhäuser, Basel, p 145-151

Kojima Y, Kägi JHR 1978 Metallothionein. Trends Biochem Sci 3:90-93

Kojima Y, Berger, C, Vallee BL, Kägi JHR 1976 Amino-acid sequence of equine renal metallothionein-1B. Proc Natl Acad Sci USA 73:3413-3417

Kojima Y, Berger C, Kägi JHR 1979 The amino acid sequence of equine metallothioneins. In: Kägi JHR, Nordberg M (eds) Metallothionein. Birkhäuser, Basel, p 153-161

Lerch K 1979 Amino acid sequence of copper-metallothionein from *Neurospora crassa*. In: Kägi JHR, Nordberg M (eds) Metallothionein. Birkhäuser, Basel, p173-179

Margoshes M, Vallee BL 1957 A cadmium protein from equine kidney cortex. J Am Chem Soc 79:4813-4814

Nelbach ME, Pigiet VP, Gerhart JC, Schachman HK 1972 A role of zinc in the quaternary structure of aspartate transcarbamylase from *Escherichia coli*. Biochemistry 11:315-327

Nordberg M, Kojima Y (eds) 1979 Report from the first international meeting on metallothionein and other low molecular weight metal-binding proteins. In: Kägi JHR, Nordberg M (eds) Metallothionein. Birkhäuser, Basel, p 41-124

Oh SH, Deagen JT, Whanger PD, Weswig PH 1978 Biological function of metallothionein. Am J Physiol 234:282-285

Olafson RW, Sim RG, Kearns A 1979 Physiological and chemical characterization of invertebrate metallothionein-like proteins. In: Kägi JHR, Nordberg M (eds) Metallothionein. Birkhäuser, Basel, p 197-204

Palmer G 1975 Iron-sulphur proteins. In: Boyer PD (ed) The enzymes. Academic Press, New York, vol 12:1-56

Piscator M 1964 Om kadmium i normala människonjurar samt redogörelse för isolering av metallothionein ur lever från kadmiumexponerade kaniner. (On cadmium in normal human kidneys together with a report on the isolation of metallothionein from livers of cadmium-exposed rabbits.) Nord Hyg Tidskr 45:76-82

Prinz R, Weser U 1975 A naturally occurring Cu-thionein in *Saccharomyces cerevisiae*. Hoppe-Seyler's Z Physiol Chem 356:767-776

Richards MP, Cousins RJ 1975 Mammalian zinc homeostasis: requirement for RNA and metallothionein synthesis. Biochem Biophys Res Commun 64:1215-1223

Sobocinski PZ, Canterbury WJ Jr, Mapes CA, Dinterman RE 1978 Involvement of hepatic metallothioneins in hypozincemia associated with bacterial infection. Am J Physiol 234:399-406

Webb M 1979 Functions of hepatic and renal metallothioneins in the control of the metabolism of cadmium and certain other bivalent cations. In: Kägi JHR, Nordberg M (eds) Metallothionein. Birkhäuser, Basel, p 313-320

Discussion

Brierley: You discussed the metallothionein of *Neurospora crassa*. Unlike the other metallothioneins, which have divalent metal ions, you say that this protein contains Cu^{II}. Can you say more about this?

Kägi: In most species, naturally occurring metallothioneins contain besides cadmium and/or zinc only small amounts of copper and exhibit a characteristic ratio of cysteinyl residues to the sum of metal ions that is close to three (Kojima & Kägi

1978). However, in metallothionein from rats receiving injections of copper salts (Bremner & Young 1976a) and in pigs (Bremner & Young 1976b), the proportion of copper exceeds that of the other metals and the ratio of cysteinyl residues to metal decreases to two. In *Neurospora* metallothionein, which contains only copper, the ratio is even lower (<1.2), indicating that compared to zinc and cadmium, copper is bound differently to the thiolate sites of the protein (Ammer et al 1978). This difference and the observation that copper is bound much more firmly than either cadmium or zinc as well as the fact that the complexes are inactive on electron-paramagnetic-resonance (e.p.r.) spectroscopy and show no electronic absorption bands in the visible region favours the suggestion that in these proteins, copper is present as Cu (I) (Rupp & Weser 1978).

Meister: Can you detect metallothioneins in the 'apo' form, i.e. without any metal? What is known about the turnover of the protein?

Kägi: There are no good assays for these proteins yet. A radioimmunoassay method has just been published (Van der Mallie & Garvey 1978) though I don't know if the apo form exists in measurable quantity. The material which we can isolate is always saturated with metal; we never find it partially saturated.

The turnover has been measured by several groups and is rather fast. The zinc-induced protein turns over in 18–20 hours (Feldman & Cousins 1976). The cadmium-induced protein turns over within about three to four days (Chen et al 1975). This difference is attributed to the firmer binding of cadmium to sulphur and, since the turnover probably occurs in lysosomes, it's probable that at the pH prevailing in the lysosomes zinc is dissociated whereas cadmium is not dissociated (Andersen et al 1978).

Whatley: If you isolate the protein in the presence of the metal, since that will have a very different charge from the apoprotein you will never see the apoprotein, even if it is there *in vivo*.

Kägi: That isn't completely true. The first chromatographic step is gel filtration, and if this protein is in its oligomeric form (which one can doubt, of course, since it has such a high cysteine content), it should show up in the same fraction. If one adds cadmium salts to that fraction, one doesn't observe any increase in the typical cadmium trithiolate absorption at 250 nm that would exceed that attributable to the displacement of zinc by cadmium from the binding sites of metallothionein.

Le Gall: Metallothioneins seem to have a lot of similarities to desulphoredoxin from *Desulfovibrio,* which has just been sequenced in our laboratory. It contains iron, and has no free sulphide; there are eight cysteines, and also the (Cys–Cys) sequence, which is very unusual.

Kägi: We have never found iron in a metallothionein, however. Attempts to find a redox function for copper-containing forms of metallothionein have also been inconclusive (Rupp & Weser 1979).

Ziegler: The metallothionein appears to be a very basic protein, since it has an unusually high basic amino acid content.

Kägi: Ignoring contributions from the ionization of the cysteinyl side-chains, there are between 7 and 9 positive and 4 and 5 negative charges on the mammalian polypeptide chains, but the charge picture is totally altered when the metal is bound. The metalloprotein has a negative charge, in fact. As measured by moving boundary electrophoresis at pH 7.0, equine metallothionein carries two negative charges (Bethune et al 1979). The polypeptide chain of the *Neurospora* metallothionein contains an equal amount of basic and acidic groups.

Mudd: What proportion of the total zinc content is bound to this protein in the liver? And what dissociation constants do you find? I am thinking of the problem of getting zinc to the active centres of enzymes.

Kägi: This is difficult to answer. The protein does not occur in all mammalian species to the same extent. Margoshes & Vallee (1957) were fortunate to choose the horse, because horse tissues contain large quantities of metallothionein. In the kidney cortex and in the liver of the horse, more than 50% of the total zinc is bound to metallothionein (Kägi & Vallee 1960, Kägi 1970); in adult human liver, between 5 and 10% (Bühler 1974). In fetal human liver up to 70% of the soluble zinc and a similarly large proportion of the soluble copper is bound to metallothionein (D.D. Ulmer, personal communication 1970). An oxidation product of this protein is probably identical with mitochondrocuprein, a copper protein described by Porter (1974).

The dissociation constant is not known. Since the ligands are displaced from the metal by protons, it is obviously pH dependent. Ignoring hydrolysis equilibria of the metal ions, we estimated for the pH region in which all thiol ligands are fully deprotonated (pH >12) the dissociation constants for zinc and for cadmium to be of the order of 10^{-21} M and 10^{-25} M, respectively (Kägi & Vallee 1961). At neutral pH, the dissociation constant must be much higher. For zinc, it is probably near 10^{-11} M (Bühler et al 1978).

Segel: Is the metal content a function of the metals in the environment? If the primary sequences are the same, and the tertiary structures are similar for the various proteins, why does the *Neurospora* protein, say, contain only copper and no zinc? Or, if you grow the organism in a medium containing both copper and zinc, would you find the protein complexed with both metals?

Kägi: It is probably more complicated than that. Both the concentration of the metal offered and the ambient chemical conditions are important. The intracellular pH for *Neurospora* is probably on the acidic side so that, in spite of the presence of zinc in the medium, the zinc derivative may not be formed. It is also of interest in this context that in various mammals the hepatic metallothioneins have a much larger zinc : cadmium ratio than the corresponding kidney proteins, although their amino acid sequences are identical (Kojima et al 1979). This can only indicate that the chemical conditions in the two organs that regulate the binding of zinc and cadmium ions to the metal-binding sites of metallothionein are considerably different.

Swaisgood: Can you artificially form mixed-metal proteins?

Kägi: Yes, this can be done (Rupp et al 1975).

Segel: Can you measure exchange between a free metal, such as zinc, and a copper protein *in vitro?*

Kägi: There are no reports of such studies on copper metallothionein. However, cadmium ions can easily displace zinc from metallothionein, and cadmium as well as zinc ions are readily replaced by silver ions and mercurials (Kägi & Vallee 1961). A limitation on the velocity of exchange is set by the stronger binding of the metal ions to the protein at higher pH. The pH dependence of the binding constant has, however, not been determined experimentally. It has been claimed that at neutral pH, zinc metallothionein can serve as supplier of the metal in the reconstitution of some zinc enzymes from their apoproteins (Udom & Brady 1979). However, no information on the velocity of dissociation of metals from the various metallothioneins at this pH is available. Judged from the estimated binding constants, the off-velocities might be rather small.

Brierley: What is the location of the protein in *Neurospora?*

Kägi: It is in the cytoplasm. It is also cytoplasmic in mammalian liver and kidney.

Meister: You mentioned a possible metal storage function for metallothioneins. Have you considered that it might serve as a storage form of sulphur, as well as of metal?

Postgate: Is that very likely, in that the protein would have to be completely degraded to make use of the sulphur?

Kägi: In spite of its high cysteine content, the total amount of sulphur present in metallothionein of equine liver and kidney is smaller than that present in glutathione by at least a factor of 100. Thus, I do not think that it is a quantitatively important storage form of sulphur. It may, however, participate importantly in intracellular cysteine turnover. The turnover of the apoprotein is not known but it is probably much faster than the turnovers of the metal-containing forms mentioned earlier.

Roy: Can you make the protein take up other metals than Zn, Cd or Cu?

Kägi: Yes; the protein will bind Co, Ag, Pb and other metals *in vitro*.

Le Gall: Can you try iron, then?

Kägi: You can get derivatives of iron *in vitro* which look very much like some iron-sulphur proteins in their absorption spectra. But they are very unstable.

References

Ammer D, Budry R, Lerch K 1978 Structure and function of a copper-metallothionein from *Neurospora crassa*. Experientia (Basel) 34:905

Andersen RD, Winter WP, Maher JJ, Bernstein IA 1978 Turnover of metallothioneins in rat liver. Biochem J 174:327-338

Bethune JL, Budreau AJ, Kägi JHR, Vallee BL 1979 Determination of the charge of horse kidney metallothionein by free boundary electrophoresis. In: Kägi JHR, Nordberg M (eds) Metallothionein. Birkhäuser, Basel, p 207-210

Bremner I, Young BW 1976a Isolation of (copper, zinc)-thioneins from the livers of copper-injected rats. Biochem J 157:517-520

Bremner I, Young BW 1976b Isolation of (copper, zinc)-thioneins from pig liver. Biochem J 166:631-635

Bühler RHO 1974 Metallothioneine, Zinkproteine aus der menschlichen Leber; Reindarstellung und Charakterisierung. Ph. D. Thesis, University of Zürich

Bühler RHO, Leuthardt F, Kägi JHR 1978 Human hepatic metallothionein: evidence for trimercaptide zinc-binding sites. Fed Proc 37:1814

Chen RW, Whanger PD, Weswig PH 1975 Synthesis and degradation of rat liver metallothionein. Biochem Med 12:95-105

Feldman SL, Cousins RJ 1976 Degradation of hepatic zinc-thionein after parenteral zinc administration. Biochem J 160:583-588

Kägi JHR 1970 Hepatic metallothionein. Abstracts 8th International Congress of Biochemistry, p 130-131

Kägi JHR, Vallee BL 1960 Metallothionein, a cadmium- and zinc-containing protein from equine renal cortex. J Biol Chem 235:3460-3465

Kägi JHR, Vallee BL 1961 Metallothionein, a cadmium- and zinc-containing protein from equine renal cortex. J Biol Chem 236:2435-2442

Kojima Y, Kägi JHR 1978 Metallothionein. Trends Biochem Sci 3:90-93

Kojima Y, Berger C, Kägi JHR 1979 The amino acid sequence of equine metallothioneins. In: Kägi JHR, Nordberg M (eds) Metallothionein. Birkhäuser, Basel, p 153-161

Margoshes M, Vallee BL 1957 A cadmium protein from equine kidney cortex. J Am Chem Soc 79:4813-4814

Porter H 1974 The particulate half-cystine-rich copper protein of newborn liver. Relationship to metallothionein and subcellular localization in non-mitochondrial particles possibly representing heavy lysosomes. Biochem Biophys Res Commun 56:661-668

Rupp H, Voelter W, Weser U 1975 Molecular biology of copper, a circular dichroism study on copper complexes of thionein and penicillamine. Hoppe-Seyler's Z Physiol Chem 356:755-765

Rupp H, Weser U 1978 Circular dichroism of metallothioneins. Biochim Biophys Acta 533:209-226

Rupp H, Weser U 1979 Structural aspects and reduction oxidation reactions of metallothionein. In: Kägi JHR, Nordberg M (eds) Metallothionein. Birkhäuser, Basel, p 231-240

Udom A, Brady FO 1979 A physiological function for metallothionein. Abstracts 11th International Congress of Biochemistry, p 168

Van der Mallie RJ, Garvey JS 1978 Production and study of antibody produced against rat cadmium-thionein. Immunochemistry 15:293-300

Diseases of sulphur metabolism: implications for the methionine–homocysteine cycle, and vitamin responsiveness

S. HARVEY MUDD

National Institute of Mental Health, Bethesda, Md. 20205, USA

Abstract Sixteen inherited human diseases are now recognized, affecting most of the major steps in sulphur metabolism. Studies of patients with three types of homocystinuria have demonstrated unequivocally the major role of cystathionine formation in degradation of homocysteine, and the importance of homocysteine remethylation. Methionine balance studies of normal subjects and of a sarcosine oxidase-deficient subject have shown the predominant role of creatine synthesis in methionine utilization and permitted assessment of the rate of oxidation of the methyl group of methionine. Together, the results demonstrate that once regulatory adjustments have been made the rate of methylneogenesis is nicely controlled so that labile methyl groups are made available in amounts just sufficient to meet the needs for methionine. When excess methionine is ingested the four-carbon moiety is diverted into cystathionine, the methyl group is oxidized via sarcosine and the flow of partially oxidized one-carbon units is diverted away from 5-methyltetrahydrofolate toward CO_2.

Studies of cystathionine synthase-deficient patients demonstrate that the capacity to respond or not to respond to pyridoxine administration is genetically controlled, probably through structural differences in mutant cystathionine synthases. However, the properties of the enzyme crucial in conferring responsiveness have not yet been identified.

The past 15–20 years have witnessed many notable advances in knowledge about inherited human diseases affecting the metabolism of sulphur-containing compounds. To give a perspective on the present status of this knowledge I have listed in Table 1 presently recognized genetic diseases and the loci of the primary abnormalities in-so-far as they are known. This table lists 16 conditions affecting the metabolism of methionine, homocysteine, cystathionine, cysteine, β-mercapto-pyruvate, sulphite, glutathione and γ glutamylcysteine. Genetic conditions are now known which affect most of the steps in the flux of sulphur from methionine to cysteine, several steps in the conversion of cysteine sulphur to inorganic sulphate, and most steps in the conversion of cysteine to glutathione, and in the utilization of the latter compound for the γ-glutamyl cycle and for oxidation–reduction reactions.

Sulphur in biology
(Ciba Foundation Symposium 72) p 239-258

TABLE 1

Known hereditary abnormalities affecting metabolism of sulphur-containing compounds in humans[a]

A. Methionine through cystathionine
1. Methionine absorption: α-hydroxybutyric aciduria
2. ATP:methionine adenosyltransferase (EC 2.5.1.6) : hypermethioninaemia
3. Cystathionine β-synthase (EC 4.2.1.22) : homocystinuria
4. 5,10-Methylenetetrahydrofolate reductase (EC 1.1.1.68) : homocystinuria
5. Cobalamin metabolism : homocystinuria
6. γ-Cystathionase (EC 4.4.1.1) : cystathioninuria

B. Cysteine through sulphite

7. Cysteine transport in intestine and? kidney : cystinuria
8. Cystinosis
9. β-Mercaptopyruvate sulphurtransferase (EC 2.8.1.2) : β-mercaptolactate-cysteine disulphiduria
10. Sulphite oxidase (EC 1.8.3.1) : sulphituria, thiosulphaturia, *S*-sulphocysteinuria

C. γ-Glutamyl cycle

11. γ-Glutamylcysteine synthetase (EC 6.3.2.2) : haemolytic anaemia
12. Glutathione synthetase (EC 6.3.2.3) : 5-oxoprolinuria
13. γ-Glutamyltranspeptidase (EC 2.3.2.2) : glutathionuria

D. Glutathione

14. Glutathione reductase (EC 1.6.4.2) : haemolytic anaemia
15. Glutathione peroxidase (EC 1.11.1.9) : ??haemolytic anaemia
16. Glucose-6-phosphate dehydrogenase (EC 1.1.1.49) : haemolytic anaemia

[a]Most of these conditions have recently been reviewed in detail (see Stanbury et al 1978). The defect in β-mercaptopyruvate sulphurtransferase was demonstrated by Shih et al (1977), and the defect in γ-glutamyltranspeptidase by Schulman et al (1975).

Some of the conditions affecting glutathione metabolism have been covered by Dr Meister (p 135-161). Dr Johnson (p 119-133) has discussed sulphite oxidase deficiency. Even with these deletions, the list in Table 1 is too extensive to attempt even cursory coverage here. I shall therefore limit my discussion to the abnormalities resulting in homocystinuria, with emphasis on three areas: (a) the manner in which study of homocystinuria has helped to refine our concepts of mammalian methionine metabolism, (b) present ideas about the physiological rates of the reactions of the methionine–homocysteine cycle and the regulation of these reactions, and (c) the interrelated problems of genetic heterogeneity and vitamin responsiveness in cystathionine synthase-deficient patients.

HOMOCYSTINURIA AND THE METHIONINE–HOMOCYSTEINE CYCLE

Homocystinuria was first discovered in 1962 when patients at three different locations were independently found to be excreting excessive amounts of this

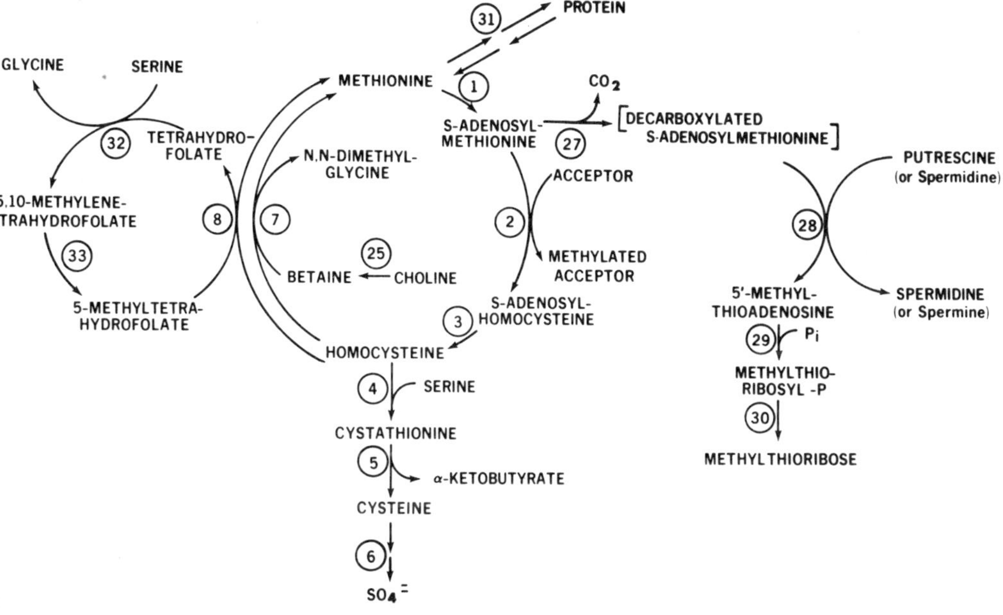

FIG. 1. Metabolic pathways relevant to methionine utilization and synthesis. See Mudd & Poole (1975) for citations to some of the original literature relevant to these reactions. (Reproduced from Mudd & Poole 1975 with the permission of the publishers.)

sulphur amino acid. It was very soon demonstrated that in addition these patients were accumulating and excreting excessive amounts of methionine (see Mudd & Levy 1978). A framework for interpreting these observations is provided by Fig. 1, which suggests that mammalian metabolism of methionine may be conceived of as a cycle in which the four-carbon moiety of methionine is successively converted to (*S*)-*S*-adenosyl-L-methionine (AdoMet), *S*-adenosyl-L-homocysteine (AdoHcy), and homocysteine. During these conversions the methyl group of methionine is transferred, forming one of a variety of methylated products. Homocysteine lies at a metabolic branch point. On the one hand, it may be condensed with serine to form the thioether cystathionine. The sulphur atom is then committed to the pathway leading to cysteine, taurine (not shown), and inorganic sulphate. The latter two compounds may be regarded as the normal end-products for methionine sulphur, accounting for approximately 20% and 80%, respectively, of the methionine intake. On the other hand, homocysteine may itself be methylated to reform methionine. At least two enzymes can catalyse this reaction. One, betaine:homocysteine methyltransferase (EC 2.1.1.5), uses as methyl donor the preformed methyl group of betaine. A second enzyme, 5-methyltetrahydrofolate:homocysteine methyltransferase (EC 2.1.1.13), utilizes the newly formed methyl group of 5-methyltetrahydrofolate. In either case, these methylation reactions complete a cycle in which the methyl of methionine has been transferred to an acceptor, but the homocysteine moiety has been conserved.

An additional important metabolic pathway for methionine (Fig. 1) is its utilization, via AdoMet and decarboxylated AdoMet, in biosynthesis of the polyamines, spermidine and spermine.

Fig. 1 is based on the results of nutritional and enzymological experiments (too numerous to cite individually here) carried out in a variety of laboratories of biochemistry over a period of almost 30 years. By the early 1960s each of the reactions in Fig. 1 (with minor exceptions) had been demonstrated *in vitro*, and the enzymic mechanisms in most were reasonably well understood. The 5-methyltetrahydrofolate-dependent methylation of homocysteine was a possible exception, for at that time the complex details of this reaction, which in mammals is catalysed by a cobalamin (B12 or Cbl)-containing enzyme and (at least *in vitro*) requires a catalytic amount of AdoMet, were just yielding to experimental elucidation. It is a tribute to these pioneering investigations that they provided a framework adequate to permit a correct prediction of the enzyme deficiency in patients accumulating abnormal amounts of both homocysteine and methionine. If it be recalled that each reaction in the methionine–homocysteine cycle (with the exception of AdoHcy hydrolase, EC 3.2.2.9), and each step in transfer of the sulphur atom from homocysteine to cysteine, is essentially irreversible under physiological conditions, one could predict that the patients in question would be deficient in cystathionine synthase activity. This deficiency would lead as a primary event to the accumulation of homocysteine. Enhanced methylation of homocysteine occurring at abnormally high concentrations of this amino acid would then lead to methionine accumulation. This prediction was soon confirmed. By now, at least 50 patients with homocystinuria accompanied by hypermethioninaemia have been subjected to enzyme studies and have invariably been found to have gross deficiencies of cystathionine synthase activity, demonstrable in liver and brain, and in cultured fibroblasts or lymphocytes (see Mudd & Levy 1978).

Returning to our understanding of methionine and homocysteine metabolism in the early 1960s, one may state that in spite of the well-advanced studies of the enzymology of the reactions in question, there was little knowledge about the rates at which these reactions proceed under various physiological conditions. Thus, the relative demands upon AdoMet made by each of the known transmethylation reactions and by the utilization of this sulphonium salt for polyamine synthesis had not been assessed. With respect to the disposition of homocysteine, authorities in this field were not convinced that the quantitative importance of cystathionine formation had been unequivocally demonstrated (V. du Vigneaud, personal communication 1964). Nor was it clear that homocysteine methylation, especially by the tetrahydrofolate pathway involving *de novo* formation of methyl groups, was physiologically significant. Thus, du Vigneaud and Rachele wrote in 1964 that 'even under conditions favoring methyl synthesis it appears that in the presence of an adequate supply of preformed methyl groups provided by the methionine in the diet the neogenesis of methyl groups is comparatively small' (du Vigneaud & Rachele 1965).

It is perhaps in the assessment of the physiological roles in humans of the reactions under discussion that studies of patients with genetic diseases have made their major contribution. Thus, it was soon possible to demonstrate that patients deficient in cystathionine synthase activity could convert the sulphur of a methionine load to inorganic sulphate only at a rate very much below normal. Conversion of cysteine sulphur to inorganic sulphate was normal (Laster et al 1965). These findings established unequivocally the major role of cystathionine synthase in degrading a large intake of methionine. Furthermore, the accumulation of methionine by cystathionine synthase-deficient patients suggested the likely importance of homocysteine methylation, at least under circumstances in which homocysteine accumulated abnormally.

Stronger evidence for the importance of homocysteine methylation was provided during the period 1969–1972 by the discovery of patients who excreted excessive homocystine (i.e. were homocystinuric) but who tended to have lower than normal blood and tissue concentrations of methionine. Investigation of patients with this constellation of abnormalities eventually revealed that they were deficient in the activity of 5-methyltetrahydrofolate:homocysteine methyltransferase, but in each case the deficient activity of this enzyme has been secondary rather than primary. One group of patients have abnormalities in cobalamin metabolism such that they do not convert the vitamin forms of Cbl to methylcobalamin, the cofactor required for 5-methyltetrahydrofolate:homocysteine methyltransferase activity (Mudd et al 1970b). Furthermore, each of the 6–8 such patients recognized to date has been unable to form the second known coenzymically active form of Cbl, i.e. adenosylcobalamin (AdoCbl). AdoCbl is the cofactor for the methylmalonyl-CoA mutase, and these patients excrete abnormally high amounts of methylmalonic acid (Mudd 1974a, Willard et al 1978). Almost surely they are defective in an early step of cobalamin metabolism, but the specific step has not yet been identified. Indeed, since such patients have been divided into two genetic complementation groups by heterokaryon analysis (Willard et al 1978), it is possible that genetic lesions at more than one locus produce this phenotype.

Among the patients with homocystinuria who tend to have lower than normal methionine concentration, a second group has been shown to be deficient in the activity of 5,10-methylenetetrahydrofolate reductase (Mudd et al 1972, Narisawa et al 1977, Rosenblatt et al 1979). Among the approximately 12 patients of this group subjected until now to enzyme studies there has been a range of residual activities of 5,10-methylenetetrahydrofolate reductase, which correlates reasonably well with the proportion of cellular folate present as 5-methyltetrahydrofolate and with the clinical severity of the disease (Rosenblatt et al 1979). The relative lack of 5-methyltetrahydrofolate in these patients explains the secondary defect in homocysteine methylation via the reaction which depends on this cosubstrate. Table 2 presents representative data illustrating that extracts of cultured fibroblasts may be used to demonstrate the enzyme deficiency in each of the three known types of homocystinuria, and showing that each of these three types occurs independently of the others.

Most relevant to this discussion is the fact that patients with abnormally low capacity for the 5-methyltetrahydrofolate-dependent methylation of homocysteine not only accumulate homocysteine, but also tend to become methionine depleted. Indeed, when one such patient was placed on a methionine intake of 655 mg/day (an intake adequate to maintain normal control subjects) he rapidly went into severe negative nitrogen balance and became clinically ill (see Mudd 1974a). These observations clearly emphasize that, at least for human beings, homocysteine methylation via 5-methyltetrahydrofolate is an ongoing reaction, probably of more importance physiologically than had hitherto been appreciated.

TABLE 2

Classification of homocystinuric subjects according to enzyme activities in fibroblast extracts

Subjects	*Enzyme specific activities*		
	Cystathionine synthase	*Methyl-THF methyltransferase*	*Methylene-THF reductase*
		(units/mg)[a]	
Controls	4-65	3.2 - 7.2	1.0-4.6
Homocystinuria–A	0-1	3.6 -7.3	3.3-5.3
Homocystinuria–B	11-24	0.06-0.60	2.0-5.1
Homocystinuria–C	23-27	3.5 -4.1	0.3-0.6

[a]Units are: nmole cystathionine/135 min per mg protein for cystathionine synthase (Uhlendorf et al 1973); nmole methionine/60 min per mg protein for 5-methyltetrahydrofolate: homocysteine methyltransferase, assayed without added MeCbl (Mudd et al 1970b); nmole formaldehyde/60 min per mg protein for 5,10-methylenetetrahydrofolate reductase (Mudd et al 1972). Since none of these enzymes were assayed at V_{max}, only vertical comparisons of the data are warranted.

THE PHYSIOLOGICAL RATE OF METHYLNEOGENESIS, HOMOCYSTEINE RECYCLING AND HOMOCYSTEINE METHYLATION

To gain a better quantitative understanding of the rate of homocysteine methylation in the young adult human, we have recently done several studies of a somewhat different design (Mudd & Poole 1975). Healthy volunteer subjects were maintained on fixed diets containing known quantities of methionine and choline. By adjustment of caloric intake each subject was brought into a metabolic steady state as indicated by lack of weight change, as well as maintenance of nitrogen and sulphur balances within their respective zones of equilibrium. Under these circumstances it may be assumed that daily total methionine utilization was just balanced by the amount of methionine made available through the diet and by methylation of homocysteine. When the creatinine and creatine excretions of these subjects were measured, it turned out that males on normal diets were using more methionine each day to synthesize these compounds (about 16.0 millimole) than they were ingesting labile methyl moieties in the forms of methionine and choline (a combined total of 13.5 millimole). This observation directly proves that these subjects must have been synthesizing at least 2.5 millimole of labile methyl group each day, presumably through the 5-methyltetrahydrofolate-dependent pathway. Of course this is a minimal estimate of the rate of methylneogenesis, which would be increased if additional pathways for methionine utilization were taken into account. Estimates of methionine utilization for the formation of as many as possible additional methylated end-products and for polyamine synthesis were made on the basis of literature values. Together, these processes appear to require approximately 2.0 millimole daily of methionine. Therefore the male subjects on normal diets must have been synthesizing daily at least 4.5 millimole of methionine methyl from more oxidized one-carbon units. For females, whose smaller muscle mass led to smaller creatinine and creatine outputs, the corresponding value for daily methylneogenesis was at least 1.7 millimole.

It was recognized that these estimates, being based in part on literature values for excretion of a variety of methylated compounds, are inexact. However, because creatine synthesis plays such a dominant role, accounting for about 80% of the estimated total methionine utilizations, this inexactness is not likely to be a source of serious error.

It was recognized, furthermore, that these estimates were minimal ones, which might be increased if additional pathways for methionine utilization could be taken into account. In particular, in making these estimates it was not possible to make a quantitative assessment of the rate of oxidation of the methionine methyl group, although a variety of evidence suggested that such a pathway might not be negligible in total methionine utilization.

Oxidation of methionine methyl groups via two major pathways recognized to occur in normal mammals proceeds by way of a common intermediate, *N*-methylglycine (i.e. sarcosine). Oxidation of sarcosine converts its *N*-methyl moiety to a one-carbon unit at the oxidation level of formaldehyde, the rate of formation of which is very difficult to measure since it may be converted rapidly to the β-carbon of serine, or to more oxidized compounds, ending in CO_2. To gain some insight into the rate of methionine methyl oxidation by humans we reasoned it might be worthwhile to study the sarcosine excretion of sarcosinuric patients. The sarcosine excreted by subjects with proven genetic defects in their capacity to oxidize sarcosine might furnish a measure of the rate at which normal humans synthesize sarcosine and subsequently oxidize this compound (Mudd & Poole 1975). So far, genetically determined sarcosinuria has been reported in 11 patients only, very few of whom are old enough to be suitable subjects for metabolic balance studies. We have recently had the opportunity to do extensive studies on one such subject, a 16-year-old female with a proven deficiency of hepatic sarcosine oxidase (EC 1.5.3.1) activity (Mudd et al 1979). For each study this subject was maintained in a metabolic steady state on a fixed diet until a constant rate of sarcosine excretion had been attained. The dietary intake of methionine, choline, choline derivative, or glycine, was then changed and thereafter kept fixed until a new steady-state excretion of sarcosine had been achieved. The most relevant results here are summarized by the statement that sarcosine formation which required methionine utilization appeared to proceed at a rate of about 2 millimole/day on a normal diet. When the intake of labile methyl moieties was decreased below normal, this rate of sarcosine formation requiring transmethylation changed little. When the intake of labile methyl moieties was increased above normal, again the rate of sarcosine formation requiring transmethylation changed little until the total labile methyl group intake came to equal 13–14 milliequivalent/day. At intakes higher than that amount, there was a marked increase in sarcosine formation from methionine, so that the incremental sarcosine formed accounted for almost all of the increased labile methyl group intake.

These data, along with appropriate assumptions, have been used to calculate the methionine balances of this subject on a normal diet, or on diets containing unusually low and high amounts of methionine (Fig. 2). On a normal diet the methionine required for the combination of creatine synthesis, formation of other methylated compounds, polyamine synthesis, and an ongoing sarcosine formation of about 2 millimoles daily, was balanced by the dietary intakes of labile methyl groups as methionine and choline, as well as by methylneogenesis of about 5 millimoles daily. When methionine intake decreased, methylneogenesis increased to balance an essentially unchanged utilization. When methionine intake was raised, and exceeded the amount required for creatine formation, methylation of other

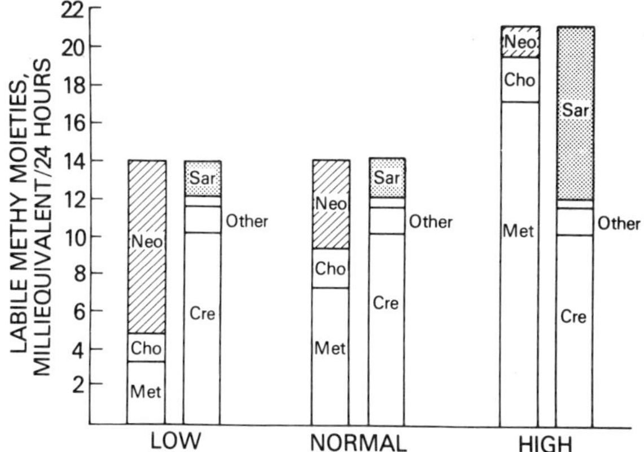

FIG. 2. Methionine balances for a 16-year-old sarcosine oxidase-deficient subject maintained on diets of low, normal, and high labile-methyl group content. A pair of bars is shown for each dietary regimen. In the left bar are indicated methionine intake (Met), choline intake (Cho), and neogenesis of methyl groups (Neo). In the right bar are indicated utilization of methionine for creatine synthesis (Cre), other transmethylation reactions (as specified in Mudd & Poole: 1975) (Other), polyamine synthesis (unlabelled), and sarcosine formation (Sar). See text and Mudd & Poole (1975) for further explanation. (Mudd et al 1979.)

compounds, polyamine synthesis, and ongoing sarcosine formation, methylneogenesis was brought almost completely to a halt and the excess methionine methyl was used largely to form sarcosine. It seems likely that the incremental sarcosine synthesis may have been due largely to methylation of glycine by AdoMet:glycine (EC 2.1.1.20) methyltransferase. The latter reaction would then normally serve as a means for removing 'excess' labile methyl groups, such removal being accomplished by the combined action of glycine methyltransferase and the sarcosine oxidizing system.

From results such as the methionine balances in Fig. 2 one may also calculate the number of times which the average homocysteine moiety cycles from methionine to homocysteine during its passage through the body, and the fraction of homocysteine which is methylated during each cycle. Such calculations for the conditions illustrated in Fig. 2 are shown in Table 3, which includes similar calculations for diets of altered choline content. With the normal diet, in this subject the average homocysteine cycled twice between methionine and homocysteine, and about 50% of the homocysteine was remethylated at each cycle. Increased methionine intake decreased methylneogenesis by decreasing homocysteine recycling. Increased choline intake decreased methylneogenesis, but in this case homocysteine recycling remained at a normal level, now presumably more predominantly due to the action of betaine:homocysteine methyltransferase. When intakes of both methionine and

TABLE 3

Effect of dietary intakes of methionine and choline on methylneogenesis, homocysteine cycling and homocysteine methylation

Dietary intake		*Minimum estimates*[a]		
Methionine	*Choline*	*Methylneogenesis*	*Cycles/Homocysteine*	*Homocysteine methylation*
(millimole/24 h)		*(millimole/24 h)*		*(%)*
7.1	2.3-1.9	4.7-5.5	2.0	49-51
17.2	2.3-1.9	1.6-2.4	1.2	17-19
7.1	7.9-6.4	−0.3-2.7	2.1-2.3	51-56
3.3	1.5-1.2	9.1-9.7	4.3-4.4	77
8.0	1.5-1.2	5.0-5.6	1.8	44-46
3.3	4.7-3.8	5.6-7.4	4.2-4.5	76-78

[a]Minimal estimates derived as explained in the text and by Mudd & Poole (1975). (Mudd et al 1979.)

choline decreased, methylneogenesis rose to 9–10 millimole daily, homocysteine cycled more than four times on the average, and the fraction of homocysteine being remethylated increased to 77%. Methionine or choline added to this labile-methyl-poor diet brought about commensurate decreases in methylneogenesis, again in the case of the methionine increment accompanied by decreased homocysteine recycling; and, in the case of the choline supplement, accompanied by an increased betaine-dependent methylation of homocysteine.

REGULATORY EVENTS SECONDARY TO CHANGES IN METHIONINE CONCENTRATION

These combined results suggest strongly that once regulatory adjustments have been made, the process of methylneogenesis is nicely controlled to provide just the amount of methionine needed for ongoing necessary transmethylation reactions and polyamine synthesis. A number of mechanisms have been reported in the literature which might contribute to such regulation. Table 4 details some of the known effects which follow an increase in methionine tissue concentration. As might be expected, in view of many of the observations already summarized here, regulatory effects following changes in tissue methionine appear to be focused on the branch point at homocysteine. Increased methionine both decreases the conversion of homocysteine back to methionine and increases the conversion of homocysteine to

TABLE 4

Some consequences in mammals of increased methionine tissue concentration

1. Increase in AdoMet concentration (Baldessarini 1975, Eloranta 1977)
(a) Methionine adenosyltransferase is normally not saturated with respect to methionine
(b) Methionine adenosyltransferase activity may increase (approx. 1.5-fold) (Finkelstein & Mudd 1967) or may not (Eloranta 1977)

2. Smaller increase in AdoHcy concentration (Eloranta 1977)

3. Increase of conversion of homocysteine to cystathionine (Laster et al 1965)
(a) Increase of homocysteine concentration would increase flux through cystathionine synthase which is normally not saturated with this substrate
(b) Cystathionine synthase activity increases (Finkelstein & Mudd 1967)
(c) AdoMet activates cystathionine synthase (approx. 2.7-fold at 1000 μM-AdoMet) (Finkelstein et al 1975)
(d) AdoHcy stimulates activity of cystathionine synthase (Finkelstein et al 1974), but effective concentration may be too high to be physiologically meaningful

4. Decrease of conversion of homocysteine to methionine
(a) K_m of 5-methyltetrahydrofolate:homocysteine methyltransferase for homocysteine is lower than is K_m of cystathionine synthase (Finkelstein 1971, Kraus et al 1978)
(b) 5-Methyltetrahydrofolate:homocysteine methyltransferase activity decreases (Kutzbach et al 1967, Finkelstein et al 1971)
(c) 5,10-Methylenetetrahydrofolate reductase is inhibited by AdoMet. Half-maximal inhibition at 2.8 μM-AdoMet (in absence of AdoHcy), and 5 μM-AdoMet (in presence of 2 μM-AdoHcy) (Kutzbach & Stokstad 1971)
(d) AdoMet activation of 5-methyltetrahydrofolate:homocysteine methyltransferase is probably fully effective at normal AdoMet concentration (K_m AdoMet = 0.2 μM) (Burke et al 1971). This activation may serve to maintain methionine concentration during periods of low intake (Krebs et al 1976)

5. Diversion of partially oxidized one-carbon units to CO_2 rather than reduction to methyl moiety
(a) As noted in 4c
(b) 10-Formyltetrahydrofolate dehydrogenase is normally not saturated with 10-formyltetrahydrofolate (Krebs et al 1976)

6. After labile methyl group intake exceeds that required for normal needs, increase of flux of methionine via sarcosine into oxidized one-carbon units (Mudd et al 1979)
(a) AdoMet:glycine methyltransferase is normally not saturated with AdoMet and is relatively insensitive to inhibition by AdoHcy (Kerr 1972)

cystathionine, thereby allowing outflow of the homocysteine moiety. The mechanism whereby an increase in methionine leads to an increase of the activity of cystathionine synthase remains to be clarified, although a change in the catalytic properties of the enzyme brought about by AdoMet could possibly explain the available observations. AdoMet is also an effector in the decreased 5-methyltetrahydrofolate-dependent methylation of homocysteine by virtue of the inhibition of 5,10-methylenetetrahydrofolate reductase by this sulphonium compound. The effector for the decrease in 5-methyltetrahydrofolate:homocysteine methyltransferase is not known. The requirement of this enzyme for activation by AdoMet is puzzling since it means that a methionine product is required for methionine formation. This effect either may be an artifact brought about by conversion of the methyltransferase to an unphysiological form during *in vitro* manipulation, or could be a means of conserving the total amount of methionine *plus* AdoMet during periods of methionine scarcity (Krebs et al 1976).

Table 4 also indicates some additional effects of methionine regulation. Krebs et al (1976) have produced evidence that when methionine is plentiful the normal flux of partially oxidized one-carbon units into 5-methyltetrahydrofolate will be diverted instead toward CO_2, because of inhibition of 5,10-methylenetetrahydrofolate reductase with consequent increased flow into 5,10-methenyltetrahydrofolate, 10-formyltetrahydrofolate, and CO_2. Furthermore, as indicated by our studies of the sarcosinuric patient, when an excess of methionine methyl is available, this group is disposed of via sarcosine and subsequent oxidation. The kinetic properties of AdoMet:glycine methyltransferase are such as to favour this effect during periods of increased AdoMet availability.

The mechanisms responsible for the changes indicated by Table 3 to occur after alterations in choline intake have not been identified. Since an increment in choline is likely to bring about an equimolar increment in methionine due to betaine-dependent homocysteine methylation, the effect of incremental choline intake may well be mediated by increase of methionine.

Finally, one may note that some of the regulatory effects under discussion are likely to come into play in the abnormal circumstances which prevail in homocystinuric patients. For example, cystathionine synthase activity in liver of 5,10-methylenetetrahydrofolate reductase-deficient patients has usually been found to be mildly to profoundly decreased (see Narisawa et al 1977 and other reports cited there). This decrease could be secondary to abnormally low tissue methionine concentrations in such patients, and partially alleviate the tendency to become methionine depleted. As another example, one may speculate that the abnormally high tissue methionine and AdoMet concentrations in cystathionine synthase-deficient patients may enhance any residual activity of the defective enzyme. Since, as will be shown below, even small increases in such residual activities may be beneficial, this effect, considered in isolation, would provide one reason for increasing the ratio of methionine to homocysteine in such patients by treatment with choline, betaine, folic acid, or some combination thereof (Mudd & Levy 1978).

PYRIDOXINE RESPONSIVENESS IN CYSTATHIONINE SYNTHASE-DEFICIENT PATIENTS AND ITS MECHANISM

I would like to turn now to a different aspect of human genetic disease and discuss the question of vitamin responsiveness. Applied to patients with inherited metabolic disorders, vitamin responsiveness usually signifies that an intake of a particular vitamin, often far in excess of the 'normal' requirement, alleviates the biochemical and, in some cases, the clinical manifestations of the disorder. More than 20 such disorders are now recognized, responsive to a wide range of vitamins (Mudd 1974b). It has been known since 1967 that cystathionine synthase deficiency is responsive to vitamin B6, usually given in the form of pyridoxine.HCl. Some, but not all, patients with this abnormality respond to high doses of pyridoxine with decreases of plasma methionine concentration to normal, and with virtual elimination of homocystine from plasma and urine (Mudd & Levy 1978). This aspect of cystathionine synthase deficiency has been studied extensively. The results obtained indicate that there is a great deal of genetic heterogeneity among patients with cystathionine synthase deficiency which may affect in complex ways the interactions between the enzyme and its cofactor. Since these insights are relevant not only to cystathionine synthase deficiency, but also possibly to a number of other genetic abnormalities, the relevant evidence will be summarized in this section. Because some of the evidence has been reviewed recently in detail (Mudd & Levy 1978), the chief conclusions will be presented here in the form of summary statements, with documentation only of newer observations: (a) Cystathionine synthase is a pyridoxal phosphate (pyridoxal-P)-dependent enzyme. (b) Responsiveness to increased pyridoxine intake is not, in general, due to overcoming of a vitamin B6 deficiency. Typically, cystathionine synthase-deficient patients metabolize vitamin B6 normally and have no deficiencies of other pyridoxal-P-requiring enzyme activities. (c) Only about half of all cystathionine synthase-deficient patients are responsive to pyridoxine. Within a given sibship all cystathionine synthase-deficient subjects are either responsive, or all are unresponsive, to pyridoxine.

When sensitive enzyme assays have been done, members of pyridoxine-responsive cystathionine synthase-deficient sibships have invariably been found to have at least traces (which may be as low as 1–2% of normal) of residual cystathionine synthase activity in their livers, whereas members of non-responsive sibships have not had detected activity (Mudd & Levy 1978). The same correlation is supported by more extensive data obtained through study of large series of cultured skin fibroblasts in two laboratories (Table 5). (d) In agreement with these indications of the presence of some residual cystathionine synthase activity in responsive patients, even on normal diets such patients have the capacity to convert

TABLE 5

Clinical responsiveness to pyridoxine and presence of detected cystathionine synthase activity

Enzyme activity[a]	*Clinical*	
	Responsive	*Not responsive*
	(Number of sibships)	
Detected	22	4
Not detected	3	13

[a]Assays were done without added pyridoxal-P. Sensitivity for detection of activity was 0.1–0.25% of mean normal. (Uhlendorf et al 1973, Fowler et al 1978.)

the sulphur of up to 8 millimole of methionine to inorganic sulphate in 24 hours, and synthesize *at least* 2.5–4.0 millimole cysteine each day from methionine. Non-responsive patients lack the latter capability. (e) Even in response to B6, cystathionine synthase-deficient patients do not have normal capabilities to metabolize homocysteine. They accumulate abnormal amounts of homocysteine after methionine loads, and are unable to convert methionine sulphur to sulphate at normal rates. (f) The latter capacity is increased about two-fold during response to pyridoxine to a maximum of about 16 millimole/24 hours. In parallel, hepatic activities of cystathionine synthase in response are increased 1.5–4 fold. (g) Normal humans have capacities to metabolize at least as much as 84 millimole methionine daily via the transsulphuration pathway, far more than the normal daily methionine intake of approximately 10 millimole.

Taking these observations together, the most reasonable explanation for pyridoxine responsiveness seems to be that responsive patients possess small residual activities of cystathionine synthase. Administration of pyridoxine brings about relatively small increases in such residual activities. Although the enhanced cystathionine synthase activities remain far below normal, they nevertheless become capable of metabolizing the normal dietary load of methionine without accumulating homocysteine to the point of spillover homocystinuria. Generally, pyridoxine-unresponsive patients possess less, or no detectable residual cystathionine synthase activity.

However, exceptional patients have occasionally been encountered in each group. No cystathionine synthase activity has been detected in fibroblasts from a few B6-responsive patients, and cystathionine synthase activities have been found in fibroblasts from an occasional non-responsive patient.

This formulation leaves several questions unresolved. First, what is the basis of the responsiveness of patients without detected cystathionine synthase in their cultured fibroblasts? Such responsiveness could be mediated by cystathionine synthase-independent mechanisms, although there is no positive evidence to support this possibility. Alternatively, residual cystathionine synthase activities may yet be detected in the *livers* of such patients, or even in their fibroblasts by use of more sensitive assays. Secondly, what is the mechanism of enhancement of hepatic cystathionine synthase activity in responsive patients during B6 administration, or is more than one mechanism responsible? To the extent that a single intrinsic molecular property of mutant cystathionine synthases is crucial in bestowing the capacity to be enhanced by B6 administration, one might hope to identify this property by demonstrating its existence in all, or most, cystathionine synthase preparations from B6-responsive patients. The argument would be strengthened if the property were not present in control cystathionine synthase, or in the cystathionine synthase activities found in exceptional B6-unresponsive patients (although neither of these conditions will necessarily be so).* So far, experiments designed to test these concepts have generally failed to reveal any property consistently present in mutant cystathionine synthases from responsive patients, yet differing from either control enzyme or mutant enzymes in non-responsive patients. With respect to the possibility that response might be due to the overcoming of an abnormally low affinity of a mutant enzyme for pyridoxal-P by increase in cofactor concentration, it has been found that the cystathionine synthase activities extracted from fibroblasts cultured from members of most B6-responsive sibships are stimulated little more than is control cystathionine synthase by the *in vitro* addition of pyridoxal-P. In fact, more marked stimulation has often been observed with extracts from those exceptional non-responsive patients who possess detectable residual cystathionine synthase activity (Table 6). Furthermore, the dissociation constants of the residual cystathionine synthases extracted from fibroblasts of B6-responsive patients overlap the dissociation constants for control

---

*The possibility that cystathionine synthase activity in normal human liver will be enhanced by doses of B6 similar to those used to elicit responses in cystathionine synthase-deficient patients remains to be explored; we are however deterred by reasonable reluctance to obtain multiple liver specimens from control human subjects. I am not aware of animal data relevant to this point. Cohen et al (1973) found no difference in hepatic cystathionine synthase activities between rats fed diets containing the equivalents of either 150–200 or 1500–2000 mg pyridoxine/70 kg man, but, as they pointed out, even the lower dose might be considered 'excessive' relative to the normal intake of 1–2 mg/70 kg man. Nor do cultured fibroblasts provide a helpful model. Increases of pyridoxine in the growth medium to 100–200 times that usually present did not increase the cystathionine synthase activities in fibroblasts from either control subjects or B6-responsive patients (Fleisher et al 1978).

The residual cystathionine synthases found in exceptional B6-unresponsive patients could possess the same molecular properties as those from responsive patients if the lack of response were due to failure of such patients to form pyridoxal phosphate normally from pyridoxine (Fowler et al 1978).

TABLE 6

Clinical responsiveness to pyridoxine and *in vitro* stimulation by pyridoxal-P

Activity (+ pyridoxal-P) / Activity (no added pyridoxal-P)	Clinical: Responsive (Number of sibships)[a]	Clinical: Not responsive (Number of sibships)[a]
1-2	16	1
2-3	3	1
3-8	3	1
8-13	-	1

[a]Only extracts with detected activity in the absence of added pyridoxal-P are shown. For control cystathionine synthase the ratio of activity when assayed with added pyridoxal-P to that when assayed without added pyridoxal-P is 1–2. (Uhlendorf et al 1973, Fowler et al 1978.)

cystathionine synthase, and for the residual activities in non-responsive patients (Table 7).

Another possibility is that responsiveness is due to stabilization of cystathionine synthase by pyridoxal-P. This is supported by the observation that control holocystathionine synthase is more stable to heat denaturation than is control apocystathionine synthase (Fowler et al 1978). However, heat inactivation studies of mutant enzymes done without prior removal of firmly bound pyridoxal-P (but without *in vitro* addition of this cofactor) have failed to demonstrate any consistent pattern. Thus, cystathionine synthases extracted from tissues of B6-responsive patients have been more (Fowler et al 1978, Fleisher et al 1978), equally (Mudd et al 1970a), or less (Fleisher et al 1978) sensitive than control enzyme. In the two cases studied, similar preparations from non-responsive patients were more sensitive than control cystathionine synthase (Fowler et al 1978). Less extensive inactivation

TABLE 7

Dissociation constants for pyridoxal phosphate[a]

Control cell lines (μM)	Cystathionine synthase deficient: B_6-responsive (μM)	Cystathionine synthase deficient: Not responsive (μM)
23	22	63
30	125	330
	130	

[a]Data taken from Fowler et al (1978).

studies with similar preparations heated in the presence of added pyridoxal-P have likewise been inconclusive. Enzymes from B6-responsive patients have been stabilized (Mudd et al 1970a, Kim & Rosenberg 1974), minimally affected (Fleisher et al 1978) or destabilized (Fleisher et al 1978) by the presence of pyridoxal-P. Taken together these observations emphasize that great genetic diversity exists between cystathionine synthase-deficient patients. It appears that premature conclusions about the mechanisms of vitamin responsiveness will be avoided only if extensive studies are made.

ACKNOWLEDGEMENT

I wish to acknowledge the many scientists whose investigations have been essential to the conclusions reached in this paper. I regret that limitation of space means that many of these contributions are cited only indirectly, through reviews and later citations.

References

Baldessarini RJ 1975 Biological transmethylation involving *S*-adenosylmethionine: development of assay methods and implications for neuropsychiatry. Int Rev Neurobiol 18:41-67

Burke GT, Mangum JH, Brodie JD 1971 Mechanism of mammalian cobalamin-dependent methionine biosynthesis. Biochemistry 10:3079-3085

Cohen PA, Schneidman K, Ginsberg-Fellner F, Sturman JA, Knittle J, Gaull GE 1973 High pyridoxine diet in the rat: possible implications for megavitamin therapy. J Nutr 103:143-151

du Vigneaud V, Rachele JR 1965 The concept of transmethylation in mammalian metabolism and its establishment by isotopic labeling through 'in vivo' experimentation. In: Shapiro SK, Schlenk F (eds) Transmethylation and methionine biosynthesis. University of Chicago Press, Chicago, p 1-20

Eloranta TO 1977 Tissue distribution of *S*-adenosylmethionine and *S*-adenosylhomocysteine in the rat. Biochem J 166:521-529

Finkelstein JD 1971 Methionine metabolism in mammals. In: Carson NAJ, Raine DN (eds) Inherited disorders of sulphur metabolism. Churchill Livingstone, London, p 1-13

Finkelstein JD, Mudd SH 1967 Trans-sulfuration in mammals: the methionine sparing effect of cystine. J Biol Chem 242:873-880

Finkelstein JD, Kyle WE, Harris BJ 1971 Methionine metabolism in mammals. Regulation of homocysteine methyltransferases in rat tissue. Arch Biochem Biophys 146:84-92

Finkelstein JD, Kyle WE, Harris BJ 1974 Methionine metabolism in mammals: regulatory effects of *S*-adenosylhomocysteine. Arch Biochem Biophys 165:774-779

Finkelstein JD, Kyle WE, Martin JJ, Pick A-M 1975 Activation of cystathionine synthase by adenosylmethionine and adenosylethionine. Biochem Biophys Res Commun 66:81-87

Fleisher LD, Longhi RC, Tallan HH, Gaull GE 1978 Cystathionine-β-synthase deficiency: differences in thermostability between normal and abnormal enzyme from cultured human cells. Pediatr Res 12:293-296

Fowler B, Kraus J, Packman S, Rosenberg LE 1978 Homocystinuria. Evidence for three distinct classes of cystathionine-β-synthase mutants in cultured fibroblasts. J Clin Invest 61:645-653

Kerr SJ 1972 Competing methyltransferase systems. J Biol Chem 247:4248-4252

Kim YJ, Rosenberg LE 1974 On the mechanism of pyridoxine responsive homocystinuria. II. Properties of normal and mutant cystathionine-β-synthase from cultured fibroblasts. Proc Natl Acad Sci USA 71:4821-4285

Kraus J, Packman S, Fowler B, Rosenberg LE 1978 Purification and properties of cystathionine β-synthase from human liver. J Biol Chem 253:6523-6528

Krebs HA, Hems R, Tyler B 1976 The regulation of folate and methionine metabolism. Biochem J 158:341-353

Kutzbach C, Stokstad ELR 1971 Mammalian methylenetetrahydrofolate reductase. Partial purification, properties, and inhibition of *S*-adenosylmethionine. Biochim Biophys Acta 250:459-477

Kutzbach C, Galloway E, Stokstad ELR 1967 Influence of vitamin B_{12} and methionine on levels of folic acid compounds and folate enzymes in rat liver. Proc Soc Exp Biol Med 124:801-805

Laster L, Mudd SH, Finkelstein JD, Irreverre F 1965 Homocystinuria due to cystathionine synthase deficiency: the metabolism of L-methionine. J Clin Invest 44:1708-1719

Mudd SH 1974a Homocystinuria and homocysteine metabolism: selected aspects. In: Nyhan WL (ed) Heritable disorders of amino acid metabolism: patterns of clinical expression and genetic variation. Wiley, New York, p 429-451

Mudd SH 1974b Vitamin-responsive genetic disease. J Clin Pathol 27, Suppl (R Coll Pathol) 8:38-47

Mudd SH, Levy HL 1978 Disorders of transsulfuration. In: Stanbury JB et al (eds) The metabolic basis of inherited disease, 4th edn. McGraw-Hill, New York, p 458-503

Mudd SH, Poole JR 1975 Labile methyl balances for normal humans on various dietary regimens. Metab Clin Exp 24:721-735

Mudd SH, Edwards WA, Loeb PM, Brown MS, Laster L 1970a Homocystinuria due to cystathionine synthase deficiency: the effect of pyridoxine. J Clin Invest 49:1762-1773

Mudd SH, Uhlendorf BW, Hinds KR, Levy HL 1970b Deranged B_{12} metabolism: studies of fibroblasts grown in tissue culture. Biochem Med 4:215-239

Mudd SH, Uhlendorf BW, Freeman JM, Finkelstein JD, Shih VE 1972 Homocystinuria associated with decreased methylenetetrahydrofolate reductase activity. Biochem Biophys Res Commun 46:905-912

Mudd SH, Ebert MH, Scriver CR 1979 Metab Clin Exp, in press

Narisawa K, Wada Y, Saito T, Suzuki H, Kudo M, Arakowa T et al 1977 Infantile type of homocystinuria with $N^{5,10}$-methylenetetrahydrofolate reductase defect. Tohoku J Exp Med 121:185-194

Rosenblatt DS, Cooper BA, Lue-Shing S, Wong PWK, Berlow S, Narisawa K et al 1979 Folate distribution in cultured human cells; studies on 5,10-CH_2-H_4PteGlu reductase deficiency. J Clin Invest 63:1019-1025

Schulman JD, Goodman SI, Mace JW, Patrick AD, Tietze F, Butler EJ 1975 Glutathionuria: inborn error of metabolism due to tissue deficiency of gamma-glutamyl transpeptidase. Biochem Biophys Res Commun 65:68-74

Shih VE, Carney MM, Fitzgerald L, Monedjikova V 1977 β-Mercaptopyruvate sulphur transferase deficiency. The enzyme defect in β-mercaptolactate cysteine disulfiduria. Pediatr Res 11:464

Stanbury JB, Wyngaarden JB, Fredrickson DS (eds) 1978 The metabolic basis of inherited disease, 4th edn. McGraw-Hill, New York

Uhlendorf BW, Conerly EB, Mudd SH 1973 Homocystinuria: studies in tissue culture. Pediatr Res 7:645-658

Willard HF, Mellman IS, Rosenberg LE 1978 Genetic complementation among inherited deficiencies of methylmalonyl CoA mutase activity: evidence for a new class of human cobalamin mutant. Am J Hum Genet 30:1-13

Discussion

Kredich: The sarcosine pathway is clearly an escape valve for excess methyl groups. Other mechanisms have been postulated for getting rid of methionine without having to go down the transsulphuration pathway. These involve conversion of methionine sulphur to sulphate without going through homocysteine. Are such pathways well established?

Mudd: Pathways have been described. The problem is to assess how much they contribute, in physiological conditions. In the patient with sarcosine oxidase deficiency sarcosine excretion appears to take off when the methyl group intake exceeds 14 millimoles; sarcosine accounts for almost all the excess methyl groups above that level. This study needs to be repeated in more patients, and we don't have them, but this finding suggests that the other pathways are not accounting for a major portion of the excess methyl groups because they virtually all appear as sarcosine, once the ongoing requirements have been satisfied.

Meister: There are situations when a drug or other compound is given that is extensively methylated, where you get an excessive amount of methylation, probably at the expense of methionine. One can get animals to live on homocysteine in place of methionine. I wonder whether, in the wide range of conditions in which humans exist, you might have a totally different partition of methyl synthesis and utilization? You could be taking in *no* dietary methyl groups.

Mudd: Yes. When one cuts down the intake of labile methyl groups, compensation occurs by forming methyl groups *de novo* through the tetrahydrofolate pathway.

Meister: I recall a human disease in which methyl mercaptan is produced, which would agree with the suggestion of another pathway for loss of excess methyl groups.

Mudd: That was reported in patients with severe liver disease (Challenger & Walshe 1955).

Rassin: Thc pathogenesis of these diseases is intriguing. Enzymic deficiencies of cystathionine synthase, 5-methyltetrahydrofolic acid:homocysteine methyltransferase and 5,-10-methylenetetrahydrofolate reductase are associated with brain dysfunction and other clinical signs and symptoms; yet when the synthesis of *S*-adenosylmethionine is blocked because of a deficiency of methionine adenosyltransferase you don't see obvious clinical symptoms. In two patients the only symptom observed is a muscle myopathy in the older patient. It is not clear what the origin of this myopathy is, because the enzyme that is defective seems to be normal in skeletal muscle. It is probably abnormal in the liver and in blood. High concentrations of methionine were observed in the muscle. It is remarkable that you can have this block with relatively few clinical symptoms, yet with the other enzyme blocks, where there are usually alternative pathways available with the various transmethylation reactions, the effects are severe.

Mudd: Is there really a deficiency of *S*-adenosylmethionine in those patients? Or do they accumulate enough methionine to flux enough of this compound through the partially deficient methionine adenosyltransferase to form a reasonably normal amount of *S*-adenosylmethionine?

I was not previously aware that the older patient with deficiency of methionine adenosyltransferase has developed myopathy. Could this be due to creatine deficiency in the muscles? As I have just shown, creatine formation consumes about 70% of the total adenosylmethionine formed under normal dietary conditions. Since guanidoacetate methyltransferase, the enzyme that forms creatine at the expense of adenosylmethionine, is chiefly, or solely, hepatic, this hypothesis would account for a muscle problem brought about by a defect limited to liver methionine adenosyltransferase. A portion of creatine ingested by human subjects may be taken into the plasma unchanged, as indicated by the urinary excretion of a greater or lesser amount of this compound (Beard 1943). Creatine formed in liver is presumably normally made available to muscle through plasma transport. Therefore creatine administration might be useful in patients with a secondary defect in hepatic creatine biosynthesis.

Rassin: In animal experiments you can change *S*-adenosylmethionine concentrations in the brain by increasing or decreasing the amount of methionine in the diet. Of course we don't know the amounts of *S*-adenosylmethionine or the activity of methionine adenosyltransferase in the brains of these patients. However, it is difficult to understand why we don't see more deleterious effects if *S*-adenosylmethionine metabolism is really disrupted.

References

Beard HH, 1943 Creatine and creatinine metabolism. Chemical Publishing Co., Brooklyn, New York
Challenger F, Walshe JM 1955 Methyl mercaptan in relation to foetor hepaticus. Biochem J 59:372-375

Similarities between cysteinesulphinate transaminase and aspartate aminotransferase

M. RECASENS and P. MANDEL

Centre de Neurochimie, INSERM, 11 rue Humann, 67085 Strasbourg Cedex, France

Abstract A method for the purification of two cysteinesulphinate transaminases, A and B (EC 2.6.1), is described. These enzymes catalyse the conversion of cysteinesulphinic acid to β-sulphinyl pyruvate. The final preparations are homogeneous by polyacrylamide gel electrophoresis, sodium dodecyl sulphate–polyacrylamide gel electrophoresis and isoelectrofocusing. The molecular weight of the subunits is 41 000 for cysteinesulphinate transaminase A and 43 400 for B. Both enzymes are unspecific, as L-aspartate, L-glutamate and L-cysteic acid serve as substrates in addition to L-cysteinesulphinic acid. Cysteinesulphinate transaminase A has a K_m of 9.8 mM for cysteinesulphinic acid and 0.25 mM for aspartic acid, whereas the B enzyme has a K_m of 6.5 mM for cysteinesulphinic acid and 1.4 mM for aspartic acid. The V_{max} values of the A and B enzymes are respectively 7.1 and 6.2 mmol h^{-1} mg^{-1} protein for aspartic acid and 45 and 9.3 mmol $h^{-1}mg^{-1}$ protein for cysteinesulphinic acid. Both enzymes exhibit maximum activity at pH 8.6. A high specific activity is found in optimal conditions for these two transaminases, the pI values being 9.06 and 5.70 for cysteinesulphinate transaminase A and B respectively. These results have been compared with those already obtained for purified aspartate aminotransferase. Similarities in the pathways of taurine and γ-aminobutyric acid (GABA) metabolism are discussed.

Taurine (2-aminoethanesulphonic acid) is a sulphur amino acid present in rather large amount in animal tissues, especially in the heart. Specific physiological roles have been suggested for taurine, in liver (bile salt conjugation, Haslewood & Wootton 1950), in heart (heart rhythm, Huxtable 1976) and in brain. For this latter tissue, a neurotransmitter role (Pasantes-Morales et al 1972), a neuromodulator role (Gruener & Bryant 1975), and a role in brain development (Sturman et al 1977) have been postulated.

In the nervous system, molecules of similar size and structure to that of taurine, generally considered to be neurotransmitters (γ-aminobutyric acid[GABA], glutamate and glycine) are widely distributed, so that it has been difficult to determine a specific role for taurine (for review, see Mandel & Pasantes-Morales 1978).

Sulphur in biology
(Ciba Foundation Symposium 72) p 259-270

These similarities, and the findings that taurine and GABA have some common inhibitory neurotransmitter properties, prompted us to study GABA and taurine metabolism in parallel.

The major pathway of GABA metabolism is known to involve its synthesis from glutamate and its degradation via succinic semialdehyde to succinate (for review, see Baxter 1976). However, the situation concerning the pathways of taurine metabolism is less clear (for review, see Awapara 1976).

Even if the cysteine S–S′ dioxide and inorganic sulphate pathways seem negligible in mammalian brain, two others appear to play major roles: the cysteinesulphinic acid (CSA) pathway and the cysteamine pathway. The synthetic CSA pathway, which we have studied, involves the oxidation of cysteine to cysteinesulphinic acid (CSA), the decarboxylation of CSA to hypotaurine and the oxidation of hypotaurine to taurine. Cysteinesulphinic acid is a key intermediate in this biosynthetic route, so its metabolism is of great interest. CSA is converted into cysteic acid by oxidation; hypotaurine by decarboxylation; and β-sulphinyl pyruvate by transamination.

Transamination and decarboxylation are the two main mechanisms (Yamaguchi et al 1973) for CSA, as for glutamate (Fig. 1).

We have found that some areas of the rat brain have very high cysteinesulphinate transaminase (CSA-T) activities (Recasens et al 1978a), about 1000 times higher than the activity of cysteinesulphinate decarboxylase (CSD). Consequently, this enzyme may play an important role by regulating the level of CSA, the precursor of taurine. However, it is not clear at present whether CSA-T (EC 2.6.1) is different from aspartate aminotransferase (L-aspartate:2-oxoglutarate aminotransferase, EC 2.6.1.1), as purified aspartate aminotransferase is able to transaminate CSA (Ellis & Davies 1961).

A similar problem appears with cysteinesulphinate decarboxylase (CSD; EC 4.1.1.29), as Blindermann et al (1978) have shown, in our laboratory, that highly purified glutamate decarboxylase (GAD; EC 4.1.1.15) uses CSA as substrate. Chatagner et al (1976) have found that purified CSD uses CSA as substrate. Moreover, Agrawal et al (1971) have shown that CSD activity parallels GAD activity.

Consequently, one might speculate that a portion of the enzymic pathway of glutamate metabolism could also be used by CSA.

By following cysteinesulphinate transminase activity we have purified to homogeneity two transaminases from rat brains. Since CSA, aspartate and cysteic acid are substrates for these two transaminases, it seems essential to learn whether these enzymes act either as cysteinesulphinate transaminases, or as aspartate aminotransferases, or as both *in vivo,* and to study the conditions governing these activities.

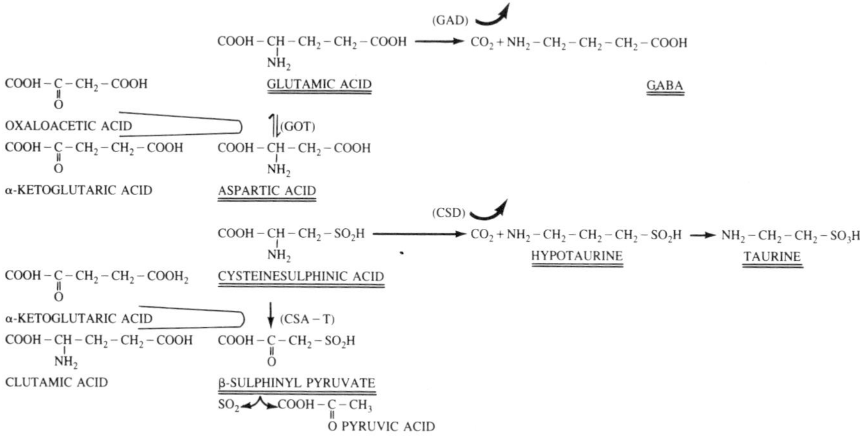

FIG. 1. Metabolic pathway of γ-aminobutyric acid (GABA) and taurine. GAD, glutamate decarboxylase; GOT, aspartate aminotransferase; CSD, cysteinesulphinate decarboxylase; CSA-T, cysteinesulphinate transaminase.

MATERIALS AND METHODS

Wistar albino rats, bred in our laboratory, were used in all experiments. The brains were quickly removed, frozen in liquid nitrogen, and homogenized in 10 mM-Tris-HCl buffer (pH 8.6) containing pyridoxal phosphate, 20 μM and magnesium sulphate, 1 mM.

CSA-T and aspartate aminotransferase activities were determined as previously described (Recasens et al 1978b).

Cysteinesulphinate transaminase A was purified by the following steps, all at 4° C:

Centrifugation of the brain homogenate at 100 000 *g*, 1 h;

Ammonium sulphate precipitation (67–85% saturation of the supernatant);

Chromatography on DEAE-Sephadex, hydroxylapatite, octylsepharose and Sephacryl.

Cysteinesulphinate transaminase B was purified as follows:

Centrifugation of the brain homogenate at 100 000 *g*, 1 h;

Ammonium sulphate precipitation (50–67% saturation) of the supernatant;

Chromatography on DEAE-Sephadex, hydroxylapatite, carboxymethyl-Sephadex and ultrogel.

Protein was determined by the method of Lowry et al (1951). Polyacrylamide gel electrophoresis was performed essentially according to Weber & Osborn (1969). Double immunodiffusion was carried out by the method of Ouchterlony (1953).

RESULTS AND DISUSSION

Typical results of the purification of cysteinesulphinate transaminase A and B are shown in Tables 1 and 2 respectively. Despite the low yield on purification, both enzymes were purified to homogeneity as checked by polyacrylamide gel electrophoresis, sodium dodecyl sulphate–polyacrylamide gel electrophoresis (Fig. 2), and analytical isoelectrofocusing.

Antibodies were raised against the A and B enzymes. Double immunodiffusion revealed that the homogenate, fractions at the different steps of purification, and the purified enzyme all gave single precipitation lines (Fig. 3). The two transaminases that we have obtained are different, as a mixture of the two purified enzymes gave two bands on SDS–polyacrylamide gel electrophoresis, with a molecular weight of 41 000 for CSA-T (A) and 43 400 for CSA-T (B) (Fig. 4). The pI values are also quite different: pI = 9.06 for CSA-T (A) and pI = 5.5 for CSA-T (B).

CSA-T (B) did not react with antibodies against CSA-T (A), and vice versa. The optimum pH (8.6) is the same for both enzymes. However, these two transaminases are not specific; in addition to CSA, aspartic acid, cysteic acid and to a lesser extent homocysteine also serve as substrates. The comparative specific activities toward CSA, aspartic acid and cysteic acid are shown in Table 3. These results provide further evidence for a difference between CSA-T (A) and (B).

The relatively low specificity of these transaminases raises the question of their role. Moreover, as aspartic acid is as good a substrate as CSA, these two transaminases could be identical to the well-known soluble and mitochondrial aspartate aminotransferases.

TABLE 1

Typical purification of rat brain CSA-T (A)

Procedure	*Specific activity (units/mg protein)*	*Yield*	*Purification (fold)*
Homogenate	0.797		
Supernatant	1.55	37.8	2
Ammonium sulphate precipitation (67-85%)	9.95	12.3	12
DEAE-Sephadex	41.1	7.4	51
Hydroxylapatite	138	4.5	172
Octylsepharose	208	2.8	260
Sephacryl (purified enzyme)	227	1.7	285

The enzyme was isolated from 100 Wistar rat brains. A unit of enzyme is defined as the amount of CSA-T which results in the production of 1 μmole of glutamate/min at pH 8.6, at 37 °C, and at a concentration of 2-oxoglutarate of 7.5 mM and cysteinesulphinic acid, 15 mM.

TABLE 2

Typical purification of rat brain CSA-T (B)

Procedure	*Specific activity (units/mg of protein)*	*Yield*	*Purification (fold)*
Homogenate	0.797		
Supernatant	1.55	37.8	2
Purified enzyme	71.4	1.3	90

For unit definition, see Table 1.

TABLE 3

Specific activities of rat brain homogenate, CSA-T (A) and (B) toward cysteinesulphinic acid, aspartic acid and cysteic acid

	Specific activity (units/mg protein, CSA as substrate)	*Specific activity (units/mg protein, Asp as substrate)*	*Specific activity (units/mg protein, CA as substrate)*	*Ratio CSA/Asp*	*Ratio CSA/CA*
Homogenate	0.797	0.522		1.53	
Purified CSA-T (A)	227	86.3	79.0	2.63	2.87
Purified CSA-T (B)	71.4	94.6	3.34	0.75	21.4

Cysteinesulphinic acid (CSA), aspartic acid (Asp) and cysteic acid (CA) concentrations in the incubation mixture are respectively 15 mM, 7.5 M and 15 mM. Specific activities are determined in optimal conditions (pH 8.6).
Unit: μmole of glutamate formed/min. For more details see Table 1.

CSA-T (A) is probably a mitochondrial enzyme, since purified mitochondria react with antibodies against CSA-T (A) whereas the 100 000 ***g*** supernatant obtained after homogenization in sucrose (0.32 M) of rat brains which have not been frozen in liquid nitrogen did not react. Hence we can say that CSA-T (A) is probably partially soluble in the mitochondria. In fact, deep-freezing of whole brains in liquid nitrogen, even if homogenization in sucrose (0.32 M) is done afterwards, is sufficient to solubilize the enzyme.

Kinetic studies showed that CSA-T (A) and (B) follow a typical pattern of a double competitive substrate inhibition involving a ping pong mechanism which is standard for transaminases, so that excess substrate results in a decrease in activity. K_m and V_{max} values are summarized in Table 4. Apparent Michaelis constants are 40 times and five times lower for aspartic acid than for CSA, respectively, for CSA-T (A) and

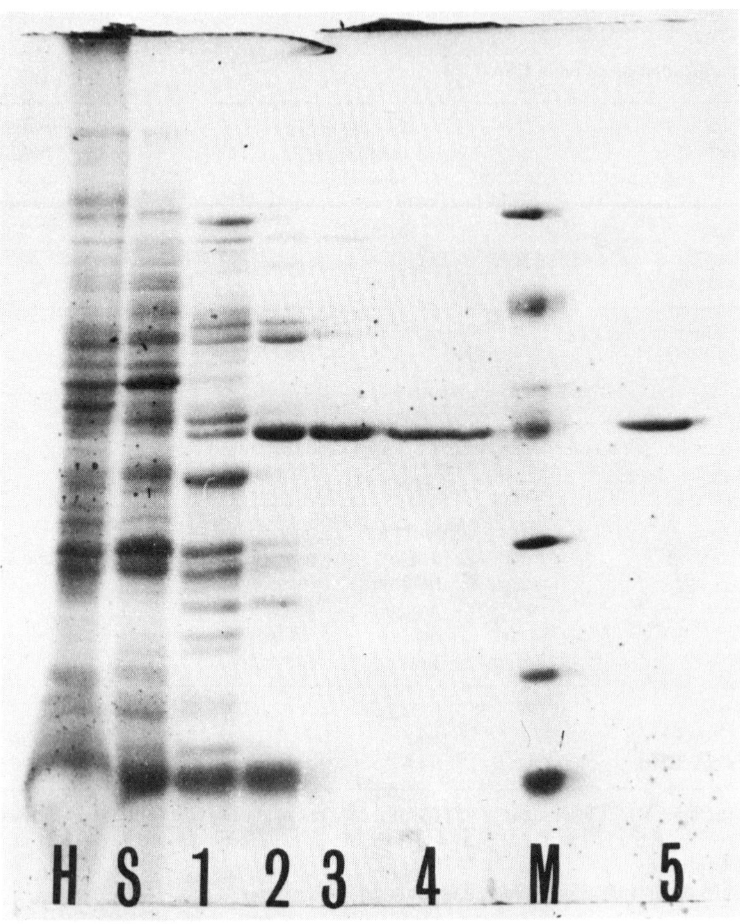

FIG. 2. SDS–polyacrylamide gel electrophoresis (gradient 10–20%) of CSA-T (A). H, homogenate: S, supernatant; 1, fraction after ammonium sulphate precipitation (67–85%); 2, fraction after hydroxylapatite; 3, fraction after DEAE-Sephadex; 4, fraction after octylsepharose; M, proteins of known weight used as references; 5, fraction after Sephacryl S_{200} (purified enzyme).

CSA-T (B), whereas the V_{max} values are respectively six and 15 times lower for aspartic acid than for CSA.

Although we have found lower Michaelis constants for aspartic acid than for CSA, in the optimal conditions used for our assay CSA is transaminated to a greater extent than aspartic acid by CSA-T (A). In order to establish the exact role of the transaminases that we have purified we need to know the cellular microdistribution of the possible substrates CSA, aspartic acid and cysteic acid and of 2-oxoglutarate.

To study the problem of the possible identity of our enzymes with aspartate

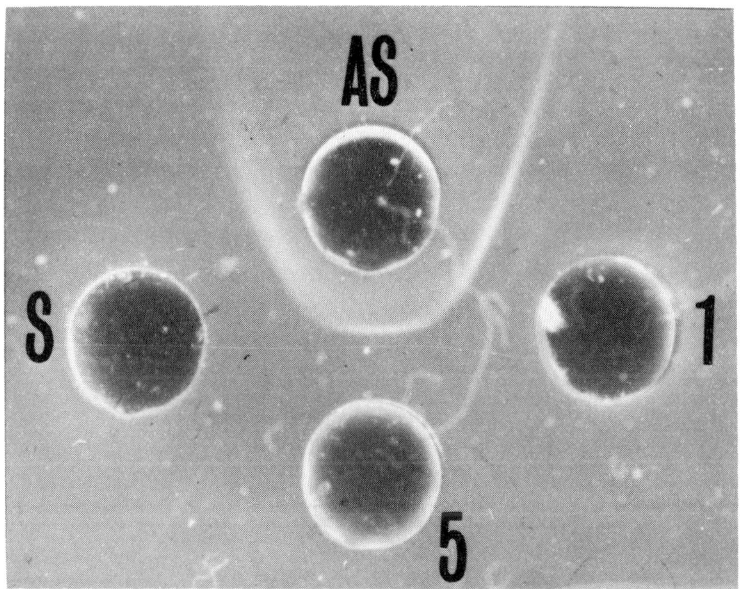

FIG. 3. Ouchterlony test. AS, antiserum to CSA-T (A); S, supernatant; 1, fraction after ammonium sulphate precipitation; 5, purified CSA-T (A).

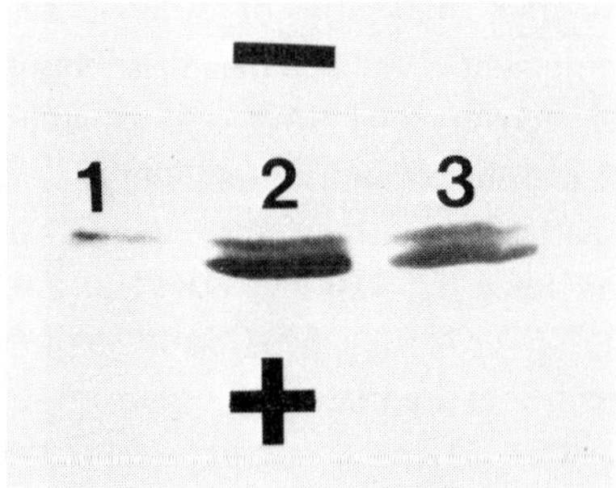

FIG. 4. SDS–polyacrylamide gel electrophoresis (gradient 10–20%). 1, purified CSA-T(B); 2–3, mixture of purified CSA-T (A) and (B).

aminotransferase we have compared the regional and subcellular distributions of CSA-T and aspartate aminotransferase activities in the adult and newborn rat nervous system (Recasens et al 1978a). When the ratios of adult:newborn CSA-T activity and of adult:newborn aspartate aminotransferase activity were studied,

TABLE 4

Apparent Michaelis constants for CSA-T (A) and (B)

Enzymes	*Michaelis constants (mM)*		V_{max} *(mmol* h^{-1} mg^{-1} *protein)*	
	Aspartate	*CSA*	*Aspartate*	*CSA*
CSA-T (A)	0.25	9.8	7.1	45
CSA-T (B)	1.4	6.5	6.2	9.3
Mitochondrial				
Aspartate aminotransferase[a] I	0.78	–	–	–
Aspartate aminotransferase[a] II	0.61	–	–	–
Aspartate aminotransferase[a] III	0.50	–	–	–
Soluble				
Aspartate aminotransferase[a]	6.7	–	–	–

[a] K_m values of isoenzymes are from Magee & Phillips (1971).

marked differences were found throughout the various brain areas examined; for instance, a seven-fold increase for CSA-T and a four-fold increase for aspartate aminotransferase in the pyriform cortex. These results suggest either that CSA-T is different from aspartate aminotransferase or that the ratio of soluble and mitochondrial enzyme changes during development. Subcellular distribution studies indicated significant differences in the recovery of CSA-T and aspartate aminotransferase activities in both supernatant and mitochondrial fractions.

A study of the development of CSA-T and aspartate aminotransferase activities in the retina and olfactory bulbs of two strains of mice (C57/Bl and C3H/He) also revealed different patterns (Gabellec et al 1978). Even though these results suggest the existence of CSA-T enzymes that are different from aspartate aminotransferase, they are not yet sufficient for us to conclude this. Critical comparison of the properties of our purified enzymes and purified aspartate aminotransferase is necessary to establish this.

Although aspartate aminotransferase has been extensively studied, most of the available information on its purification and properties has been derived from studies on liver and heart. A single report by Magee & Phillips (1971) describes the molecular properties of rat brain aspartate aminotransferase isoenzymes. Their enzyme assays were done at pH 7.4 with an appropriate amount of enzyme varying between 0.5 and 10 m units, which corresponds to 2–40 μg of enzyme, substrate concentrations being 7 mM for 2-oxoglutarate and 125 mM for aspartate. In our assay we work at the optimal pH, 8.6, using 0.003 μg of enzyme, 7.5 mM-2-oxoglutarate and 7.5 mM-aspartate. At pH 7.4, CSA-T(A) and aspartate aminotransferase are very low. Soluble aspartate aminotransferase has a lower molecular weight than the mitochondrial enzyme (Magee & Philips 1971), while we found CSA-T (B) to have a higher molecular weight than mitochondrial CSA-T (A). However, the molecular weights are in the same range (about 85 000). As we found for our CSA-T (A) and (B), the soluble aspartate aminotransferase isolated by Magee & Phillips was not immunologically reactive against antisera prepared against the purified mitochondrial aspartate aminotransferase. Some differences appear between our results and those of Magee & Phillips (1971) for Michaelis constants (Table 4). They might be due to the different enzymic assay, the different method of purification, and/or to the different strains of rat used.

At present we are unable to say whether or not our transaminases (A) and (B) are different from the soluble and mitochondrial aspartate aminotransferase.

CSA-T, like cysteinesulphinate decarboxylase, is able to use cysteic acid, glutamate and aspartate as substrates, so we have no evidence for a specific enzyme catabolizing CSA. As the CSA pathway may be considered to be the major one for the biosynthesis of taurine (Bradford et al 1976), it appears strange, to say the least, that two putative inhibitory neurotransmitters, GABA and taurine, should be synthesized by the same enzyme. Rassin & Gaull (1975) have suggested that general metabolic functions of the transsulphuration enzymes are associated with the soluble component, whereas the synthesis of neurotransmitters is associated with the particulate component. This might support the hypothesis that if CSA and glutamate are substrates of the same enzymes, the membrane microenvironment of the enzyme determines the choice of the substrate and consequently compensates for the lack of specificity of the enzyme itself. The relative concentration of substrates near the membrane could also play a significant role.

CONCLUSION

Purified cysteinesulphinate decarboxylase, like glutamate decarboxylase, utilizes cysteic acid, glutamic acid, cysteine and sulphinic acid as substrates. By following cysteinesulphinic acid activity we have purified two transaminases which also use aspartic acid, glutamic acid and cysteic acid as substrates. These findings on enzymes involved in GABA and taurine metabolism raise questions about the existence of specific pathways, regulatory mechanisms and/or cellular compartments for these amino acids.

ACKNOWLEDGEMENT

This work was supported by a grant from the Institut National de la Santé et de la Recherche Médicale (INSERM), contract no. 79.5.354.6.

References

Agrawal HC, Davison AN, Kaczmarek LK 1971 Subcellular distribution of taurine and cysteine sulphinate decarboxylase in developing rat brain. Biochem J 122:759-763

Awapara J 1976 The metabolism of taurine in the animal. In: Huxtable R, Barbeau A (eds) Taurine. Raven Press, New York, p 1-19

Baxter CF 1976 Some recent advances in studies of GABA metabolism and compartmentation. In: Roberts E et al (eds) GABA in nervous system function. Raven Press, New York

Blindermann JM, Maitre M, Ossola L, Mandel P 1978 Purification and some properties of L-glutamate decarboxylase from human brain. Eur J Biochem 86:143-152

Bradford HF, Davison AN, Wheler GHT 1976 Taurine and synaptic transmission. In: Huxtable R, Barbeau A (eds) Taurine. Raven Press, New York, p 303-310

Chatagner F, Lefauconnier JM, Portemer C 1976 On the formation of hypotaurine in various tissues of different species. In: Huxtable R, Barbeau A (eds) Taurine. Raven Press, New York, p 67-72

Ellis RJ, Davies DD 1961 Glutamic-oxaloacetic transaminase of cauliflower. 1. Purification and specificity. Biochem J 78:615-623

Gabellec MM, Recasens M, Mandel P 1978 Developmental pattern of cysteine sulphinic acid transaminase activity in some areas of mouse nervous system. Life Sci 23:1263-1270

Gruener R, Bryant HJ 1975 Excitability modulation by taurine: action on axon membrane permeabilities. J Pharmacol Exp Ther 194:514-521

Haslewood GAD, Wootton V 1950 Comparative studies of 'Bile salts' 1. Preliminary survey. Biochem J 47:584-597

Huxtable R 1976 Metabolism and function of taurine in the heart. In: Huxtable R, Barbeau A (eds) Taurine. Raven Press, New York, p 99-119

Lowry OH, Rosebrough NJ, Farr AL, Randal RJ 1951 Protein measurement with the folin phenol reagent. J Biol Chem 193:265-275

Magee SC, Phillips AT 1971 Molecular properties of the multiple aspartate aminotransferases purified from rat brain. Biochemistry 10:3397-3405

Mandel P, Pasantes-Morales H 1978 Taurine in the nervous system. Neuroscience 3:158-193

Ouchterlony O 1953 Antigen antibody reaction in gels. Acta Pathol Microbiol Scand 32:231-240

Pasantes-Morales H, Klethi J, Urban PF, Mandel P 1972 The physiological role of taurine in retina: uptake and effect on electroretinogram (ERG). Physiol Chem Phys 4:339-348

Rassin DK, Gaull GE 1975 Subcellular distribution of enzymes of transmethylation and transsulphuration in rat brain. J Neurochem 24:969-978

Recasens M, Gabellec MM, Austin L, Mandel P 1978a Regional and subcellular distribution of cysteine sulphinate transaminase in rat nervous system. Biochem Biophys Res Commun 83:449-456

Recasens M, Gabellec MM, Mack G, Mandel P 1978b Comparative study of miscellaneous properties of cysteine sulphinate transaminase and glutamate oxaloacetate transaminase in chick retina homogenate. Neurochem Res 3:27-35

Sturman JA, Rassin DK, Gaull GE 1977 Taurine in development. Life Sci 21:1-22

Weber K, Osborn M 1969 The reliability of molecular weight determinations by dodecyl-sulphate–polyacrylamide gel electrophoresis. J Biol Chem 244:4406-4412

Yamaguchi K, Sakakibara S, Asamizu J, Ueda I 1973 Induction and activation of cysteine oxidase of rat liver. II. The measurement of cysteine metabolism *in vivo* and the activation of *in vivo* activity of cysteine oxidase. Biochim Biophys Acta 297:48-59

Discussion

Meister: Have you done any experiments with β-sulphinyl pyruvate? Have you prepared it, or shown that it is formed? This compound has been in the text-books for about 30 years but so far as I know no one has ever prepared it.

Recasens: I have not done experiments with β-sulphinyl pyruvate, as this is outside the aim of my work.

On the second point I must say no, but again the necessary experiments lie beyond the aim of my work, which deals with the possible regulation of the cysteinesulphinic acid (CSA) level by cysteinesulphinate transaminase. What I can say is that CSA is transaminated to form pyruvic acid, whether by production of β-sulphinyl pyruvate which gives spontaneously and/or rapidly SO_3^{2-} and pyruvic acid, or by direct production of pyruvic acid during the transamination mechanism. (Good evidence for CSA transamination *in vivo* is given by Yamaguchi et al 1973.) When CSA has been transaminated it can no longer serve as a precursor for taurine biosynthesis. This is the main point, and this may be an important regulatory mechanism for taurine metabolism *in vivo*.

Meister: My feeling is that cysteinesulphinate binds to the transaminase to give the usual aldimine Schiff base which then tautomerizes to the ketimine. This eliminates sulphite, so that you end up with the ketimine corresponding to pyruvate. When that is hydrolysed you get pyruvate and the pyridoxamine form of the enzyme, which can transaminate back to the pyridoxal form. There is no evidence for β-sulphinyl pyruvate formation and it may not be an intermediate. This enzyme is a transaminase, but the reaction may be more complicated than usual because a desulphination occurs.

Recasens: This question involves an interesting theoretical point on the mechanism of CSA transamination. We cannot say at present whether CSA, like aspartate, is transaminated by a classical 'ping-pong' bi-bi mechanism or by a more complicated mechanism as you suggest, although kinetic studies have shown classical 'ping pong' patterns of aminotransferase whether CSA or aspartate is used as substrate. However, since β-sulphinyl pyruvate has not been isolated, the exact mechanism of transamination remains unknown.

Rassin: It's important to look at these isolated sulphur pyruvic acid compounds, because there is some evidence that the mercaptopyruvate may be a potent inhibitor of decarboxylation. Allylglycine has been used as an inhibitor of glutamate decarboxylase but the keto-acid derivatives of cysteine and homocysteine are much more potent (Reingold & Orlowski 1978).

Meister: You get a ketimine, whose structure is analogous to oxaloacetate in which oxygen is replaced by nitrogen. Such a compound would be expected to eliminate sulphite rapidly, perhaps on the enzyme, so you may not get a free keto acid (β-sulphinyl pyruvate) at all.

Chatagner: If you start with cysteic acid as substrate you can easily isolate sulphopyruvic acid, so that is good evidence that β-sulphinic pyruvic acid exists, but is immediately and non-enzymically desulphinated.

Meister: With the sulphonic acid corresponding to cysteine (cysteic acid), transamination gives sulphopyruvic acid.

Dodgson: Could one use that well-known 'suicide' substrate for aspartate aminotransferase, developed in our laboratories — serine sulphate? When this sulphate ester is incubated with aspartate aminotransferase an analogous reaction to that described by Dr Meister leads to the elimination of sulphate. The other product is an acrylic acid derivative which inactivates the enzyme by reacting with a lysine residue. You should try this substrate on both your enzymes.

You may even manage to distinguish between the two, because one might attack the 'suicide' substrate and the other might not. Dr Robert John, who has done this work (John & Fassella 1969, John et al 1973), could provide you with the substrate.

Ziegler: Have you done the amino acid analysis on the two forms of the enzyme and, if so, how different are they?

Recasens: No; we have not done that yet.

References

John RA, Fasella P 1969 The reaction of L-serine *O*-sulfate with aspartate aminotransferase. Biochemistry 8:4477-4482

John RA, Bossa F, Barra D, Fasella P 1973 Active-site labelling of cytoplasmic aspartate aminotransferase from pig heart by L-serine *O*-sulphate. Biochem Soc Trans 1:862-864

Reingold DF, Orlowski M 1978 Inhibition of human and mouse brain glutamate decarboxylase by the α-keto analogs of cysteine and homocysteine. Biochem Pharmacol 27:2567-2570

Yamaguchi K, Sakakibara S, Asamizu J, Ueda I 1973 Induction and activation of cysteine oxidase of rat liver. II. The measurement of cysteine metabolism *in vivo* and the activation of *in vivo* activity of cysteine oxidase. Biochim Biophys Acta 297:48-59

Taurine in development and nutrition

GERALD E. GAULL[*,+] and DAVID K. RASSIN[*,‡]

*Department of Human Development and Nutrition, New York State Institute for Basic Research in Mental Retardation, Staten Island, NY 10314 and Departments of Pediatrics[+] and Pharmacology,‡ Mount Sinai School of Medicine of the City University of New York, NY 10029, USA

Abstract Taurine is an amino acid that is widely distributed in the fluids and tissues of man. In mammals, taurine is a major end-product of methionine metabolism. Taurine is found in most mammalian tissues but is only present in trace amounts in many plants.

During fetal development of the brain in man and other mammals taurine is present in high concentrations and declines to lower, adult concentrations during neonatal life. However, during this time there is a net accumulation of taurine when the amount per brain rather than per gram of tissue is calculated. In man, taurine is apparently an essential nutrient, unlike in other animals which have a much greater capacity to synthesize this compound. The human infant, in particular, needs a dietary supply of taurine to synthesize the bile salt taurocholate.

Thus, taurine appears to be an important component of the developing brain and must be supplied to man in the diet.

BIOLOGICAL ASPECTS OF TAURINE

Taurine has been known to be a constituent of living organisms since its identification in ox bile early in the 19th century (Tiedemann & Gmelin 1827). It has been thought to have existed before the origin of life (Choughuley & Lemmon 1966). By the end of the 19th century taurine was known to be widely distributed in animals but to have a rather limited occurrence in plants (Jacobsen & Smith 1968). It is found in some algae and fungi, but its distribution in higher plants is not widespread.

In animals it is present in both invertebrates and vertebrates. It is often found in large quantities in molluscs. Its presence in marine invertebrates, but not in freshwater animals, suggests a role in osmolar regulation (Simpson et al 1959). It is found in large concentrations in many animal tissues, and the concentration of taurine in a given tissue may vary widely from species to species. There are also

Sulphur in biology
(Ciba Foundation Symposium 72) p 271-288

large differences amongst species in capacity to synthesize taurine (Jacobsen & Smith 1968).

Despite its ubiquitous and abundant distribution in animal tissue, taurine takes part in few known biochemical reactions. From a quantitative point of view the only biochemical reaction in which it takes part to any considerable extent is the conjugation with bile acids in the liver. Limited deamination to isethionic acid may take place in some organs (Peck & Awapara 1967), but even some of these reports have not been confirmed (Applegarth et al 1977, Fellman et al 1977, Hoskin & Kordik 1977). It can be converted to inorganic sulphate by intestinal microflora but not by mammalian tissue (Boquet & Fromageot 1965). Taurine is not a constituent of any known protein, although it has been identified as a constituent of some peptides (Reichelt & Kvamme 1973, Reichelt & Edminson 1974). The latter compounds are of little quantitative importance, and their physiological significance is uncertain.

Although there are numerous proposals for alternative pathways for the biosynthesis of taurine (Cf. Jacobsen & Smith 1968), the enzyme immediately responsible for this synthesis in physiologically significant amounts is thought to be cysteinesulphinate decarboxylase (L-cysteine-sulphinate carboxy-lyase, EC 4.1.1.29). There are as large differences amongst species in the *in vitro* capacity of this enzyme as there are in the concentration of taurine itself. For example, as we measure it, the activity of cysteinesulphinate decarboxylase is a thousand-fold higher in adult rat liver than it is in adult human liver (Gaull et al 1977). In addition, adult man has only a limited ability to synthesize taurine from an oral load of L-cysteine *in vivo* (Swan et al 1964). In contrast, the rat has a considerable capacity to convert dietary cysteine to taurine (Sturman & Cohen 1971). A fundamental puzzle in the whole problem of the biological function of taurine is that it is found in such large concentration in various tissues when many of its presently known or proposed functions would require only a fraction of that known to be present.

During the last decade interest in taurine has increased as a result of accumulating evidence that it is involved in neuronal and retinal function. It is one of the major constituents of the free amino acid pools in all mammalian neural tissue, usually exceeded in concentration by glutamate only (Jacobsen & Smith 1968). It is structurally similar to γ-aminobutyrate (GABA), glycine and choline. A number of biochemical and electrophysiological studies of taurine have led to a suggestion that it may have a role as a modulator or transmitter in neural tissue, especially in the visual system and perhaps in the olfactory system. The subcellular distribution of taurine and of cysteinesulphinate decarboxylase suggests that both are enriched in nerve-ending particles and that taurine may be bound to the vesicular protein, at least in some species (Rassin et al 1977). Furthermore, there are specific mechanisms in nerve-ending particles and retina for uptake of taurine (Starr & Voaden 1972,

Schmid et al 1975). More recently it has been shown that taurine undergoes rapid axonal transport with selective enrichment in the nerve-ending particles in the contralateral tectum after injection into the goldfish optic system (Ingoglia et al 1976). The presence of taurine and of cysteinesulphinate decarboxylase in photoreceptor cells, the photoreceptor cell degeneration in taurine-deficient cats, and the changes in the electroretinogram induced by taurine deficiency and taurine repletion in the kitten suggest an important function and possibly a structural role for taurine in the visual system (Pasantes-Morales et al 1976, Macaione et al 1976, Hayes et al 1975). Finally, the high uptake of taurine by olfactory bulb, the preferential sparing of taurine in olfactory bulb of the taurine-deficient cat, and its axonal transport in the olfactory system (Sturman et al 1978, F.L. Margolis, personal communication) suggest that taurine may also have a special function in the olfactory bulb. In addition, taurine has been implicated in the epilepsies, in both man and animal (Van Gelder & Courtois 1972, Van Gelder et al 1975), and it has been found in either increased or decreased amounts in certain neurological disorders (Perry et al 1975, 1977).

In addition, it has been proposed that taurine may be involved in the regulation of excitable membranes. It is thought to stabilize membrane calcium under conditions in which there is a depletion of electrolytes (Huxtable & Bressler 1973). Another interesting feature of taurine is its presence in large amounts in the adrenal gland, where it has been reported to stabilize the adrenal medullary granules against stress-induced disruption (Nakagawa & Kuriyama 1975). Taurine is also found in very large quantities in blood platelets and leucocytes. Its role in these two blood elements is not known, but there has been considerable interest in the uptake systems for taurine by the blood platelets. The platelet uptake systems resemble those found in brain (Ahtee et al 1974). Taurine is also found in large concentrations in the pituitary and pineal glands. In pineal glands in organ culture, taurine increases the rate of production of *N*-acetylserotonin and melatonin from tryptophan by stimulating the activity of the *N*-acetyltransferase. The stimulation is blocked by L-propanolol, suggesting that the mechanism of action is via beta-adrenergic receptors. The rate of release of [^{14}C]taurine from pineal gland was increased by L-noradrenaline and by cyclic nucleotides. It is possible that taurine could modulate the known adrenergic stimulation of the pineal gland by competing with noradrenaline for beta-adrenergic receptors (Wheler et al 1979). These results open up new possibilities for the study of the central nervous system function of taurine.

In summary, taurine is found widely distributed in the animal kingdom but not in the plant kingdom. It is found in large amounts where there are bioelectric membranes and contractile proteins.

TAURINE IN DEVELOPMENT

The most striking generalization that can be made about taurine during development is that in the brain of most mammalian species at birth taurine is the ninhydrin-positive compound present in greatest concentration. In man, it is the free amino acid present in highest concentration during the second trimester of pregnancy. During this period of gestation the taurine concentration of brain is decreasing, and this decrease correlates with the increasing crown–rump length of the fetus. Since the third trimester of human gestation is essentially unavailable for systematic study, we turned to the monkey, whose sulphur metabolism, in many ways, is close to that of man (Gaull et al 1972, Sturman et al 1976). The taurine concentration of occipital lobe in rhesus monkey brain is also much higher in the fetus than it is in the adult (Sturman & Gaull 1975). No correlation between the brain taurine concentration and the gestational age of the monkeys was found; however, later stages of gestation of the monkey were examined than were for man. The concentration of taurine in the occipital lobe of the newborn monkey decreased steadily after birth, reaching concentrations found in the occipital lobe of the adult monkey at about 7–9 months after birth. The concentration of taurine in fetal monkey liver is higher in every case than it is in mature monkey liver. The same relationship is true in human liver; however, after birth the taurine concentration of liver decreases rapidly. Concentrations of taurine equal to those found in adult monkey liver are reached within the second week of postnatal life.

The concentration of taurine in the retina of the rat, the gerbil and the cat increases during development; that of the monkey retina remains constant from late gestation to maturity (cf. Sturman et al 1977b). It is of interest that the rat, the gerbil and the cat are born with their eyes closed, whereas the monkey is born with its eyes open. The activity of cysteinesulphinate decarboxylase in retina of the chick and of the rat increases during development (Pasantes-Morales et al 1976) in a manner similar to that found in rat brain.

When the taurine concentrations of neonatal monkey brain and of neonatal rat and rabbit brain are plotted as a function of the weaning time, they decrease in a linear fashion to reach concentrations at the time of weaning equal to those found in the adult. The high concentration of taurine in fetal and newborn brain and its slow decrease postnatally suggest that it may have a role in brain development, *per se,* in addition to any functional role that it may have in mature brain. The concentrations of many amino acids are greater in fetal tissues and fluids than they are in tissues and fluids of the mature animal of the same species (Agrawal & Himwich 1970) but none shows the four- to five-fold difference found for brain taurine. In fact some amino acids, such as glutamate and cystathionine, are present in even lower concentrations in fetal brain than they are in mature brain (Sturman et al 1976).

The origin of the very high concentration of taurine in developing brain is uncertain and perplexing. The activity of cysteinesulphinate dexcarboxylase is low early in the development of rat brain (Agrawal et al 1971). It is increasing at a time when the concentration of taurine in brain is decreasing. The same situation is found in the fetal rhesus monkey: taurine is present in high concentrations at a time when only minimal cysteinesulphinate decarboxylase is available to catalyse its synthesis. By contrast GABA, a carboxylic acid analogue of taurine, and the analogous enzyme of synthesis, glutamate decarboxylase, have a parallel pattern of development.

Thus GABA is apparently present as a result of the enzyme responsible for its synthesis while the biosynthesis of taurine via cysteinesulphinate decarboxylase seems unlikely to account for the large concentrations found in fetal and newborn brain (Rassin et al 1979b).

An efficient and highly selective transport system for attaining and maintaining high intracellular concentrations of taurine in brain during development has been suggested, and such a mechanism might account for the large concentration in newborn brain (Agrawal et al 1971). Experiments with slices of brain from seven-day-old rats suggest that they can sustain intracellular concentrations of taurine better than those of adult rats in dilute incubation medium. It was striking in these experiments that the change in taurine concentration in the intracellular water of the brain slices was slow, and no steady state was reached during incubation for up to 150 minutes (Oja 1971). Considerable work has been done on high and low affinity uptake systems for taurine in various types of preparations of central nervous system. They may be more relevant to the question of the possible neurotransmitter or neuromodulator role of taurine than they are to development (cf. Rassin & Gaull 1978).

Our own studies have shown that [^{35}S]taurine injected into the pregnant rat enters the brain of the fetus as rapidly as it enters the liver of the fetus, maximum values in fetal liver being reached after 12 hours (Sturman et al 1977a). In contrast, labelled taurine *in vivo* enters the brain of the mother more slowly than it enters the liver, maximum values in maternal brain being reached after 5–7 days (Sturman et al 1977a). The rapid exchange of [^{35}S]taurine observed in fetal brain apparently changes to a much slower exchange after birth. No such differences in exchange rates are apparent between fetal liver and mature liver. Although these experiments cannot be interpreted unequivocally, because it is impossible accurately to determine the precursor specific activity, the decreases in specific activity in the brain of the newborn pup suggest a half-life in brain of approximately 44 days under these conditions as against a half-life of 22 days in the brain of the dam. Experiments in which [^{14}C]taurine was injected intracisternally into mature rats and rates of disappearance of [^{14}C]taurine from brain were measured suggested a considerably shorter half-life for taurine in mature brain (Collins 1974). The explanation for these differences is not immediately apparent, and it would be useful for half-life measurements from both the intraperitoneal and intracerebral routes to be made in the same laboratory using the same methods. Further systematic study of this problem is of considerable importance since the bulk of the present data suggest that the turnover rate of taurine in brain is considerably slower than that of protein.

Further interpretation of the data obtained during the postnatal decrease in taurine concentration indicates that there is actually an increase in the total brain pool of taurine when this pool is calculated as micromoles/brain rather than as micromoles/gram of brain. This postnatal increase in the total brain pool of taurine in the face of a decrease in brain concentration of taurine implies that a dilutional effect occurs during early postnatal development. In the rat this is the time of glial maturation. However, the hyperplasia and hypertrophy of various cellular elements in brain development occur at widely different times relative to birth in different species, whereas brain taurine concentrations decrease after birth in all species. An interesting correlation, however, is that the total amount of taurine per brain reaches a maximum and then levels off at the time (about 20 days) in postnatal growth of the rat brain when there is a sharp decrease in the rate of gain in brain weight and a levelling off of the *in vivo* exchange of [^{35}S]taurine. This is also the time at which neuronal proliferation is decreasing and glial proliferation is increasing (Brizzee et al 1962). We have suggested, therefore, that the accumulation of taurine in developing brain is mainly neuronal and is associated with axonal and dendritic proliferation, i.e. the proliferation of membranes. This is compatible with the increase in the relative amount of taurine contained in the synaptosomal fraction during development both in the rat (Rassin et al 1977) and in the monkey (Rassin et al 1979b). Thus, taurine is important in early development for functions other than nerve transmission. Later in development the synaptosomal, neurotransmitter or neuromodulator pool is protected from the overall decrease in taurine concentration.

An indication of the function of the large soluble pool of taurine in brain may be extrapolated from experiments demonstrating the axonal transport of taurine in the goldfish optic system (Ingoglia et al 1976). Current concepts of axonal flow suggest that the transport of taurine along the axon is probably accomplished in association with the movement of macromolecules. Such a taurine–macromolecule association may explain why such a large pool of soluble taurine of unexplained function is found in the brain. The soluble taurine presumably was released from binding to macromolecular structures during tissue preparation, thereby obscuring the possible important structural and functional association of taurine within the neurons of the brain. It is interesting to speculate that the large soluble pool of taurine is important for the maintenance of the internal structure of the neuron that controls the axonal movement of organelles and macromolecules (Gaull & Rassin 1979). This postulated function is consistent with the apparent lack of catabolism and the long half-life of taurine. One such group of macromolecular structures are the microtubules, and it is an interesting correlate to the above speculation that microtubular protein has a similar pattern of development to that of taurine. In the neonatal mouse brain there is twice as much tubulin (colchicine-binding protein) as in the adult (Koehn & Olsen 1978). Furthermore, tubulin from neonatal brain is more stable than that of the adult preparation.

Finally, it seems that there is a dietary requirement for taurine in the rapidly growing human infant. In studies designed to investigate the effects on preterm and full-term infants of feeding formulas of different quantity and quality of protein and comparing these infants with infants fed human milk, an interesting pattern of plasma and urine amino acid concentrations was observed. The amino acid concentrations in plasma and urine generally were either increased or unchanged when compared with those infants fed human milk. The striking exception was taurine. The concentration of taurine in the urine of preterm infants fed the formulas was lower from the first week of study than that of infants fed human milk. The plasma taurine decreased steadily and by the fourth week of study was significantly lower in the plasma of infants fed formulas than it was in infants fed the human milk (Gaull et al 1977). Preliminary studies in term infants show generally similar results (Rassin et al 1979a). Furthermore, as we measure it, the concentration of taurine in plasma and urine of breast-fed term infants is considerably higher than that of preterm infants fed pooled human milk at a fixed volume. The reason for this difference is not, as yet, clear. In effect, however, the differences in plasma and urine taurine concentration between the infants fed formulas and those fed human milk are greater in term infants than they are in preterm infants.

There is evidence for the relative inability of man to synthesize taurine, and the pattern of deficiency of taurine in the plasma and urine led us to examine the milk of various species, as well as the formulas, in these studies. Taurine is a major constituent of the free amino acid pool of milk in a number of species (Rassin et al 1978). Human milk contains a considerable amount of taurine while bovine milk, from which most of the formulas are prepared, contains only minimal amounts of taurine (Gaull et al 1977). The formulas in this study of preterm infant feeding reflected the lack of taurine in the bovine milk from which they were prepared. The three major commercial infant formulas available in the United States contained only small amounts of taurine, and soy-based formulas contained none at all. For most species studied, taurine is a major constituent of the free amino acid pool of milk and it has a greater concentration early in lactation.

It is likely that the effects of feeding a taurine-deficient diet to a human infant may be moderated by his capacity to conjugate bile acids with glycine, although he conjugates bile acids solely with taurine at birth. Infants fed human milk remain predominantly taurine conjugators of bile acids, whereas those fed taurine-deficient formulas become predominantly glycine conjugators of bile acids (Brueton et al 1978, Watkins et al 1979).

Thus, taurine appears to be important during development of the nervous system and it must be supplied to the human neonate in the diet. The functions of this compound are not fully understood, but it is associated with so many important physiological systems that its properties deserve continued study.

ACKNOWLEDGEMENT

This manuscript could not have been completed without the gracious and capable assistance of Ms Lee Antonucci.

References

Agrawal HC, Himwich WA 1970 Amino acids, proteins and monoamines of developing brain. In: Himwich WA (ed) Developmental neurobiology. Thomas, Springfield, Illinois, p 287-310

Agrawal HC, Davison AN, Kaczmarek LK 1971 Subcellular distribution of taurine and cysteinesulphinate decarboxylase in developing rat brain. Biochem J 122:759-763

Ahtee L, Boullin DJ, Paasonen MK 1974 Transport of taurine by normal human blood platelets. Br J Pharmacol 52:245-251

Applegarth DA, Remtulla M, Williams IH 1977 Does isethionic acid occur in heart and brain tissue? Clin Res 24:646A

Boquet PL, Fromageot P 1965 Sur l'origine de la taurine excrétée par le rat irradié. Bull Inf Sci Tech (Paris) 93:7-18

Brizzee KR, Vogt J, Kharetchko X 1962 Postnatal changes in glia/neuron index with a comparison of methods of cell enumeration in the white rat. Prog Brain Res 4:136-149

Brueton MJ, Berger HM, Brown GA, Ablitt L, Iyangkaran N, Wharton BA 1978 Duodenal bile acid conjugation patterns and dietary sulphur amino acids in the newborn. Gut 19:95-98

Choughuley AWU, Lemmon RM 1966 Production of cysteic acid and taurine and cystamine under primitive earth conditions. Nature (Lond) 210:628

Collins GGS 1974 The rates of synthesis, uptake and disappearance of [^{14}C]taurine in eight areas of the rat central nervous system. Brain Res 76:447-459

Fellman JH, Roth ES, Fujita TS 1977 Is taurine metabolized to isethionic acid in mammalian tissue? Trans Am Soc Neurochem 8:90

Gaull GE, Rassin DK 1980 Taurine and brain development: human and animal correlates. In: Meisami E, Brazier MAB (eds) Neural growth and differentiation. Raven Press, New York (IBRO Symp 5) p 461-477

Gaull GE, Sturman JA, Räihä NCR 1972 Development of mammalian sulfur metabolism: absence of cystathionase in human fetal tissues. Pediatr Res 6:538-547

Gaull GE, Rassin DK, Räihä NCR, Heinonen K 1977 Milk protein quantity and quality in low-birth-weight infants. 3. Effects on sulfur amino acids in plasma and urine. J Pediatr 90:348-355

Hayes KC, Carey RE, Schmidt SY 1975 Retinal degeneration associated with taurine deficiency in the cat. Science (Wash DC) 188:949-951

Hoskin FCG, Kordik ER 1977 Hydrogen sulphide as a precursor for the synthesis of isethionate in the squid giant axon. Arch Biochem Biophys 180:583-586
Huxtable R, Bressler R 1973 Effect of taurine on a muscle intracellular membrane. Biochim Biophys Acta 323:573-583
Ingoglia NA, Sturman JA, Lindquist TD, Gaull GE 1976 Axonal migration of taurine in the goldfish visual system. Brain Res 115:535-539
Jacobsen JG, Smith LH Jr 1968 Biochemistry and physiology of taurine and taurine derivatives. Physiol Rev 48:424-511
Koehn JA, Olsen RW 1978 Microtubule assembly in developing mammalian brain. Arch Biochem Biophys 186:114-120
Macaione S, Tucci G, DeLuca G, DiGiorgio RM 1976 Subcellular distribution of taurine and cysteine sulphinate decarboxylase activity in ox retina. J Neurochem 27:1411-1415
Nakagawa K, Kuriyama K 1975 Effect of taurine on alteration in adrenal functions induced by stress. Jpn J Pharmacol 25:737-746
Oja SS 1971 Exchange of taurine in brain slices of adult and 7-day-old rats. J Neurochem 18:1847-1852
Pasantes-Morales H, Lopez-Colombe AM, Salceda R, Mandel P 1976 Cysteine sulphinate decarboxylase in chick and rat retina during development. J Neurochem 27:1103-1106
Peck EJ Jr, Awapara J 1967 Formation of taurine and isethionic acid in rat brain. Biochim Biophys Acta 141:499-506
Perry TL, Bratty PJA, Hansen S, Kennedy J, Urquhart N, Dolman CL 1975 Hereditary mental depression and Parkinsonism with taurine deficiency. Arch Neurol 32:108-113
Perry TL, Currier RD, Hansen S, MacLean J 1977 Aspartate-taurine imbalance in dominantly inherited olivopontocerebellar atrophy. Neurology 27:257-261
Rassin DK, Gaull GE 1978 Taurine and other sulfur-containing amino acids: their function in the central nervous system. In: Fonnum F (ed) Amino acids as chemical transmitters. Plenum Publishing, New York, p 571-597
Rassin DK, Sturman JA, Gaull GE 1977 Taurine in developing rat brain: Subcellular distribution and association with synaptic vesicles of [^{35}S]taurine in maternal, fetal and neonatal rat brain. J Neurochem 28:41-50
Rassin DK, Sturman JA, Gaull GE 1978 Taurine and other free amino acids in milk of man and other mammals. Early Hum Dev 2:1-13
Rassin DK, Järvenpää A-L, Räihä NCR, Gaull GE 1979a Breast feeding versus formula feeding in full-term infants: effects on taurine and cholesterol. Pediatr Res 13:406
Rassin DK, Sturman JA, Gaull GE 1979b Source of taurine and GABA in the developing rhesus monkey. Trans Am Soc Neurochem 10:144
Reichelt KL, Edminson PD 1974 Biogenic amine specificity of cortical peptide synthesis in monkey brain. FEBS (Fed Eur Biochem Soc) Lett 47:185-189
Reichelt KL, Kvamme E 1973 Histamine-dependent formation of *N*-acetylaspartyl peptides in mouse brain. J Neurochem 21:849-859
Schmid R, Siegart W, Karobath M 1975 Taurine uptake in synaptosomal fractions of rat cerebral cortex. J Neurochem 25:5-9
Simpson J, Allen K, Awapara J 1959 Free amino acids in some aquatic invertebrates. Biol Bull (Woods Hole) 117:371-381
Starr MS, Voaden MJ 1972 The uptake, metabolism and release of ^{14}C-taurine by rat retina *in vitro*. Vision Res 12:1261-1269
Sturman JA, Cohen PA 1971 Cystine metabolism in vitamin B_6 deficiency: evidence of multiple taurine pools. Biochem Med 5:245-268
Sturman JA, Gaull GE 1975 Taurine in the brain and liver of the developing human and monkey. J Neurochem 25:831-835
Sturman JA, Gaull GE, Niemann WH 1976 Cystathionine synthesis and degradation in brain, liver and kidney of the developing monkey. J Neurochem 26:457-463
Sturman JA, Rassin DK, Gaull GE 1977a Taurine in developing rat brain: maternal-fetal transfer of [^{35}S]taurine and its fate in the neonate. J Neurochem 28:31-39

Sturman JA, Rassin DK, Gaull GE 1977b Taurine in development. Life Sci 21:1-22

Sturman JA, Rassin DK, Hayes KC, Gaull GE 1978 Taurine deficiency in the kitten: exchange and turnover of [^{35}S]taurine in brain, retina, and other tissues. J Nutr 108:1462-1476

Swan P, Wentworth J, Linksweiler H 1964 Vitamin B_6 depletion in man: urinary taurine and sulfate excretion and nitrogen balance. J Nutr 84:220-228

Tiedemann F, Gmelin L 1827 Einige neue Bestandtheile der Galle des Ochsen. Ann Physik Chem 9:326-337

Van Gelder NM, Courtois A 1972 Close correlation between changing content of specific amino acids in epileptogenic cortex of cats and severity of epilepsy. Brain Res 43:477-484

Van Gelder NM, Sherwin AL, Sacks C, Andermann F 1975 Biochemical observations following administration of taurine to patients with epilepsy. Brain Res 94:297-306

Watkins JB, Järvenpää AL, Räihä N, Szczepanik Van-Leween P, Klein PD, Rassin DK, Gaull G 1979 Regulation of bile acid pool size: role of taurine conjugates. Pediatr Res 13:410 (abstr)

Wheler GHT, Weller JL, Klein C 1979 Taurine: stimulation of pineal *N*-acetyltransferase activity and melatonin production via a beta-adrenergic mechanism. Brain Res 166:65-75

Discussion

Ziegler: Are there any specific inhibitors for cysteine dioxygenase or cysteine-sulphinate decarboxylase?

Rassin: Not so far as I know. We should like to have some specific inhibitors of taurine synthesis. The compounds now being experimented with may be agonists rather than antagonists.

Chatagner: Yamaguchi and his colleagues recently reported that in rats treated with glucagon there is a marked decrease in the level of hepatic cysteine dioxygenase (Hosokawa et al 1978). We looked to see whether there is a change in cysteinesulphinate decarboxylase and in the taurine level, but we found that, under similar conditions, there is no change at all (C. Portemer & F. Chatagner, unpublished results).

Meister: I think Dr Ziegler is asking whether there is another way of making taurine that we don't know about. I was interested in another point you mentioned. You noted that isethionic acid may *not* be formed from taurine, as had been thought. Did you also say that aspartate is decarboxylated in one of your systems? I am not aware that aspartate decarboxylase occurs in mammalian tissues.

Rassin: When we were setting up the rat liver decarboxylase assay we compared aspartate, glutamate and cysteinesulphinate as substrates. We could measure no glutamate decarboxylation in liver. There was some apparent aspartate decarboxylation, as measured by CO_2 release, about 10% of the CO_2 released from cysteinesulphinate.

Meister: This may be explained by transamination of aspartate followed by decarboxylation of the resultant oxaloacetate. Unless you can find β-alanine, I would be doubtful about genuine decarboxylation of aspartate.

Rassin: I agree.

Meister: We have noted, in the course of analysing tissue culture media of human fibroblasts, a gradual increase in taurine content. This seems to be a consequence of cell growth, and suggests that these cells are making taurine. Is this a system in which one could study taurine formation?

Rassin: We have examined changes in the amino acid content of media from cultured cells while looking for ways to study inborn errors of amino acid metabolism. We didn't see much in the way of changes in the amino acids, other than a decrease in glutamine, which is probably a result of hydrolysis during incubation at 37°C. There were no changes in taurine concentration.

Kredich: Have you done taurine balance studies, comparing intake and output, in adults? Is there evidence that there is little taurine synthesis in the adult human? And secondly, is there taurine in a vegan diet?

Rassin: I don't know whether vegans, eating no animal products, get any taurine in their diet. The dogma is that such a complete vegetarian diet should not provide much taurine. I have only ever measured taurine in the plasma of one vegetarian, not a vegan, and the taurine concentration was normal. We haven't done any balance studies in adults; all our studies have been in infants.

Roy: There have been reports of taurine being formed from 3′-phosphoadenosine 5′-phosphosulphate (PAPS) by an unknown mechanism. Is this true? I was approached by physiologists complaining that their cats were going blind, presumably because of taurine deficiency, yet there are reports that cat liver can synthesize taurine from PAPS and so presumably from SO_4^{2-}.

Rassin: Several reports suggest that taurine can be synthesized from sulphate via a mechanism involving PAPS (Martin et al 1974). Certainly in the cat, giving sulphate doesn't help the blindness, nor is radioactive sulphate incorporated into taurine in whole-animal experiments (Knopf et al 1978).

Chatagner: Some years ago it was reported that incorporation of sulphate into taurine was seen in the chick embryo, but only in this tissue (Machlin et al 1955).

Postgate: Is there any clinical expression of taurine deficiency which results from formula feeding as opposed to breast feeding?

Rassin: There are no clinical symptoms specifically associated with taurine deficiency in man at the moment. The situation is complicated, however, because any clinical effects associated with formula feeding cannot be specifically related to taurine, as increases in other components such as tyrosine and phenylalanine in plasma have been observed in some formula-fed infants. These infants, particularly those fed high protein casein-predominant formulas, go through prolonged periods of acidosis. It isn't yet possible to separate the effects of the individual biochemical changes observed. The only direct effect of taurine deficiency in infants is the change in the taurocholate/glycocholate ratio in the duodenal bile.

Rose: We must be the species most guilty of feeding our young on the milk of other animals and formula preparations, and yet we seem to get away with it. Is it possible that so far as fundamental central nervous system functions are concerned, the taurine depletion resulting from feeding infants on cow's milk and so on does not matter too much in the long run, because the cells which require taurine for such processes might be able to 'hold on' to enough of it to last them until the diet becomes more varied?

Rassin: That may or may not be true. There are other biochemical changes associated with formula feeding which indicate that that is an inappropriate conclusion. These formula-fed infants also become cholesterol deficient, as measured by their serum cholesterol content, and the combination of cholesterol deficiency with taurine deficiency and the changes in taurocholate/glycocholate ratios make one wonder whether their lipid uptake and synthesis is adequate. The bulk of the evidence suggests that we ought to be considering ways to improve formulas for when human milk is not available to feed infants. The biochemical changes we have observed suggest that formula feeding is biochemically inappropriate, even if infant blindness or seizures are not observed in formula-fed infants. We have more sophisticated ways of measuring behaviour and brain function now than formerly and when we re-examine formula- and human milk-fed infants at seven or eight years old, we shall perhaps detect differences between them that we were unable to measure before.

Dodgson: Sulphur metabolism is rather strange in cats, I believe, and perhaps ought to be investigated further. The cat makes little use of glucuronide conjugation in metabolizing drugs and xenobiotics but it does conjugate phenols with sulphate. Professor R.T. Williams and his colleagues once investigated the metabolism of xenobiotics in lions (French et al 1974). He found the same thing there. Another report (Rambaut & Miller 1965) suggested that the cat has no requirement for exogenous sulphur-containing amino acids.

Meister: The cat and related mammals make a sulphur-containing amino acid, felinine ($HOOC.CH(NH_2).\ CH_2.S.C(CH_3)_2.CH_2.CH_2OH$), which is not found in other species (Westall 1953, Trippett 1957).

Roy: There have also been reports (Rambaut & Miller 1965) that the cat doesn't need methionine.

Idle: We have been studying another aspect of taurine metabolism which also concerns drug metabolism.

In drug metabolism in general there are three reactions which involve sulphur: firstly, the sulphation of phenols and alcohols; secondly, glutathione conjugation, which ultimately produces mercapturic acids that are of great interest now because certain carcinogens and other toxic agents may have an important interaction with glutathione; and finally the conjugation of aromatic acids with taurine. Aromatic acids arise in three different ways *in vivo*. They can be exogenous, in the diet; for example benzoic acid. They can come from intermediary metabolism, like phenylacetic acid and some of its derivatives; or they may be given as drugs.

Aromatic acids are metabolized by two pathways in mammals, firstly by conjugation with sugars, chiefly glucuronic acid, and secondly by conjugation with the amino acids glycine, glutamine and taurine. The way the 'decisions' are made by the animal in selecting a conjugation pathway seems to depend upon two factors. The structure of the aromatic acid determines largely whether the compound is conjugated with glucuronic acid or an amino acid, and the species determines which amino acid is used for conjugation. For example, glutamine conjugation seems to occur largely in primates and taurine conjugation in carnivorous species. The conjugates are readily excreted and are eliminated in the urine.

Taking glycine conjugation of benzoic acid as an example, an acyl adenylate is formed which is transferred to CoA by a mitochondrial enzyme system forming a benzoyl-CoA derivative. A second enzyme system also in the mitochondria transfers glycine onto benzoic acid, forming hippuric acid (Schachter & Taggart 1954).

There is an analogous system with glutamine for which Dr Meister demonstrated the enzymology (Moldave & Meister 1957). Very little is known about the enzymic mechanism responsible for taurine conjugation. One can only assume that since glycine and glutamine conjugation is handled by the CoA system in mitochondria, taurine probably is as well. However, it must be pointed out that this is not the same enzyme system which makes taurocholate or glycocholate, which is essentially a non-mitochondrial one.

Of the various mammalian species that have been examined, in the rat, aromatic acids largely form glycine conjugates, with a few (like diphenylacetic acid) forming glucuronic acid conjugates for reasons of chemical structure which make them incompatible with the glycine-conjugating enzymes. There is relatively little if any taurine conjugation in the rat (Idle et al 1978).

In the ferret, a mustelid carnivore, which conjugates bile acids by forming taurocholate, the major metabolite for the elimination of aromatic acids is the taurine conjugate (Idle et al 1978). The rest is made up largely with glycine. Here, the species determines which amino acid is used for conjugation. I am sure this is for dietary reasons. The ferret, like the cat and dog, is a carnivore with high levels of taurine in the liver (10 μmol/g wet wt.) compared to the rat, guinea-pig or rabbit, which have a maximum of 1 μmol taurine per g wet liver.

In the cat, taurine conjugation is less important than in the ferret but with some aromatic acids it can be the predominant metabolite. The dog is similar to the cat.

To summarize, in species like cat, dog and ferret taurine conjugation of aromatic acids represents an important pathway in the elimination of these substances from the body, but we know nothing about the enzymology of this process and, moreover, we don't know what is the endogenous substrate for the aromatic acid taurine-conjugating enzyme.

Meister: Many people look upon reactions of this sort as being significant in relation to administered compounds; for example, administered phenylacetate and benzoate are excreted as the corresponding derivatives of glutamine and glycine. Phenylacetylglutamine, however, arises in the normal metabolism of phenylalanine, so reactions of this type need not invariably be viewed as a consequence of giving a foreign compound. Dr Gregory Thompson in our laboratory has been looking at the formation of hippurate and has obtained evidence indicating that benzoate is formed in the course of normal metabolism. Dr Thompson's studies were done in germ-free rats that were fed a chemically defined diet that did not contain benzoate; his animals excreted hippurate in their urine. This supports similar conclusions based on studies on conventionally reared rats and on humans (Stein et al 1954, Grumer 1961, Olson et al 1963, Napier et al 1964, Olson 1966).

Idle: It is important to try to understand why these reactions occur in the first place, because adding an amino acid to what is already a polar grouping isn't any particular advantage, except that there is some evidence that the resulting conjugates are more readily secreted by the renal tubules. The conjugates are not much more acid and not much more polar, except in the case of taurine conjugates, where you lower the pK to around one. However, the species which produce predominantly taurine conjugates eliminate the drug from the body more *slowly* than those species that make glucuronides, so it doesn't seem to be conferring an advantage on the animal. Substances like hippurate may have a more subtle biochemical role than previously has been understood.

Dodgson: It might be worthwhile reminding ourselves, since we have talked about bile salts, that when you move down in the evolutionary scale to the amphibia and the fishes the amino acids glycine and taurine are replaced by sulphate. You have a steroid nucleus and an alkyl side-chain with either a primary or secondary hydroxyl group which is sulphated via the agency of a sulphotransferase enzyme. In the same organisms a second sulphotransferase exists that transfers sulphate to simple primary or secondary alcohols such as pentanol or 3-hexanol. During the course of evolution the capacity to produce bile alcohols and their corresponding sulphates has been lost; mammals synthesize bile acids and conjugate them with glycine or taurine. However, the capacity to sulphate simple primary and secondary alcohols still persists in mammals, presumably indicating that the system has a function — as yet unknown — to play in mammalian metabolism.

Rassin: The spiny lobster has a receptor on its antennulae specific for taurine and β-alanine which does not respond to α-amino acids. Seawater has a background concentration of α-amino acids which does not cause electrical activity in these receptors. However, the spiny lobster's food of molluscs almost always contains taurine as one of the four or five most concentrated amino acids. Fuzzessery et al (1978) have suggested that the spiny lobster uses its taurine receptors to detect its food supply.

Dodgson: What is the basic retinal lesion in the taurine-deficient cat?

Rassin: There is a structural change in the photoreceptor cells. The membranes deteriorate. The tapetum (which reflects light) also undergoes structural changes, although it is not known what effect these have on vision.

Dodgson: But chemically speaking we don't really know what is going on?

Rassin: We don't know whether taurine is a neurotransmitter or something else.

Idle: Taurine is an unusual compound for a neurotransmitter. Unlike other known neurotransmitters it doesn't have a metabolic deactivation mechanism. The other central and peripheral neurotransmitters are taken up by storage mechanisms but are also metabolized. If taurine is being synthesized at a relatively constant rate, how does it leave the brain? It's not building up; surely it can't diffuse across the blood–brain barrier, or it would get in from the outside that way. There must be some metabolic deactivation.

Rassin: Not necessarily. Classical neurochemistry began investigations of neurotransmitters with acetylcholine, and acetylcholinesterase was the catalyst for the very rapid breakdown of the transmitter. Most of the evidence for other neurotransmitters is that metabolic breakdown is a minor component in their inactivation. Catechol-*O*-methyltransferase cannot inactivate noradrenaline (norepinephrine) and dopamine rapidly enough to account for the termination of their action. There are no known inactivating enzymes for enkephalins or serotonin. Most transmitters are inactivated by a high affinity re-uptake mechanism at the nerve ending. The unanswered question is what happens to taurine after it has been taken up. Taurine may move in and out of the brain, although one cannot alter the brain concentrations by peripheral loads of the compound.

Meister: Dr Idle has a point, however, because none of the 'putative' neurotransmitters is a dietary essential. It seems rather careless of evolution to have developed an important organ such as the nervous system by using a casual compound that you have to get from the environment and can't make yourself.

Rassin: But transmitters are perhaps all under some dietary regulation. We just do not yet fully understand the control mechanism. Choline intake apparently controls acetylcholine concentrations in brain; and serotonin is very finely attuned to dietary tryptophan supply. The K_m of tryptophan hydroxylase is about the same as the plasma tryptophan concentration. Serotonin concentrations can be altered by changing the tryptophan content of the diet (see review by Cooper et al 1978). As we come to understand the relationship of dietary supply and brain neurotransmitter concentrations, I suspect that this will be seen as a more rather than less important basis for understanding brain function.

Idle: I take the point about noradrenaline and dopamine. I am glad to hear that they *are* metabolized slowly because they are circulating hormones, as opposed to acetylcholine which is locally acting. You are talking about a function of taurine which is extremely local. If what you say is right, there must be a very rapid uptake and storage process, after it has had its effect, because if taurine is not metabolized, then at the point of release it would build up in concentration. After all, it is still being synthesized.

Rassin: Taurine like all putative neurotransmitters has been found to have high affinity uptake mechanisms at the synapse. These are systems by which a small amount of neurotransmitter might be very rapidly removed from the synapse. These high affinity uptake systems have also been found for compounds like proline. The presence of these uptake systems alone is not a sufficient criterion to identify a neurotransmitter.

Kägi: You say that the taurine content of the brain is correlated with the number of neurons. Are there certain neurons which are specialized for taurine?

Rassin: Nobody has been able to demonstrate a region of the brain that is particularly associated with taurine function, as has been done for catecholamines. We do not yet have the immunofluorescent-antibody techniques with which to localize taurine, nor specific inhibitors, comparable to 6-hydroxydopamine, used to destroy catecholamine-containing neurons, for the chemical destruction of taurine neurons.

Dodgson: What is the charge on taurine at physiological pH?

Rassin: Taurine is a zwitterion with a very low pK_a (1.5) for the sulphonic group and a high pK_b (8.7) for the amino group. At physiological pH it is apparently neutral.

Meister: I would have thought that it would be slightly acidic (pK_i, about 5). It comes out early on the amino acid analyser.

Chatagner: There may be some reaction between the NH_2^+ group and the SO_3^-; it is not acidic.

Idle: Taurine can coil around and then the sulphonic acid can influence the ionization of the amine group internally.

Whatley: Is there evidence for retinal damage in taurine deficiency in any other species besides the cat?

Rassin: No. This finding has, to date, been specific to the cat. Rats don't become blind on taurine-deficient diets and other mammals haven't been thoroughly investigated.

Whatley: I am looking for clues as to whether taurine is actually a constituent of the membrane of the retina. I am thinking of the similarity between the structure of the membrane systems of the photoreceptors and the sulpholipid-containing membrane systems of chloroplasts.

Rassin: We have been pursuing a possible relationship between taurine and cholesterol in membranes, but it's hard to imagine that the breakdown in such a general mechanism would only make itself evident in one species and one tissue, the cat retina. We need more experiments in other species.

Idle: Professor Dodgson pointed out another interesting feature of the cat, that it is unable to make glucuronic acid conjugates (p 282). The lion is similar in this respect and the civet and genet, also feline carnivores. One can imagine that dietary aromatic acids, unable to form glucuronic acid derivatives in these species, may mop up taurine and glycine and start to deplete the taurine pool, if there is a lot of acid and very little dietary taurine. In other species the acids would be handled by glucuronic acid conjugation.

Kredich: If so, some toxic metabolite that can't be properly conjugated might be causing the cat to go blind. It may have nothing to do with lack of taurine in the retina but rather the accumulation of a toxic compound.

Rassin: It's hard to imagine where a toxic component is coming from, however. The prevention or reversal of the retinal lesion is specific for taurine. The diet that the cats get is a purified casein. Cysteine, methionine, sulphate and vitamin B6 will not function as taurine replacements.

Roy: Is the taurine concentration decreased in the urine on a taurine-deficient diet? If taurine is being used to get rid of something produced endogenously it might not be very different from normal.

Rassin: In general the first biological response of a mammal to a taurine-deficient diet is to drastically reduce the excretion of taurine in the urine. This response has been well characterized in man and other mammals.

Kägi: It is known that the tapetum lucidum of the dog and fox eye contains large amounts of zinc cysteinate of unknown function (Weitzel et al 1955). In the cat eye, there is a zinc metalloprotein with a rather high cysteine content which bears some similarity to the metallothionein I discussed earlier (p 223-237, Croft 1972). Is there any possible relationship between the occurrence and function of these compounds and taurine metabolism?

Rassin: This is why Dr H. Wisniewski and Dr G. Wen in our Institute have been looking at tapetum structure. Their micrographs of the changes in morphology in taurine deficiency show a drastic disruption of the tapetum's organized structure.

References

Cooper JR, Bloom FE, Roth RH (eds) 1978 The biochemical basis of neuropharmacology (3rd edn). Oxford University Press, New York

Croft LR 1972 Isolation of a zinc protein from the tapetum lucidum of the cat. Biochem J 130:303-305

French MR, Bababmuni EA, Golding RR, Basir O, Caldwell J, Smith RL, Williams RT 1974 The conjugation of phenol, benzoic acid, 1-naphthyl acetic acid and sulphadimethoxine in the lion, civet and genet. FEBS (Fed Eur Biochem Soc) Lett 46:134-137

Fuzzessery ZM, Carr WES, Ache BW 1978 Antennular chemosensitivity in the spiny lobster, *Panuliris argus:* Studies of taurine sensitive receptors. Biol Bull (Woods Hole) 154:226-240

Grumer HD 1961 Formation of hippuric acid from phenylalanine labelled with C-14 in phenylketonuric subjects. Nature (Lond) 189:63-64

Hosokawa Y, Yamaguchi K, Kohashi N, Kori Y, Ueda I 1978 Decrease of rat liver cysteine dioxygenase (cysteine oxidase) activity mediated by glucagon. J Biochem (Tokyo) 84:419-424

Idle JR, Millburn P, Williams RT 1978 Taurine conjugates as metabolites of arylacetic acids in the ferret. Xenobiotica 8:253-264

Knopf K, Sturman JA, Armstrong M, Hayes KC 1978 Taurine: an essential nutrient for the cat. J Nutr 108:773-778

Machlin LJ, Pearson PB, Denton CA 1955 The utilization of sulfate sulfur for the synthesis of taurine in the developing chick embryo. J Biol Chem 212:469-475

Martin WG, Truex CR, Tarka M, Hill KJ, Gorby WG 1974 The synthesis of taurine from sulfate. VIII. A constitutive enzyme in mammals. Proc Soc Exp Biol Med 147:563-565

Moldave K, Meister A 1957 Enzymic acylation of glutamine by phenylacetic acid. Biochim Biophys Acta 24:654-655

Napier EA, Jr, Kreyden RW, Henley KS, Pollard HM 1964 Coenzyme Q and phenylketonuria. Nature (Lond) 202:806-807

Olson RE 1966 Biosynthesis of ubiquinones in animals. Vitam Horm 24:551-574

Olson RE, Bentley R, Aiyar AS, Dialameh GH, Gold PH, Ramsey VG, Stringer CM 1963 Benzoate derivatives as intermediates in the biosynthesis of enzyme Q in the rat. J Biol Chem 238:3146-3148

Rambaut PC, Miller SA 1965 Studies of sulfur amino acid nutrition in the adult cat. Fed Proc 24:373

Schachter D, Taggart JU 1954 Glycine *N*-acylase: purification and properties. J Biol Chem 208:263-275

Trippett S 1957 J Chem Soc, p 1929

Weitzel G, Buddecke E, Fretzdorff A-M, Strecker F-J, Roester U 1955 Struktur der im *Tapetum lucidum* von Hund und Fuchs enthaltenen Zinkverbindung. Hoppe-Seyler's Z Physiol Chem 299:193-213

Westall RG 1953 The amino acids and other ampholytes of urine. 2. The isolation of a new sulphur-containing amino acid from cat urine. Biochem J 55:244-248

Stein WH, Paladini AC, Hirs CHW, Moore S 1954 Phenylacetylglutamine as a constituent of normal human urine. J Am Chem Soc 76:2848-2849

General discussion

RHODANESE IN THIOBACILLI AND HIGHER ORGANISMS

Postgate: In this final discussion we shall try to cover topics that we have neglected so far in the symposium. The enzyme rhodanese is one such topic. It is apparently ubiquitous in thiobacilli; it was originally found in mammalian liver. It reacts with cyanides and thiosulphates.

Kelly: The thiobacilli metabolize polythionates to thiosulphate and then to sulphate. The complete pathway is not fully understood (Kelly 1978). One problem is the initial step in catalysing thiosulphate metabolism. It is probably a cleavage, separating the two sulphur atoms. Rhodanese is an enzyme that can do this (thiosulphate sulphurtransferase; EC 2.8.1.1). If this is the effective enzyme, what is the acceptor for the sulphane-sulphur of the thiosulphate? And has there been any progress in studying the mammalian enzyme? The mechanism of the mammalian enzyme might throw light on the thiobacillus enzyme. Lipoic acid has been shown to be an acceptor for the sulphur by the mammalian enzyme, and this is also used by the bacterial enzyme, so possibly lipoate persulphides are intermediates (Silver & Kelly 1976). Is it known what rhodanese does physiologically in the vertebrate liver, or in higher plants, where it also occurs? And does its presence or absence have any clinical significance — is there any disease known where rhodanese is absent?

Rassin: Hoskin & Kordik (1977) have suggested that rhodanese is active in the pathway leading from cysteine to isethionic acid and this is why taurine is by-passed in the squid axon.

Kelly: So it would be generally involved in cyanide metabolism? I suspect that the fact that cyanide is a favoured substrate in the way the enzyme is normally assayed is artificial, because in the thiobacilli, which have considerable rhodanese activity, you never see any cyanide.

Segel: There are cyanogenic glucosides that might enter into a human or animal diet. Rhodanese might be a way of detoxifying them. Apparently, cigarette smoking elevates rhodanese levels in the liver, which is interesting because here we have an inducible enzyme that is involved in detoxifying the cyanide!

Kelly: Thiobacillus denitrificans showed a three-fold increase in rhodanese concentration when cyanide (40 μM) was added to the culture medium (Bowen et al 1965).

Whatley: It is said that we may ingest the equivalent of two or three times the lethal dose of cyanide each day in vegetables; the difference between taking a single lethal dose as a KCN pill and taking the equivalent dose in vegetables is that it is all released at once from the pill, whereas it is released slowly from the amygdalin-like compounds in vegetables and so can be successfully detoxified.

METHIONINE METABOLISM

Postgate: May we turn now to some further aspects of methionine metabolism, particularly the question of what is known about pathways of methionine degradation in microorganisms and animals, and about the relative contributions to the various metabolic pathways of methionine in mammals.

Rassin: We have been interested for some time in the dynamics of methionine cycling. It has never been completely clear what the relative fates of methionine are, because it has so many possible products — to homocysteine and then recycled back to methionine; to *S*-adenosylmethionine and then to polyamines; to protein; and to cysteine. We don't know why metabolic blocks at particular sites in the pathway result in particular accumulations. For instance, if cystathionine synthase is blocked, why does homocysteine build up, whereas if the block is one step further on, at cystathionase, you don't see any homocysteine accumulation? There is no build-up of polyamines in homocystinuria.

Mudd: I tried to tackle this problem in my paper (p 239-258). If we are talking about an adult in a metabolic steady state, with no growth and so no net protein synthesis or breakdown, the net intake of methionine is reflected by the excretion of an equivalent amount daily of inorganic sulphate (and a relatively small amount of taurine), and the conversion of the four-carbon moiety to α-ketobutyrate and its metabolites. The methyl group comes out largely as creatine, accounting for at least 80% of the estimated net methionine utilization. The polyamines account for about 0.5 mmol/day. If you give a labelled tracer amount of methionine, 90% of it goes into protein, and the fate of the label does not reflect the net metabolism of methionine.

Rassin: What complicates the situation is that in some tissues you don't have complete transsulphuration, for instance in brain, where you have essentially a block in the further metabolism of cystathionine to cysteine because of extremely low cystathionase activity (Rassin & Gaull 1975). Only minimal amounts of cystathionine are converted to cysteine in the brain, and it is hard in that situation to understand what the relative fate of methionine is.

Mudd: The fate of homocysteine differs from tissue to tissue. Judged by relative enzyme activities, in some tissues (for example, small intestine) all the homocysteine is used for cystathionine synthesis. Other tissues such as testis, heart and lung may recycle all the homocysteine to methionine. Enzyme activities upon which this conclusion is based are reviewed in Mudd (1974). In liver, where the quantitatively dominant part of homocysteine metabolism occurs, in normal dietary circumstances at any moment on the average homocysteine is flowing roughly in equal amounts to cystathionine and to methionine, as I showed (p 239-258).

Bright: I am interested to know how methionine is broken down when cysteine is not made, as in plants and bacteria.

Meister: Methionine, in animals, can undergo transamination to α-keto-γ-methiolbutyrate. The keto acid can be converted to 3-methylthiopropionate, which in turn may lose CH_3SH to form propionate. Another possibility is conversion of the keto acid to α-ketobutyrate and CH_3SH. Benevenga's group has done recent work in this area (see Stipanuk 1977).

Segel: If you give a large dose of methionine to *Penicillium*, the culture smells differently. I suspect the organism makes methyl sulphide as a sulphur overflow product.

Mudd: In higher plants, we studied the cleavage of cystathionine, which in animals gives cysteine and α-ketobutyrate. In higher plants it goes the other way to give homocysteine and pyruvate (Giovanelli & Mudd 1971).

Postgate: To what extent are the pathways that Dr Mudd described represented in bacteria?

Jones-Mortimer: In *Escherichia coli,* methionine sulphur never gets into cysteine (Roberts et al 1955, Jones-Mortimer 1967).

Dodgson: E. coli won't grow on methionine. What does it mean, then, when a bacterium will grow on methionine?

Jones-Mortimer: Probably sulphate contamination of the medium (Roberts et al 1955, Bohinski & Mallette 1965). Or, of course, that a pathway from methionine to cysteine exists in that organism.

Rassin: In any of these plant or microbial systems do you see free pools of homocysteine?

Mudd: In higher plants like *Chlorella* and *Lemna* there is a little free homocysteine (of the order of 1 nmol/μmol of protein methionine: Giovanelli et al 1979).

TRANSPORT OF SULPHATE AND POLYTHIONATES IN BACTERIA AND EUKARYOTES

Kelly: May we consider again the question of sulphate transport? Any fundamental problems related to sulphate transport must presumably also occur for uptake into thiobacilli of thiosulphate, tetrathionate and other polythionates, as all are divalent sulphur oxyanions. Does anyone have any views on the mechanism or the energy demand for such transport, given that the thiobacilli have a problem in just growing on the energy available from the oxidation of these compounds?

Hamilton: With the thiobacilli and the dissimilatory sulphate reducers, one compound is taken in (sulphide or thiosulphate with the thiobacilli, sulphate with the sulphate reducers) but another has to come out (sulphate from the thiobacilli, sulphide from the sulphate reducers). One can think therefore in terms of a potential exchange mechanism. With assimilatory sulphate reduction, on the other hand, we are dealing with a unidirectional flux and the mechanism might be quite different.

Dodgson: Sulphate is transported through the gut wall of mammals and one unconfirmed report has involved 3′-phosphoadenosine 5′-phosphosulphate (PAPS) in the process (Astudillo et al 1964). Is there in some bacteria a second sulphate-transporting system with a much higher K_m?

Jones-Mortimer: There is only one sulphate transport system normally in *E. coli* and *Salmonella* (but cf. Howarth 1958).

Kredich: The sulphate permease system in *E. coli* and *S. typhimurium* transports thiosulphate and a *cysA* mutant doesn't utilize thiosulphate unless you use very high concentrations. You can push it past the permease, as it were, but so far as I know, you can't do that with sulphate.

Whatley: There are two K_m's for sulphate uptake into higher plant tissues. When plants take in ions from the environment they may transfer most of them into the vacuole, and an enormous concentration change may result. The concentration of the ion in the cytoplasm itself is difficult to determine experimentally. It is hard to know whether the two K_m values are associated with the vacuole (tonoplast) membrane and the external membrane of the cell respectively.

Dodgson: We have been studying the induction of a primary alkylsulphatase enzyme in a pseudomonad, using alkane sulphonates as inducers. We tried to correlate the specific activity of the enzyme produced with the quantity of inducer added to the culture medium. We obtained curious inducer concentration/specific enzyme activity curves with an initial peak, a subsequent fall and then a further increase. We thought that perhaps we were observing the effects of two transport systems, the first and most efficient one being inhibited by excess inducer.

Chatagner: Could you change the medium, to see the second phase only?

Dodgson: Using the initial slope of the first part of the inducer concentration/specific enzyme activity curves, you can plot K_{inducer} constants analogous, for example, to K_m. The K_{inducer} values obtained for alkylsulphonates of different chain length, when plotted against chain length, give a straight-line relationship. So a binding phenomenon of some sort is operating.

Whatley: In *Paracoccus* there is no evidence for a carrier protein being involved in sulphate transport. Protons are pushed out into the environment as a result of respiration and each sulphate ion taken in is accompanied by three or four protons — a proton symport mechanism, the energy for which is derived by respiration.

Segel: In *Penicillium* the stoichiometry suggests that one proton and two other positive charges (either one calcium ion or two potassium ions) must bind to the carrier before sulphate binds. It behaves as a rapid equilibrium 'random A B ordered C' ter-reactant enzyme for the net transport of sulphate into the cell.

Postgate: Can thiosulphate or tetrathionate be taken up by similar systems?

Segel: In fungi, thiosulphate is taken up by the sulphate transport system. In a sulphate permease-negative strain of *P. notatum* tetrathionate was taken up by a saturable mechanism. So, I assume that tetrathionate is taken up by a permease other than the sulphate permease.

Hamilton: These transport systems are energy-linked through a chemiosmotic type of mechanism and therefore are integral to the whole system of energy production. There is an alternative possibility, namely not to take the material into the cell. Is thiosulphate necessarily taken into the cell? Iron oxidation can occur outside the cell, and a completely different mechanism operates (see later discussion, p 299).

Kelly: Trudinger (1965) showed that thiobacilli weren't freely permeable to thiosulphate, but it obviously entered the cell somehow. If one knew the pathway of oxidation one could make deductions about the enzymology. Some of the enzymes are membrane bound, but about half of the rhodanese activity is in the cytosol. Some of the other enzymes involved are soluble, which suggests that thiosulphate does get into the cell.

Hamilton: What is the pH range at which these thiobacilli grow?

Kelly: On inorganic sulphur compounds it is broad, from pH 1.9 to about 10, depending on the species.

Whatley: We have to be careful about assuming that all organisms take up the same substance in the same way or that they must actually take up a substance simply because they use it. Nitrate is used by *E. coli* as a terminal electron acceptor in dissimilatory nitrate metabolism by a reaction on the outer surface of the cell membrane. In *Paracoccus*, nitrate must enter before it is used for that purpose.

METABOLISM OF INSOLUBLE MINERAL SULPHIDES

Kelly: We discussed the uptake of certain insoluble substrates earlier (p 80), but has anyone any ideas about how to approach this problem experimentally? Are insoluble mineral sulphides taken up by bacteria in the same way as elemental sulphur, relying on the dissolution/ionization of the substrate (or evaporation of the substrate), perhaps by close physical association with the substrate? How does one study this experimentally?

Postgate: The classical way has been to show an effect of particle size on the reaction rate (the smaller the particle size, the faster the substrate is utilized).

Brierley: A decrease in particle size does not increase the rate of substrate utilization when the particle approaches the size of a bacterium. This was demonstrated by Pinches et al (1976) during the leaching of chalcopyrite by *Thiobacillus ferrooxidans*. Decreasing the mineral particle size gave rise to an increased surface area for leaching with a corresponding increase in copper released. However, when the particle size was greatly reduced, leach rates diminished. This was attributed to geometrical effects or to an interference in the bacterial–mineral interaction.

Kelly: One idea has been to separate the organisms from the substrate by means of a semipermeable membrane, in order to investigate the problem.

Brierley: Using semipermeable membranes to separate the substrate from the bacteria has limited application when experimenting with acidophilic microorganisms. The acid environment disrupts the integrity of the membrane.

Postgate: In terms of the size of ions, even hydrated ions, semipermeable membranes provide enormous distances for diffusion. Their kinetic effect on reaction rates would be huge.

Kelly: Has anyone studied other insoluble substrates, say the ingestion of solid particles, or the breakdown of materials like lignins by fungi? Presumably secreted enzymes operate; there is no evidence for these in the breakdown of minerals. The proteins would have to be acid stable.

Le Gall: The enzyme for cellulose digestion is secreted by *Clostridium thermocellum*.

Postgate: Hydrocarbons and sulphur have a vapour pressure in air and also in any other environment, so they will be among water molecules even if they are not hydrated. I think minerals are a different problem, however.

Whatley: There has been a lot of unresolved argument about how plants take up cations, not from the soil solution but from clay particles on which they are adsorbed. The root hairs, which pick up ions, grow to conform in shape to the clay particles. It has been suggested that there is a direct hydrogen:potassium or hydrogen:calcium exchange at the surface without the intervention of an intermediate solution phase.

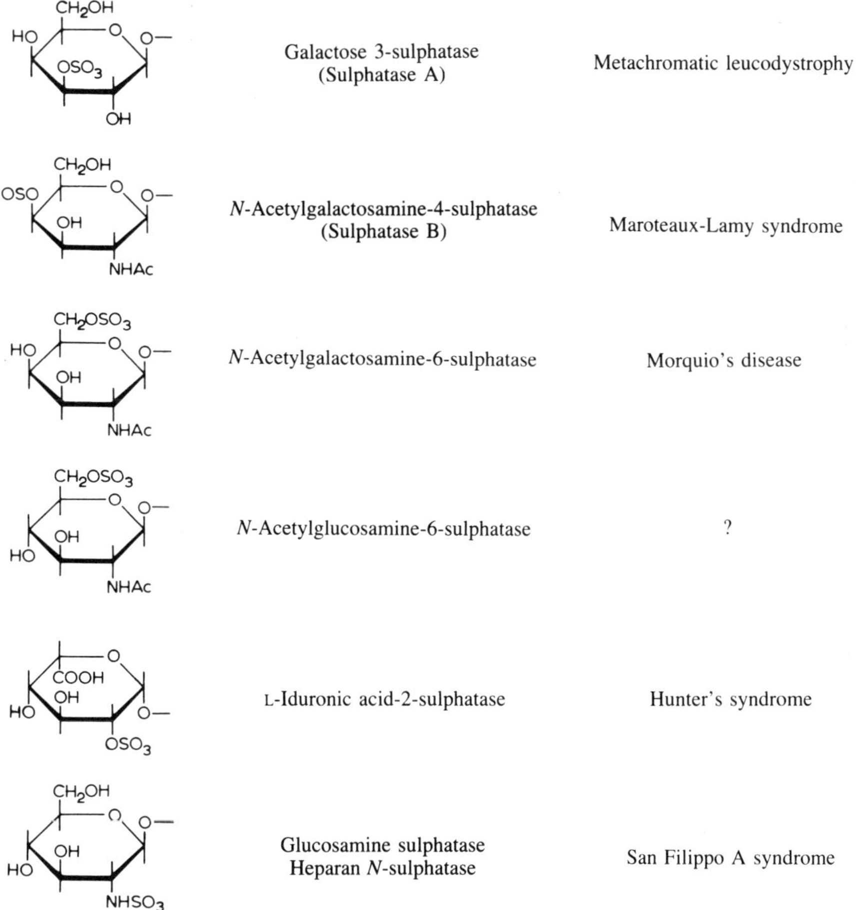

FIG. 1 (Roy). Inherited diseases caused by deficiencies of lysosomal sulphatases in man.

INBORN ERRORS OF SULPHATE METABOLISM

Roy: There are now six known inborn errors of metabolism, each caused by a defect in a specific lysosomal sulphatase; these are summarized in Fig. 1. Individually these diseases are rather rare but taken together their occurrence is far from insignificant, especially when account is taken of another condition, placental sulphatase deficiency, in which there is a lack of the microsomal arylsulphatase/steroid sulphatase complex. Despite the name, this deficiency is not restricted to the placenta but is also apparent in the skin and fibroblasts of the affected children, who also show an X-linked ichthyosis.

An interesting point that has emerged is that there exist apparently normal heterozygotes for metachromatic leucodystrophy whose tissues have only about 10% of the normal level of sulphatase A (Fluharty et al 1978a), close to that found in the condition itself. This suggests that the normal level of sulphatase A is greatly in excess of the body's requirements. If this is the case, the implications for potential therapy are considerable.

As well as the six individual conditions listed in Fig. 1 there is another condition, multiple sulphatase deficiency, in which all six enzymes are defective, together with the microsomal sulphatases that are deficient in placental sulphatase deficiency. This means that up to a total of nine sulphatase activities may be affected, although not necessarily all to the same extent (Austin 1973). To complicate matters further, Fluharty et al (1978b) have recently shown that while the enzyme defects can be shown in fibroblasts cultured in MEM/CO_2, when the cells are cultured in MEM/Hepes there is an increase in the level of sulphatase A, but not sulphatases B or C nor iduronate sulphatase. The level of sulphatase A in the fibroblasts cultured in the latter medium was close to normal, suggesting that the genome for sulphatase A must exist in fibroblasts from cases of multiple sulphatase deficiency. Much obviously remains to be learned about this most interesting condition.

Meister: An enzyme might be deficient in one tissue and not in another. This situation could develop if the enzyme was unstable and subject to degradation, and if the degradative mechanism were more active in one tissue than another. For example, consider an enzyme in red blood cells that is unstable. That enzyme cannot be replaced during the life of the peripheral red cell, whereas the same enzyme in the liver or kidney could be replaced. One might see phenotypically a deficiency of the erythrocyte enzyme and not the liver enzyme. This kind of explanation might account for situations in which there are differences in activity in different tissues. The loss of nine enzymes at once suggests that some fundamental co-factor is required for all of them.

Roy: Nobody has ever found this, or a common inhibitor.

Mudd: An explanation of the father with 10% of the normal sulphatase A activity who was himself normal is simply that 10% is not zero. It is found with many enzymes, including cystathionine synthase, that the normal amount of enzyme is in vast excess of the capacity needed to deal with the normal amount of substrate. For example, on a normal diet one has to convert about 10 mmol/day of methionine sulphur to sulphate. With a load of methionine an adult can handle 100 mmol/day without any problem, if he is normal, so about 10% of the normal enzyme level should be enough. We also know that there is a great difference to a patient whether he has no detectable activity or as little as 1–2% of normal activity. This can make a big clinical difference between accumulating homocysteine or not. Around 5%

might be sufficient with a normal load. Again, in Lesch-Nyhan disease, 0.1% of normal enzyme activity makes a big difference to the clinical well-being of the patient. Perhaps the excess capacity of these various enzymes keeps the substrate down to a lower level than would otherwise be the case.

Rassin: You get different patterns of enzyme activity in heterozygotes; one doesn't often see the 50% that one would theoretically expect. Yet these heterozygotes in general do not have clinical problems.

Rose: One of the conditions Dr Roy mentioned is placental sulphatase deficiency, in which the microsomal enzyme arylsulphatase C is missing. This disease is probably more prevalent than might be realized. The enzyme (arylsulphate sulphohydrolase, EC 3.1.6.1) is almost certainly an oestrogen sulphohydrolase and is probably active towards a number of steroid sulphates that are important as metabolic intermediates (see Dodgson & Rose 1975). The disease is an X-linked genetic disorder in which the enzyme is missing from the placenta in humans (France & Liggins 1969), hence the name placental sulphatase deficiency.

This enzyme is thought to be important for producing free oestrogen from sulphoconjugates towards the end of pregnancy, with the result that oestrogen excretion goes up and delivery of the child proceeds normally. When the enzyme is missing, extra oestrogen is not produced, the mother has a difficult labour and problems arise. Only one or two cases were known until recently but in collaboration with Dr R.E. Oakey at Leeds University we have studied about 20 in the last two years, obtained from various parts of the country, where steroid excretion in pregnancy is being watched carefully. The true incidence of this condition may be quite sizeable and although it is not a sulphur deficiency as such it is an example of how the important role played by sulphur in metabolism is gradually assuming prominence.

Jones-Mortimer: To go back to the enzyme that hydrolyses APS (adenosine 5′-phosphosulphate sulphatase), a possible role for it is in the prevention of selenate toxicity. I suspect that the selenate ion as such is not very toxic, but if it gets reduced, then it is. This enzyme could be a means of stopping selenate analogues of APS being produced.

Kelly: How would it work? The enzyme destroys APS. How stable is APSe? Is this the route for the formation of things such as seleno-cysteine?

Segel: Has anyone ever seen a selenate ester?

Roy: No. Nissen & Benson (1964) looked in plants and couldn't find any. There are reports of the chemical synthesis of alkyl selenates, which are obviously very unstable (Meyer & Wagner 1922).

Postgate: You find selenate going through to seleno-amino acids in plants, so presumably the initial steps of selenium uptake are similar to those of sulphur uptake in plants.

Whatley: We were unable to detect PAPSe although we could find APSe quite readily in extracts of *Paracoccus* (Burnell & Whatley 1975).

Postgate: Don't forget that monofluorophosphate is also a close structural analogue of selenate which could be used to examine uptake and assimilation pathways (see p 45).

Roy: The reduction of selenate has been said (Dillworth & Bandurski 1977) to depend on the formation of APSe, which is reduced (non-enzymically?) by thiols to selenite.

Whatley: That is acceptable, but I don't think PAPSe exists.

Kredich: It must exist! If you grow *E. coli* on selenate the colonies turn orange from the production of selenium.

Roy: But why must you have PAPSe?

Kredich: The reduction of sulphate in *E. coli* and *Salmonella* proceeds through PAPS, as far as we know, and I would assume that selenate reduction also requires the formation of PAPSe in those organisms.

ENERGY GENERATION IN SULPHUR METABOLISM

Hamilton: I would like to take up and extend a point we have already touched on. It is the question of thinking in terms of chemiosmotic mechanisms when considering problems of energy generation.

One can consider cellular energetics solely in terms of thermodynamics. Where two reactions are coupled, there must be release of sufficient free energy in one reaction to drive the second. One can quantify this free energy in terms of joules or, with redox reactions, millivolts. Where coupling involves a chemiosmotic mechanism, as appears to be the case with redox reactions, free energy is conserved as the protonmotive force comprised of transmembrane gradients of potential and of pH.

Thermodynamic considerations, however, can only define the possibility of any particular reaction or series of reactions taking place. It tells one nothing directly about the mechanism of the reaction. Indeed, there are mechanisms which allow apparently impossible thermodynamic events to occur. Sometimes such considerations can highlight areas of unsolved, or even unrecognized, problems and suggest particular experimental approaches. I want to cite two instances in which the chemiosmotic hypothesis has fulfilled a particular heuristic role in this respect.

John Ingledew and his associates at Dundee (Ingledew et al 1977, Cox et al 1979) have been working with *Thiobacillus ferrooxidans*. This is an acidophilic organism that can grow at pH 2 and obtain its energy from the oxidation of ferrous iron. The intracellular pH is around 6.0, so that in chemiosmotic terms there is already a protonmotive force of 240 mV in the form of a pH gradient of 4.0 units,

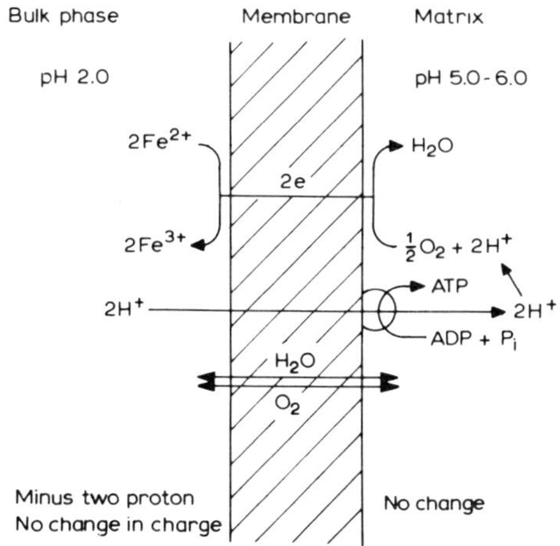

FIG. 1 (Hamilton). Scheme for oxidative phosphorylation, at the expense of the acid environment, in *Thiobacillus ferrooxidans*. (From Ingledew et al 1977, with permission of Elsevier/North-Holland Biomedical Press)

inside alkaline. This serves to explain the ability of the cells to derive sufficient energy for ATP synthesis from the very unfavourable oxidation of ferrous iron.

In fact the oxidation takes place outside the cell and serves simply as a source of electrons which are passed across the membrane to the cytoplasmic side via a copper-containing protein, rusticyanin, and cytochromes *c* and *a*. There, these electrons combine with protons in the reduction of oxygen to water. The cells possess a normal F_0F_1-ATPase and the protons enter the cell through this mechanism, driven by the pH gradient and generating the synthesis of ATP (Fig. 1).

In this organism, therefore, there is a variant of the standard chemiosmotic mechanism in which the transmembrane flux of electrons serves to combine with the protons in the reduction of oxygen and so prevent the dissipation of the pH gradient which in these extreme acid conditions is the driving force for ATP synthesis.

I should also like to draw attention to a suggestion put forward by Paul Wood (1978) for the structural and functional organization of the energy-generating reactions in the sulphate-reducing bacteria (Fig. 2, p 300). Hydrogenase and cytochrome c_3 are located in the periplasm, reactions such as pyruvate:ferredoxin oxidoreductase are in the cytoplasm, and the membrane contains normal electron transport components. From Wood's scheme one can see how the oxidation of hydrogen or pyruvate, for example, can generate the necessary gradients of potential and pH across the membrane.

This highlights the problem already referred to with respect to the entry of sulphate to the site of APS and sulphite reductases in the cytoplasm, and the subsequent efflux

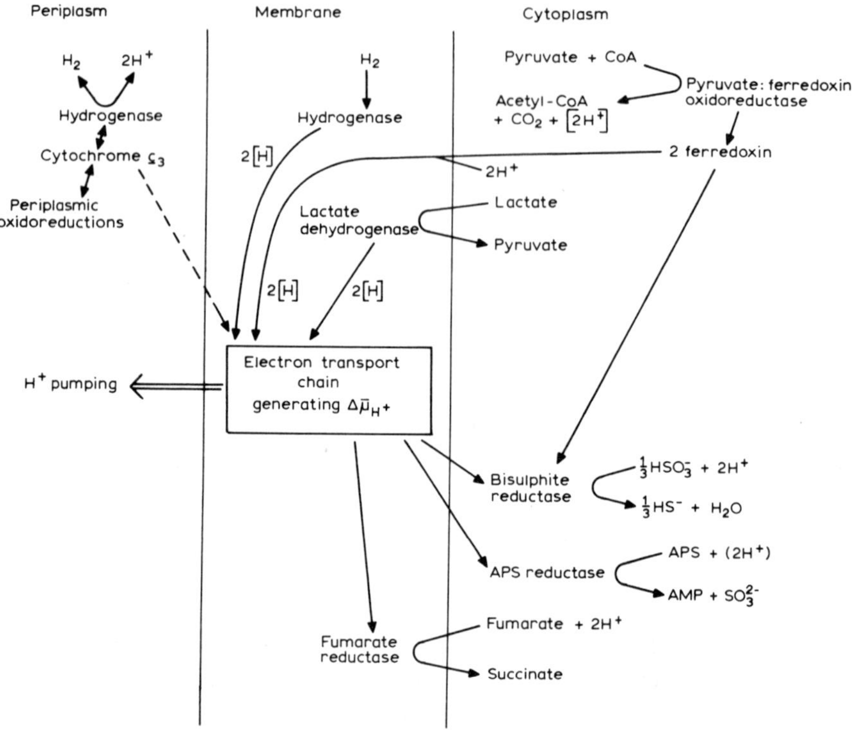

FIG. 2 (Hamilton). Proposal for electron transport in *Desulfovibrio*. A scheme for coupling various oxidations and reductions to the generation of a proton gradient across the cytoplasmic membrane is shown. The proton numbers in brackets are those for the complete reactions. (From Wood 1978, with permission of the Federation of European Biochemical Societies.)

of sulphide from the cell. That is to say, a conceptual model of this character allows one to ask more direct questions, and so far as I know is something which has not featured strongly in energetic considerations in these organisms.

Segel: I take it that these organisms will not grow at pH 6.5?

Hamilton: Not on iron, but these thiobacilli can oxidize sulphur compounds and grow at pH values above 2.

Kelly: Would it be logical that if you calculated the apparently available free energy for the iron oxidation, you would expect the apparent efficiency of the utilization of the free energy to be very high, if that is the mechanism?

Hamilton: I suppose you would. I would wonder, however, what the meaning of such a calculation was.

Kelly: We have done experiments on *Thiobacillus ferrooxidans* growing in chemostat culture on Fe^{2+} and they approximate to 100% efficiency, from the free energy change, which is surprising (Kelly 1978). The other point is that if you think of the oxidation of sulphide or thiosulphate by this type of organism, both oxidations require two moles of oxygen for each mole of sulphate oxidized, so the electron transport is equivalent for the two substrates. Just on a P/O ratio calculation one can say that if the electron transport pathway is the same, you should get the same energy and hence same growth yield. In fact, both the yield and efficiency of conservation of energy from sulphide oxidation are only about 60% of those from thiosulphate.

Hamilton: There is the possibility that reduced ferredoxin can donate electrons to the sulphide reductase without going through the membrane-bound electron transport chain; this would be a 'wasted' oxidation, energetically speaking. As I suggested to Dr Le Gall earlier (p 81), although his cytochrome c_3 can give oxidation, perhaps energy generation is only possible if other electron transport components allow the build up of transmembrane gradients of potential and pH.

Le Gall: What are the 'periplasmic oxidoreduction' reactions defined by Dr Wood? What is the evidence for their existence, and what is the nature of the dotted line in Fig. 2 from cytochrome c_3 to 'electron transport chain generating $\Delta\bar{\mu}_{H^+}$'?

Hamilton: This is a theoretical paper of Wood's. He is simply suggesting that hydrogenase and other periplasmic oxidoreductases might be present.

Le Gall: Omitting the 'periplasmic oxidoreduction' reactions, what we know about the electron transfer chain could agree with that picture; all you have to do is to allow some c_3 to enter the membrane, and then you have hydrogenase outside, c_3 halfway between and ferredoxin on the other side. We know that ferredoxin interacts with c_3 and hydrogenase (Bell et al 1978) and that hydrogenase is inactive towards ferredoxin in the absence of c_3.

Postgate: Wood's paper seems consistent with present knowledge but I don't see any obvious experiments leading from it.

Hamilton: It is very difficult to get measurements of phosphorylation in bacterial systems; with subcellular preparations one can get values of around 0.1 which are meaningless. It is possible to measure proton efflux, however, from whole-cell systems and there is much information relating the stoichiometry of that reaction to energy-coupling sites. Also, vesicles might be prepared with different sidedness and their activity determined with different permeant and impermeant oxidants and reductants. Such experiments would yield information on the functional organization of the redox components within the membrane, and the potential for each oxidoreduction to generate useful metabolic energy.

SULPHIDE OXIDATION IN ANIMAL TISSUES

Kelly: What is known about sulphide oxidation in animal tissues? I have not seen anything on this in the past 10 years.

Dodgson: We studied this in 1971/1972 (Curtis et al 1972). Earlier work on the site of the detoxification/oxidation of sulphide was somewhat conflicting and had implicated the plasma on the one hand (Haggard 1921) and the erythrocyte on the other (Lovatt-Evans 1967). Using [^{35}S]sulphide and the rat as experimental animal we were able to show that sulphide detoxification occurs in both plasma and red cells, primarily as a result of binding to blood proteins, particularly albumin in the case of plasma. The subsequent rate of oxidation of the bound sulphide to sulphate by blood was much too low to account for the rate of oxidation in the whole animal. However, the bound sulphide had only a transient existence in blood and was rapidly oxidized in the tissues. The liver and kidneys were certainly very active in this respect, as evidenced by perfusion experiments. We never saw any radiolabel in the brain unless we had administered a lethal dose of sulphide. We then took a preliminary look at the localization of the oxidative process within the liver cell. The mitochondrial fraction, although very sensitive to sulphide concentration, oxidized sulphide to thiosulphate and, in the presence of reduced glutathione, to sulphate.

Kelly: Is there any good evidence for enzymic sulphide oxidation in animal tissues or is it just a chemical process?

Dodgson: It requires a viable mitochondrial preparation.

Ziegler: Why is H_2S so toxic?

Le Gall: H_2S makes a breach in the haem molecule and you get a sulphur derivative. This is irreversible, but I don't know if it explains the toxicity. Its first action is on the olfactory nerve; people are killed because they don't smell the H_2S.

Roy: Is H_2S bound preferentially to albumin, or is this binding simply because there is more albumin in plasma than globulin?

Rose: Just over 50% of the protein-bound sulphur was associated with the albumin fraction in our experiments; this could reflect the relative concentrations of the various plasma proteins rather than any specific binding.

IRON-SULPHUR PROTEINS AND SULPHOLIPIDS

Postgate: The iron-sulphur proteins have not been mentioned much so far (p 83), but at least in some microbes one knows that their iron-sulphur clusters are termini of sulphur assimilation, just like sulphur-containing amino acids and certain vitamins. It is surprising that many physiologists have not noticed their relevance, though we have known about the iron-sulphur proteins for more than 10 years now and they must account for 10 to 30% of the sulphur in some anaerobic bacteria. The form in which the sulphur is incorporated is known precisely for the bacterial (Fe_4S_4) ferredoxins and related enzymes. The structure of some ferredoxins is known from both X-ray and sequence data. The Fe_2S_2 proteins are less well understood. Another new group of sulphur compounds which ought to have occupied us more is the sulpholipids (see p 15, 68).

Whatley: There really is a lot of sulphur in the sulpholipids, which are found in all green parts of higher plants, associated with the light-harvesting lamellae of chloroplasts; they may well have an important structural function. About one-third of the sulphur in the leaf is present as sulpholipid, a further third is in the protein of the chloroplast, and the remainder is in the rest of the cytoplasm.

Dodgson: J.L. Harwood is now looking at the biosynthesis of chloroplast sulpholipids. He holds the view that they may have both structural and metabolic roles to play in the chloroplast, but firm proof of either role is not yet available (Harwood & Nicholls 1979).

Le Gall: As far as cysteine-containing proteins are concerned, I never realized before how tricky the picture is. You have cysteine-containing proteins that form disulphide bridges, for redox functions and for structural reasons; you also have cysteine-containing proteins for detoxification and for storage of heavy metals. Finally, you have the metal-containing proteins; in this last category the formation of disulphide bridges has to be prevented, so the structure in the neighbourhood of the cysteine residues is very specialized.

Whatley: When the iron-sulphur clusters are formed they normally connect different arms of the protein via cysteine–cysteine links, so they have some structural significance as well as redox functions.

Postgate: I put great emphasis on the iron-sulphur proteins because they have such a variety of biochemical and physiological functions, which are well understood; they are involved in the regular respiratory pathways in man, and in APS reductase, sulphite reductase, xanthine oxidases and nitrogenase. But aside from that there are proteins which just have cysteine, not sulphide, binding iron, like the rubredoxins, of unknown function, and the remarkable proteins which Dr Kägi described (p 223-237).

But Dr Kelly is interested in the significance of metalloproteins and whether they could be exploited for extracting metals.

Kelly: A vital question in the chemical industry is how to extract metals from dilute solutions, ideally by some natural method. Could one persuade *Neurospora* to produce large quantities of the specific protein and hence to concentrate copper from, say, seawater? What is the proportion of the total protein of *Neurospora* that can be tied up in that protein? Is it significant, and is it subject to induction by having copper in the growth medium?

Kägi: It is not a significant amount of the total protein of *Neurospora crassa*. The protein is formed only in the stationary phase of growth. This coincides with the time when copper is taken up from the medium. However, the amount of copper found in the copper-metallothionein is only 10% of the total copper tied up by *Neurospora crassa*. It is possible that a part of the copper unaccounted for is present as a polymerized and insoluble form of metallothionein, as has been observed for copper-metallothionein from newborn and fetal bovine and human liver (Porter 1974, Rydén & Deutsch 1977).

Dr Lerch of our laboratory has also studied the dependency of the formation of *Neurospora* copper-metallothionein on copper supply. *Neurospora* is not very sensitive to copper ions. Over a large concentration span one gets normal growth. When the copper concentration is increased further (>4 mM), however, a point is reached where the potential for intracellular copper binding is exhausted and, as a consequence, the intracellular concentration of free copper ions starts to increase and growth is inhibited, signalling toxicity. Interestingly, *Neurospora* copper-metallothionein is formed in large quantity only at a relatively high copper concentration (> 0.5 mM). If the organism is supplied with insufficient sulphur the picture is different, in that a smaller amount of copper-metallothionein is formed and the sensitivity to copper increases, with the metal becoming toxic at a lower concentration. Thus, there seems to be a clear connection between the formation of this protein in large quantity and the onset of copper toxicity (K. Lerch, personal communication 1979).

Whatley: What is the copper content at the point immediately before it becomes toxic?

Kägi: As the toxicity varies with the composition of the nutrient medium, it is difficult to give figures.

Brierley: There are many microorganisms that accumulate metals (Norris & Kelly 1979). Scanning electron microscopy and surface analyses by X-ray dispersion show electron-dense sites of metal accumulation (Gale & Wixson 1979). Is it possible that these observed electron-dense areas may be rich in metallothionein protein and this protein may, in fact, be responsible for metal accumulation?

Le Gall: I believe that when you give copper to yeasts they start making a lot of cysteine and there is a desulphydration mechanism; copper sulphide is formed and stays on the cell wall.

Kägi: If one cultivates yeast at a concentration of 0.2 mM-copper, a protein is produced which has a molecular weight below 10 000, contains 24% cysteine, and is believed to be a metallothionein (Prinz & Weser 1975).

Postgate: We should at least briefly touch on the economic importance of the sulphur-metabolizing organisms.

Brierley: Microorganisms are used commercially to extract uranium and copper from low-grade ores. At present, nearly 12% of the total copper production of the United States can be attributed to activity of bacteria belonging to the genus *Thiobacillus* (Brierley 1978). Biological processes are beginning to play a key role in the removal of dilute concentrations of trace elements from industrial effluents. These processes involve the adsorption of metals onto mixed algal populations (Gale & Wixson 1979) and the precipitation of metallic sulphide compounds mediated by the anaerobic bacterial generation of hydrogen sulphide (Jackson 1978). We (C.L. Brierley & J.A. Brierley, unpublished work) are currently investigating the role of microorganisms and algae in removing trace amounts of uranium, selenium and molybdenum from uranium mine waters. These processes are effective in the restoration of industrial effluents and are energy conservative and do not require the installation of costly equipments. The construction of holding ponds or meandering stream systems is all that is required.

Could you elaborate on the potential for industrial exploitation of the microorganisms under discussion?

Postgate: There are a number of important economic aspects of microbial sulphur metabolism that we have not talked about very much. The sulphate-reducing bacteria are involved in all sorts of economic areas (see Postgate 1979), such as the anaerobic corrosion of metals, which costs millions of dollars a year. The thiobacilli and sulphate reducers between them cause corrosion of metals and stonework. Sulphate-reducing bacteria cause the penultimate stage in terrestrial and aqueous organic pollution, giving rise to problems such as H_2S in water, discoloration of paint, damage to metal installations, and spoilage of foods by sulphides; they cause great problems in oil technology, and underground gas stores get contaminated with H_2S produced by sulphate-reducing bacteria. They have even been thought of as having been involved in the generation of oil; they are certainly involved in the coalescence of oil deposits and in damage to oil-well machinery.

The sulphate-reducing bacteria affect agriculture: they contribute to the nutrition of sheep because they are present in the rumen and assist the sulphur nutrition of their animal hosts. They are economically expensive organisms, but in compensation they are involved in the formation of most of the world's sulphur supplies and a large proportion of our sulphide minerals.

Hamilton: The aspects you mention are all negative, other than the formation of sulphur deposits.

Le Gall: So long as we are not sure about what kind of sulphur is essential for the equilibrium of the global sulphur cycle in the atmosphere, we should be careful about saying this. These bacteria may contribute a lot to this equilibrium.

* * *

Postgate: It would be impossible to summarize such an extremely wide-ranging symposium, and I shall not try. In this final discussion we have tried to draw some threads together, but the garment that we have knitted has enormous lacunae in it. The conclusion that I come to is that we should hold this sort of symposium much more often. As I said at the beginning, in the area of nitrogen research, regular international symposia have been funded and have stimulated the progress of research spectacularly. If we could get something like that started in the sulphur area, we could hold symposia of this kind more frequently.

The final discussion has revealed how aspects of sulphur metabolism, particularly microbial sulphur metabolism, have not only medical and nutritional importance but also considerable consequences for industry and in the natural environment. Research in many of these areas is under-supported, perhaps because negative topics — corrosion, pollution, deterioration and so on — fall between the research responsibilities of industry and government. Yet the scientific problems posed, and the pay-offs in potential applications, are considerable. Industry as well as government should put money into supporting meetings such as ours, which would be a valuable way of promoting the science; as essential as the direct support of research itself.

References

Astudillo MD, Espliguero MS, Zumel CL Sanz F 1964 Transport of sulphate ion through intestine. Rev Esp Fisiol 20:113-120

Austin JH 1973 Studies in metachromatic leukodystrophy. Multiple sulfatase deficiency. Arch Neurol 28:258-264

Bell GR, Lee JP, Peck HD, Le Gall J 1978 Reactivity of *Desulfovibrio gigas* hydrogenase toward artificial and natural electron donors or acceptors. Biochimie (Paris) 60:315-320

Bohinski RC, Mallette MF 1965 Behavior of *Escherichia coli* B in sulfate-limited medium. Can J Microbiol 11:663-669

Bowen TJ, Happold FC, Taylor BF 1965 Some properties of the rhodanese system of *Thiobacillus denitrificans*. Biochem J 97:651-657

Brierley CL 1978 Bacterial leaching. CRC Crit Rev Microbiol 6:207-262

Burnell JN, Whatley FR 1975 A new, rapid, and sensitive assay for adenosine 5′-phosphosulphate (APS) kinase. Anal Biochem 68:281-288

Cox JC, Nicholls DG, Ingledew WJ 1979 Transmembrane electrical potential and transmembrane pH gradient in the acidophile *Thiobacillus ferro-oxidans*. Biochem J 178:195-200

Curtis CG, Bartholomew TC, Rose FA, Dodgson KS 1972 Detoxication of sodium ^{35}S-sulphide in the rat. Biochem Pharmacol 21:2312-2321

Dillworth GL, Bandurski RS 1977 Activation of selenate by adenosine 5′-triphosphate sulphurylase from *Saccharomyces cerevisiae*. Biochem J 163:521-529

Dodgson KS, Rose FA 1975 Sulfohydrolases. In: Greenberg DM (ed) Metabolic pathways, vol 7: Metabolism of sulfur compounds. Academic Press, New York, p 359-431

Fluharty AL, Stevens RL, Kihara H 1978a Cerebroside sulfate hydrolysis by fibroblasts from a metachromatic leukodystrophy parent with deficient arylsulfatase A. J Pediatr 92:782-784

Fluharty AL, Stevens RL, Davis LL, Shapiro LJ, Kihara H 1978b Presence of arylsulfatase A in multiple sulfatase deficiency disorder fibroblasts. Am J Hum Genet 30:249-255

France JT, Liggins GC 1969 Placental sulfatase deficiency. J Clin Endocrinol Metab 29:138-141
Gale NL, Wixson BG 1979 Removal of heavy metals from industrial effluents by algae. Dev Ind Microbiol, in press
Giovanelli J, Mudd SH 1971 Transsulfuration in higher plants: partial purification and properties of β-cystathionase of spinach. Biochim Biophys Acta 227:654-670
Giovanelli J, Mudd SH, Datko AH 1979 Sulphur amino acids in plants. In: Miflin BJ, Stumpf PK, Conn EE (eds) The biochemistry of plants: a comprehensive treatise, vol 5: Amino acids and derivatives. Academic Press, London & New York
Haggard HW 1921 J Biol Chem 49:519
Harwood JL, Nicholls RG 1979 The plant sulpholipid – a major component of the sulphur cycle. Biochem Soc Trans 7:440-447
Hoskin FCG, Kordik ER 1977 Hydrogen sulfide as a precursor for the synthesis of isethionate in the squid giant axon. Arch Biochem Biophys 180:583-586
Howarth S 1958 Suppressor mutations in some cystine-requiring mutants of *Salmonella typhimurium*. Genetics 43:404-418
Ingledew WJ, Cox JC, Halling PJ 1977 A proposed mechanism for energy conservation during Fe^{2+} oxidation by *Thiobacillus ferro-oxidans:* chemiosmotic coupling to net H^+ influx. FEMS (Fed Eur Microbiol Soc) Lett 2:193-197
Jackson TA 1978 The biogeochemistry of heavy metals in polluted lakes and streams at Flin Flon, Canada, and a proposed method for limiting heavy-metal pollution of natural waters. Environ Geol 2:173-189
Jones-Mortimer MC 1967 Sulphur metabolism in micro-organisms. D Phil Thesis, University of Oxford
Kelly DP 1978 Bioenergetics of chemolithotrophic bacteria. In: Bull AT, Meadow PM (eds) Companion to microbiology. Longman, London, p 363-386
Lovatt-Evans C 1967 The toxicity of hydrogen sulphide and other sulphides. Q J Exp Physiol Cogn Med Sci 52:231-248
Meyer J, Wagner J 1922 Organic derivatives of selenic acid. Ber Dtsch Chem Ges 55:1216-1222
Mudd SH 1974 Homocystinuria and homocysteine metabolism: selected aspects. In: Nyhan WL (ed) Heritable disorders of amino acid metabolism: patterns of clinical expression and genetic variation. Wiley, New York, p 429-451
Nissen P, Benson AA 1964 Absence of selenate esters and 'selenolipid' in plants. Biochim Biophys Acta 82:400-402
Norris PR, Kelly DP 1979 Accumulation of metals by bacteria and yeasts. Dev Ind Microbiol 20:299-308
Pinches A, Al-Jaid FO, Williams DJA 1976 Leaching of chalcopyrite concentrates with *Thiobacillus ferrooxidans* in batch cultures. Hydrometallurgy 2:87-103
Porter H 1974 The particulate half-cystine-rich copper protein of newborn liver. Relationship to metallothionein and subcellular localization in non-mitochondrial particles possibly representing heavy lysosomes. Biochem Biophys Res Commun 56:661-668
Postgate JR 1979 The sulphate-reducing bacteria. Cambridge University Press, Cambridge
Prinz R, Weser U 1975 A naturally occurring Cu-thionein in *Saccharomyces cerevisiae*. Hoppe-Seyler's Z Physiol Chem 356:767-776
Rassin DK, Gaull GE 1975 Subcellular distribution of enzymes of transmethylation and transsulphuration in rat brain. J Neurochem 24:969-978
Roberts RB, Abelson PH, Cowie DB, Bolton ET, Britten RJ 1955 Studies of biosynthesis in *Escherichia coli*. Carnegie Institute of Washington Publication 607, Washington DC, p 369-371
Rydén L, Deutsch HF 1978 Preparation and properties of the major copper-binding component in human fetal liver; its identification as metallothionein. J Biol Chem 253:519-524
Silver M, Kelly DP 1976 Rhodanese from *Thiobacillus* A2. J Gen Microbiol 97:277-288
Stipanuk M 1977 Doctoral Dissertation. Interrelationships of methionine and cysteine metabolism, University Microfilms International, Ann Arbor, London
Trudinger PA 1965 On the permeability of *Thiobacillus neapolitanus* to thiosulphate. Aust J Biol Sci 18:563-568
Wood PM 1978 A chemiosmotic model for sulphate respiration. FEBS (Fed Eur Biochem Soc) Lett 95:12-18

Index of contributors

Entries in **bold** *type indicate papers; other entries refer to discussion contributions*

Indexes prepared by M.R. Hardeman

Subject index